Intelligent Mechatronics
Edited by Ganesh R. Naik

Published by InTech
Janeza Trdine 9, 51000 Rijeka, Croatia

Publishing Process Manager Katarina Lovrecic

Technical Editor Teodora Smiljanic

Cover Designer InTech Design Team

Additional hard copies can be obtained from orders@intechopen.com

Intelligent Mechatronics, Edited by Ganesh R. Naik
p. cm.
ISBN 978-953-307-300-2

INTELLIGENT MECHATRONICS

Edited by **Ganesh R. Naik**

For EU product safety concerns:
IN TECH d.o.o., Prolaz Marije Krucifikse Kozulić 3, 51000 Rijeka, Croatia,
info@intechopen.com or visit our website at intechopen.com.

Meet the editor

Ganesh R. Naik received his B.E. degree in electronics and communication engineering from the University of Mysore, Mysore, India, in 1997, M.E. degree in communication and information engineering from the Griffith University, Brisbane, Australia, in 2002, and PhD degree in the area of digital signal processing from RMIT University, Melbourne, Australia, in 2009. He is currently an academician and researcher at RMIT University. As an early career researcher, he has authored more than 60 papers in peer reviewed journals, conferences and book chapters over the last five years. His research interests include pattern recognition, blind source separation techniques, audio signal processing, biosignal processing, and human–computer interface. Dr. Naik was the Chair for the IEEE Computer Society CIT08 Conference, Sydney and a member of the organising committee for IEEE BRC2011 conference, Vitoria, Brazil. He was a recipient of the Baden–Württemberg Scholarship from the University of Berufsakademie, Stuttgart, Germany (2006–2007). Recently Dr. Naik was awarded with ISSI overseas fellowship from skilled Institute Victoria, Australia.

Contents

Preface XI

Part 1 Intelligent Robotics 1

Chapter 1 **A Mechatronic Perspective on Robotic Arms and End-Effectors 3**
Pinhas Ben-Tzvi and Paul Moubarak

Chapter 2 **A Torque Cancelling System for Quick-Motion Robots 21**
Daigoro Isobe

Chapter 3 **Locomotion Control for Legged Robot by Virtual Contact Impedance Method 41**
Fumiaki Takemori

Chapter 4 **Development of a Simulation Environment Applied to the Study of Fault-Tolerant Control Systems in Robotic Manipulators. Theoretical and Practical Comparisons 51**
Claudio Urrea and John Kern

Chapter 5 **Kinematic Task Space Control Scheme for 3DOF Pneumatic Parallel Robot 67**
Luis Hernández, Eduardo Izaguirre, Ernesto Rubio, Orlando Urquijo and Jorge Guerra

Part 2 System Interfacing, Instrumentation and Control 85

Chapter 6 **Blind Source Separation Based Classification Scheme for Myoelectric Prosthesis Hand 87**
Ganesh R. Naik and Dinesh Kumar

Chapter 7 **Feedback Control and Time-Optimal Control about Overhead Crane by Visual Servo and These Combination Control 103**
Yasuo Yoshida

Chapter 8 **Intelligent Methods for Condition Diagnosis of Plant Machinery 119**
Huaqing Wang and Peng Chen

Part 3 Physical System Modelling and Real Time Applications 141

Chapter 9 **Methodology for Reusing Real-time HiL Simulation Models in the Commissioning and Operation Phase of Industrial Production Plants 143**
Sebastian Kain, Frank Schiller, and Sven Dominka

Chapter 10 **Hybrid Planning for Self-Optimization in Railbound Mechatronic Systems 169**
Philipp Adelt, Natalia Esau, Christian Hölscher, Bernd Kleinjohann, Lisa Kleinjohann, Martin Krüger and Detmar Zimmer

Chapter 11 **An Evidence Accrual Data Fusion Technique for Situational Assessment 195**
Stephen C. Stubberud and Kathleen A. Kramer

Chapter 12 **Intelligent Mechatronic System for Automatically Evaluating the Training of the Laparoscopic Surgeon 219**
Minor A., Lorias D., Ortiz Simon and Escamirosa F.

Chapter 13 **Reliability of Authenticated Key Establishment Protocols in a Complex Sensor System 229**
Kalvinder Singh and Vallipuram Muthukkumarasamy

Preface

Background and motivation

Over the last decade there has been an exponential growth in Mechatronics and intelligent systems activity, a growth that has lead to the development of exciting new products used in every day life. The discipline of Mechatronics is enormous in magnitude. Ideally, it combines mechanics, electronics, software engineering, information systems, communication, control and artificial intelligence.

Mechatronics is defined as the field of study involving the analysis, design, synthesis, and selection of systems that combine electronic and mechanical components with modern controls and microprocessors. Mechatronics is an engineering field that refers to mixed systems' tight integration. Currently, this integration can be viewed as based on digital computer monitoring and control, but it cannot be denied that integration can be based on any other signal processing system and any form of raw power that can be modulated and transferred to the mixed system in accordance with the output of the digital signal processor. Mechatronics refers to monitoring, control and integration not only of lumped parameters systems, but also of distributed parameters systems. This interdisciplinary approach is valuable to students because virtually every newly designed engineering product is a Mechatronic system.

Intended Readership

This book is intended for both mechanical and electronics engineers (researchers and graduate students) who wish to get some training in smart electronic devices embedded in mechanical systems. The book is partly a textbook and partly a monograph. It is a textbook as it provides a focused interdisciplinary experience for undergraduates that encompasses important elements from traditional courses as well as contemporary developments in Mechatronics. It is simultaneously a monograph because it presents several new results and ideas and further developments and explanation of existing algorithms which are brought together and published in the book for the first time. Furthermore, the research results previously scattered in many scientific journals and conference papers worldwide, are methodically collected and presented in the book in a unified form. As a result of its twofold character the book is likely to be of interest to graduate and postgraduate students, engineers and scientists working in the fields of Mechanical engineering, communication, electronics, computer science, optimisation, and neural networks. Furthermore, the book may also be of interest to researchers working in different areas of science, as a number of results and concepts have been included which may be useful for their further research. One can read the book through sequentially but it is not necessary since each chapter is essentially self-contained, with as few cross references as possible. Therefore, browsing is encouraged.

Apart from the technical side, I would like to express my thanks to Prof. Aleksandar Lazinica and Prof. Katarina Lovrecic, of InTech publishing, for their continuous help and support. Last but not least, I thank all the authors who have put in enormous efforts for the publication of this work.

Dr. Ganesh R Naik
RMIT University,
Melbourne,
Australia

ganesh.naik@rmit.edu.au

Part 1

Intelligent Robotics

1

A Mechatronic Perspective on Robotic Arms and End-Effectors

Pinhas Ben-Tzvi and Paul Moubarak
Robotics and Mechatronics Laboratory
Department of Mechanical and Aerospace Engineering
The George Washington University
United States of America

1. Introduction

The robotic industry has constantly strived towards developing robots that harmoniously coexist with humans. Social robots, as they are often dubbed, differ from their industrial counterparts operating in assembly lines by almost all aspects except the adjective "robotic". Social robots are often classified as robots that interact with humans, suggesting that they must possess a human-like morphology in order to fit this designation. A broader definition of the term social robots, however, encompasses any robotic structure coexisting in a society, capable of bringing comfort or assistance to humans. These robots can range from housekeeping wheeled rovers to bipedal robots, prosthetic limbs and bionic devices.
The distinction between industrial robots and social robots stems from the different environments in which they operate. The nature of the interaction with humans and the surroundings in an urban environment imposes a new stream of requirements on social robots, such as mobility, silent actuation, dexterous manipulation and even emotions. Unlike industrial robots where these constraints are alleviated in favor of strength and speed, the development of social robots for an urban environment is associated with more extreme specifications that often relate to engineering challenges and social considerations, including public perception and appeal. The robot will either be accepted by society or rejected due to unattractive or unfamiliar features. Many of these considerations are sometimes ignored by researchers although they are critical to the integration of these robots in the society as an adjunct to human faculty.
In the context of robotic manipulation related to social robots operating in an urban environment, which constitutes the scope of this chapter, the progress achieved in this field in terms of hardware implementation is remarkable. Recent developments feature manipulator arms with seven degrees of freedom and robotic hands with twenty four joints that replicate the dexterity of a human hand. This level of dexterity is appealing to the end-user because it brings familiarity to the general conception of robotic limbs, thus making the technology more acceptable from a social standpoint especially when it comes to bionic integration and prosthetic rehabilitation.
However, the cost of this technology is high due to hardware complexity and size. Other urban applications, such as search-and-rescue or police operations, favor higher payload capabilities of the arm and end-effector over a higher level of manipulation and dexterity.

Choosing between payload capabilities and dexterity is a decision a user has to make when selecting a robotic system. With the current actuators technology, these two parameters seem to be inversely proportional, with systems providing one or the other, but seldom both.

The social perception of a robotic arm or hand is also affected by the level of autonomy it can provide. In general, the complexity of the kinematics associated with these systems makes their real-time control complicated when operated in closed loop with sensor feedback. A sensor network including tactile sensors, slip sensors, proximity sensors, and encoders is often incorporated into the arm and hand structure in order to execute a desired control scheme. Conversely, bionic devices such as prosthetic hands take advantage of electromyographic (EMG) signals generated by the operator's neural system to control the motion of the prosthetic limb. A complete sensor network in this case is often not required as the operator relies on his senses - including vision - to achieve the desired manipulation. The challenge however resides in the development of a robust pattern-recognition method capable of decoding the original signal in order to control the limb functions.

In this chapter, the major contributions made in the field of robotic arms and end-effectors are evaluated and venues for prospective research outlook are identified. Due to the multi-disciplinary nature of this field and the broad range of possible applications, a comprehensive introduction of the topic requires the coverage of all aspects of the technology including sensors, actuators and automation schemes. Thus, by evaluating the state of the technology from a mechatronic perspective, we can synthesize the multi-disciplinary nature of this field in a chapter that brings together an understanding of the current challenges and advocates for subsequent developmental opportunities.

2. Sensing technology

Sensors play a critical role in the development of robotic arms and end-effectors. In the human anatomy, the skin provides sensorial information to the brain via a variety of nerve endings that react to physical stimulations such as changes in temperature and pressure. This sensorial information can be broadly classified into three major categories: proprioception, haptic perception and exteroception. Proprioception provides feedback on the position of body parts, such as the angular position of the arm's elbow and wrist. Haptic perception enables the recognition of objects via the sense of touch, while exteroception allows the perception of changes in physical variables in reaction to external stimuli. In robotic applications, there exists no single sensor with sensing capabilities comparable to the human skin. In most applications, a dedicated sensor must be integrated in the system in order to measure each and every desired variable.

2.1 Proprioception

Proprioception, such as joints position measurements, is often achieved using encoders technology for robot arms and end-effectors. These can be either absolute or incremental and can measure linear position, as well as angular position of the joints. Linear and angular velocity can be extracted from encoders' data by differentiating the position measurements with respect to time. Resistive, capacitive, optical and magnetic encoders have been studied for this purpose with each principle possessing distinctive properties (Tobita et al., 2005). For end-effector applications however, a unique challenge arises with respect to the integration of encoders on the joints. This is due to the tightness of the available space, especially in the fingers. Thus in this case, miniature encoders fabricated using MEMS-CMOS technology are desirable with sensor footprint of less than 5 x 5 mm^2 (Nakano et al., 2005).

2.2 Haptic perception

Haptic perception is achieved using tactile and force sensors. This perception is essential for handling objects, providing feedback on the amount of force or grip applied on the objects. In the most simplistic form, a tactile sensor measures the pressure exhibited by an object on a membrane which deflects proportionally to the applied pressure or force. Many techniques exist to convert the deflection of the membrane into an electrical signal. These are often implemented using piezoelectric or piezoresistive materials such as Zinc Oxide or Lead Zirconate Titanate (PZT). Membrane deflection also affects the capacitance between the substrate and the membrane. Thus, another method of implementing tactile sensors is through capacitance measurement (Castelli, 2002). These transduction principles of operation are illustrated conceptually in Figure 1.

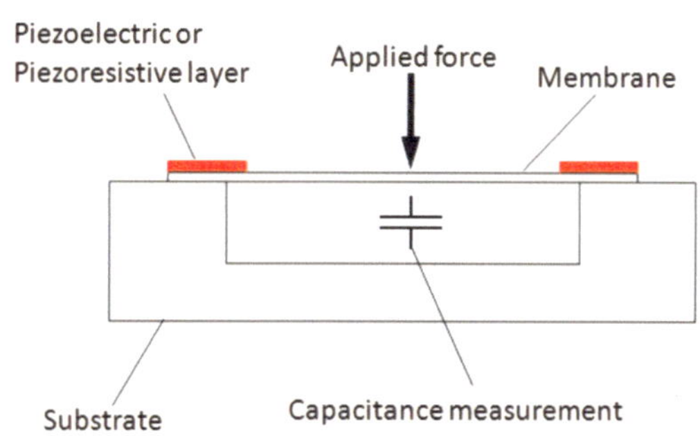

Fig. 1. A conceptual illustration of the operation principle of common tactile sensors

In general, detection of normal loads as well as shear loads is desirable in robotic end-effector applications. Normal load measurements provide information on the griping force exerted on the object, while shear load measurements can detect whether or not the object is slipping during handling maneuvers. Capacitive tactile sensors are most sensitive to normal loads, as their mode of operation requires the deflection of a membrane. Conversely, piezoelectric and piezoresistive materials can be employed to detect normal loads as well as shear loads generated by the surface traction between the object and the sensor face during slippage (Cotton et al., 2007) .

These two components of the applied load can be equally detected using other technologies such as strain gages and optical devices. Load measurements through strain gages integrated in a Wheatstone bridge is a well established procedure, and thus is more cost effective in comparison to piezoelectricity and piezoresistivity (Hwang et al., 2007). Optical measurements on the other hand can provide significant accuracy in the readings (Sato et al., 2010). However this technology requires the implementation of a camera in the structure of the sensor and the incorporation of image processing techniques.

A single tactile sensor is unable to detect the haptic perception of all fingers of a robotic end-effector. In reality, arrays of individual sensors, referred to as tactels, are incorporated together in a distributed structure constituting the tactile sensor. Tactels can be thought of as image pixels, each being sensitive to external loads. Similar to digital imaging, the resolution of a distributed tactile sensor defines the number of tactels on a given surface of the sensor, which consequently dictates the overall sensitivity of the sensor.

2.3 Exteroception

Exteroception on robotic arms and end-effectors is implemented using dedicated sensors. Most commonly, parameters such as temperature and humidity are relevant to robotic applications. These can often be sensed by incorporating appropriate sensors in the structure of the hand, most notably in the fingers. The integration of exteroceptive sensors within the structure of tactile sensors is a common practice gaining more momentum in the field. In some cases, the same physics that govern an exteroceptive parameter also govern a different haptic parameter. For instance, a capacitive sensor with top electrodes in a comb-like structure can detect the proximity of an object to the fingers (exteroceptive), as well as the collision of the object with the fingers (haptic). This is achieved by monitoring the fringe capacitance of two adjacent electrodes as a function of the changes in the dielectric constant influenced by the proximity of the object to the electrodes (Lee et al., 2009). The principle of operation is shown in Figure 2. Other techniques, such as tactile and thermal feedback provided by a single sensor, have also been successfully demonstrated (Yang et al., 2006).

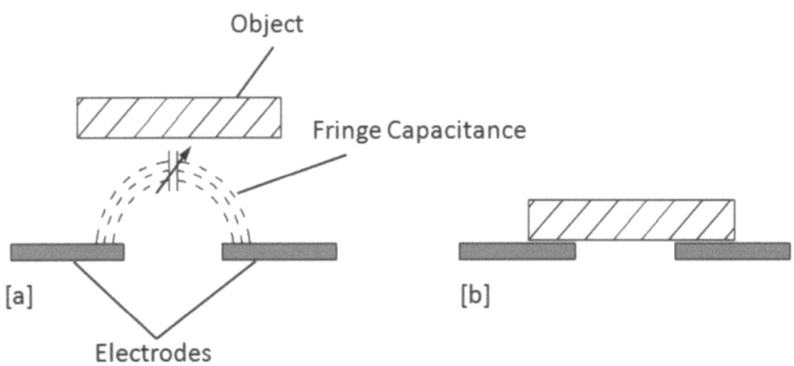

Fig. 2. A dual proximity-tactile sensor for exteroceptive and haptic feedback. [a] Proximity mode. [b] Contact haptic mode

3. Actuation technology

Actuators occupy the largest space in the structure of robotic arms and end-effectors. Although in most cases the same actuation principles that are adopted to actuate a robotic manipulator are also employed to actuate the fingers and joints of an end-effector, the constraints involved in both applications are quite different. Therefore, in order to make the content more meaningful, the two topics are separated and the discussion on the actuation of manipulator arms is carried separately from the discussion on the actuation of end-effectors. For end-effectors, we further distinguish between three categories: highly dexterous end-effectors, self-contained end-effectors and a combination of both. Each of these categories possesses inherent characteristics related to structural complexity and payload capability. Thus, treating their unique aspects separately becomes necessary.

3.1 Actuation of manipulator arms

Electrical motors constitute the most common technology to actuate the joints of manipulator arms. In most cases, the torque generated by the motor is amplified through a gearbox assembly coupled to the motor output shaft. Every motor is capable of actuating one joint at a time. Thus, in manipulator arms with no redundant joints, the number of motors equals to the number of joints. A typical spatial manipulator for a humanoid robot possesses seven independent joints similar to a human arm. These joints provide shoulder, elbow and wrist rotation. In some applications however, the exact replication of the kinematic characteristics of human arms is not desirable. For instance, industrial robotic manipulators often require the incorporation of prismatic joints that allow one link to slide inside the other. On the other hand, mobile robots intended for military applications, such as the one shown in Figure 3, may possess manipulator arms with only two or three actuated joints. A complex manipulator arm on a mobile robot is usually not advantageous due to issues related to ease of use and battery power. Since mobile robots normally operate on limited battery power, reducing the complexity of the arm joints translates into a reduction in power consumption, which ultimately extends the range of operation of the mobile robot (Ben-Tzvi et al., 2008; Ben-Tzvi, 2010; Moubarak et al., 2010).

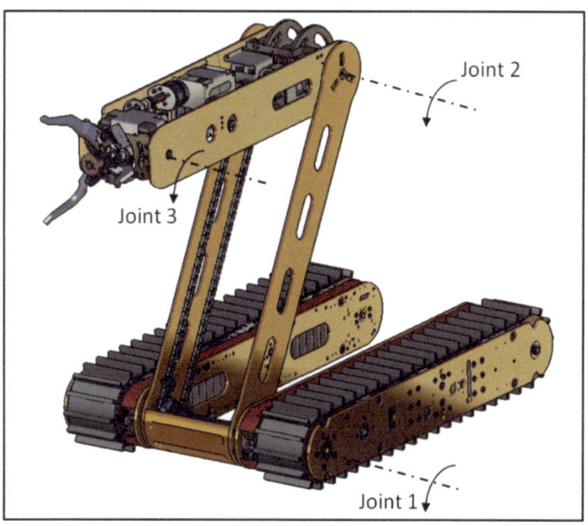

Fig. 3. A mobile military robot with a manipulator arm containing three joints

Hyper-redundant manipulator arms have also been developed using electrical motor technology. A manipulator is dubbed hyper-redundant when it possesses more than the necessary number of actuated degrees of freedom to execute a specific task. These manipulators can provide maneuverability levels analogous to elephant trunks, and are ideal for operations inside tight and narrow environments, such as inside the rubbles of a collapsed building in the aftermath of an earthquake (Chirikjian, 2001). In general, building hyper-redundant manipulator arms using electrical motors results in a discrete non-continuous articulated structure. A more compliant and continuous design shown in Figure 4 can be developed using flexible composite materials such as the Nickel-Titanium alloy (NiTi). NiTi alloys are generally used in the development of shape memory alloys (SMA) and exhibit prehensile characteristics. Thus, by running actuated tendons inside a hollow cylinder of NiTi alloy, it is possible to create a hyper-redundant continuum manipulator with adjustable flexibility dictated by the tension of the tendons (Camarillo et al., 2008).

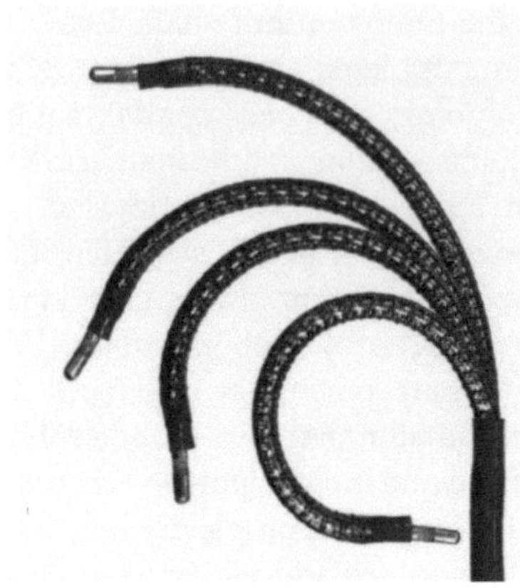

Fig. 4. A continuum manipulator with tendon actuation

The combination of tendons or cable-drive technology and electrical motor power enables the development of manipulators exhibiting a more natural motion of the joints analogous to the human arm. Normally, cable-driven arms consist of three serially connected links with a 3-DOF shoulder joint, a 1-DOF elbow joint and a 3-DOF wrist joint. All joints are driven by cables actuated by electrical motors. Unlike joint actuation achieved by electrical motors, which requires direct coupling to the joint, cable-drive allows the relocation of the motors to the base of the arm and the transmission of the motor power to the joint via cables and pulleys. The position of the motors at the base of the arm reduces the overall weight of the links, which offers the advantage of increasing the overall payload capabilities of the arm (Mustafa et al., 2008; Ben-Tzvi et al., 2008). A commercial product of this technology known as the WAM™ arm has already been developed.

3.2 Actuation of robotic end-effectors

Table 1 classifies a family of selected robotic end-effectors into three major categories:

a. Highly dexterous end-effectors
b. Self-contained end-effectors
c. Combination of both aspects 1 and 2

The first category relates to end-effectors capable of providing dexterity levels comparable to the human hand without constraining the size and weight of the eventual structure. These hands often include four actuated fingers and a thumb and are capable of providing integrated wrist motion. The second category relates to end-effectors that contain all hardware necessary to operate the joints within the hand's structure. Normally, these end-effectors compromise the dexterity for the self-containment aspect of the structure. The third category combines the benefits of both dexterity and self-containment.

3.2.1 Actuation of highly dexterous end-effectors

Dexterous robotic anthropomorphic hands are mechanical end-effectors that possess a structural compliance comparable to the human hand. The structure of these hands includes four fingers and an opposable thumb mounted on a carpal frame or palm, with each of the

Category	Robot Hand	Structural Features	Actuation Mechanism	Joints/DOF
Highly Dexterous	UB Hand III	4 fingers 1 thumb	Brushed motors with pulling tendons	20/16
	Shadow Hand	4 fingers 1 thumb Wrist motion	Air Muscles with pulling tendons	24/20
	Robonaut Hand	4 fingers 1 thumb Wrist motion	Electrical Brushless motors with flex shafts	22/14
	DIST Hand	3 fingers 1 thumb	Electrical Brushless motors with pulling tendons	16/16
Self-Contained	Barrett Hand	3 fingers Fingers spread motion	Electrical Servo-motors	8/4
	GWU Hand-I	3 fingers Wrist motion	Electrical Brushless motors with worm gears	5/3
Highly Dexterous And Self-Contained	DLR Hand II	3 fingers 1 thumb Curling Palm	Electrical Brushless motors with tooth-belt gear	17/13
	GIFU Hand-II	4 fingers 1 thumb	Electrical Servo-motors	20/16
	Ultralight Hand	4 fingers 1 thumb Wrist motion	Flexible Fluidic Actuators	18/13

Table 1. Comparison of structural characteristics for selected robotic end-effectors

fingers representing a serial linkage mechanism connected by four joints. The advantage of these dexterous hands resides in the ability of the fingers to grasp objects of different shapes and sizes. This enables the restoration of fine motor skills in prosthetic hands, or the accomplishment of delicate tasks requiring a high level of precision, such as remote tele-operated surgery. For this family of anthropomorphic robotic hands, geometrical constraints, such as the overall size of the hand, are often traded for the dexterity level of the fingers. This results in a complicated mechanical structure where the overall weight and size are not taken into account during the design stages.

In most applications however, the top joint of the fingers, which connects the distal phalange to the intermediate phalange, is coupled mechanically to the third joint. This technique moderately simplifies the structural complexity by reducing the number of degrees of freedom for each finger by one. Consequently, the number of joints in the hand will exceed the number of degrees of freedom, making the independent actuation of the coupled joints impossible. The UB-II hand shown in Figure 5 is an anthropomorphic robotic end-effector with a total of 16 degrees of freedom (20 joints), where some of the joints have been mechanically coupled to others in order to reduce the complexity of the overall mechanism. The unique feature of the UB-II hand resides in the tendon-actuation of the joints enabled by servo-motors located in the forearm. The continuum compliance of the fingers in this case is achieved using helical springs mounted inside the shell of each finger. These springs enable the phalanx of the fingers to bend in a continuous fashion, while concurrently restoring the shape of the fingers to the original non-flexed configuration when the tension in the tendons is eliminated (Lotti et al., 2005).

Fig. 5. The UB-II anthropomorphic hand with 16-DOF

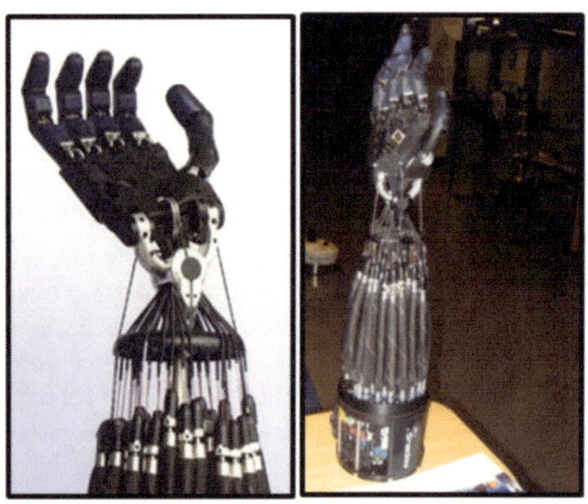

Fig. 6. The Shadow hand actuated with air muscles

The wrist joints of the UB hand are not integrated within the structure of the hand; rather, the wrist motion is achieved independently by the manipulator arm carrying the hand on the end-link. Shadow hand kinematics differs from the UB-hand kinematics by the addition of four joints and four degrees of freedom (24 joints and 20 DOF's). One actuated joint is appended to the thumb while the other is added to the little finger, both located inside the metacarpal frame. These two actuated joints allow the palm to curl inwards in a fashion similar to the human carpus. The remaining two joints are appended to the wrist and provide the flexion/extension and adduction/abduction motions of the wrist.

Actuation of the 20 degrees of freedom of the Shadow hand is achieved by pneumatic air muscles mounted on the forearm as shown in Figure 6. Emulating the biological characteristics of a human hand, the actuation of the air muscles is coupled to the joints via tendons routed through the carpus and metacarpus. The volume of every muscle is controlled by the inside air pressure. Each tendon is connected to a pair of antagonistic air muscles, which pull the tendon in one direction or the other in order to achieve clockwise or counter-clockwise rotation of the corresponding joint. The actuation of the 20 degrees of freedom of the Shadow hand therefore requires a total of 40 actuators or air muscles.

3.2.2 Actuation of self-contained end-effectors

The objective of this technology is to develop universal robotic end-effectors that can be mounted on a variety of manipulator arms without significantly modifying the structure of the arm. In the design of highly dexterous anthropomorphic hands, the forearm is used to house most of the hardware, such as servo-motors and pneumatic actuators. Conversely, in the self-contained design, the hardware required to operate the hand is located inside the hand structure itself in an attempt to reduce the overall size of the end-effector. However, the space available inside the palm is relatively small. This leads to a trade-off between the size of the actuators and the number of actuators that can be housed inside the carpal frame. The size of the actuators dictates the payload capability of the end-effector, while the number of actuators determines the level of dexterity the end-effector can exhibit.

For the self-contained end-effectors category, the payload capability is favored over the level of dexterity. This characteristic is desirable for applications that require the manipulation of heavy objects with minimal level of dexterity, such as field robotic and military operations. The Barrett Hand shown in Figure 7 provides a payload capability of 6kg and a total of three fingers, with each finger containing two links connected together by a servo-actuated joint. Fingers F1 and F2 (Figure 7) each contain an extra joint that allows them to rotate peripherally around the wrist to reconfigure the spread angle with respect to the stationary finger F3. With no wrist motion, the Barrett hand contains a total of 8 joints with only 4 degrees of freedom.

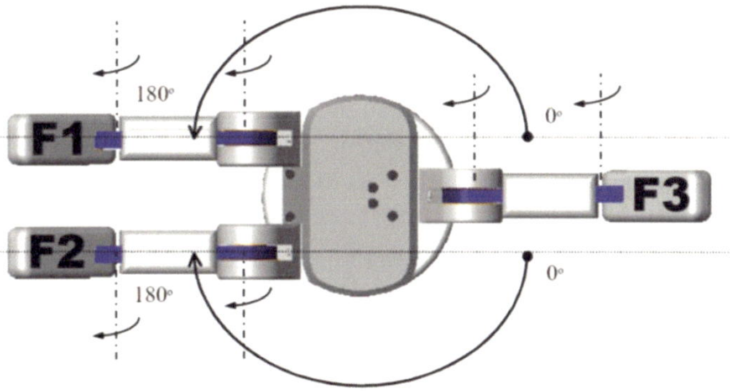

Fig. 7. The Barrett Hand with three fingers showing peripheral spread motion

Another self-contained robotic hand designed at the George Washington University (GWU) provides a payload capacity of 50kg and an integrated 2-DOF wrist motion, allowing flexion/extension, opening/closing of the fingers, and pronation/supination maneuvers as shown in Figure 8. In order to maintain the compact size of the overall structure, the high dexterity level of the fingers is traded for a high payload capability. Thus, each finger contains one joint actuated via a central worm and a brushless motor located inside the wrist. The fingers spread 110° from the closed configuration. The 2-DOF motion of the wrist on the other hand is driven by two separate motors integrated inside the structure. Wireless data communication between the finger sensors and the end-effector processor, as well as between the end-effector processor and the robot processor, allows the accomplishment of endless rotation around the wrist joints. Similar to the Barrett hand, the GWU-Hand-I integrates all hardware inside the end-effector structure including motor drivers, battery power and RF-modules. (Moubarak et al., 2010).

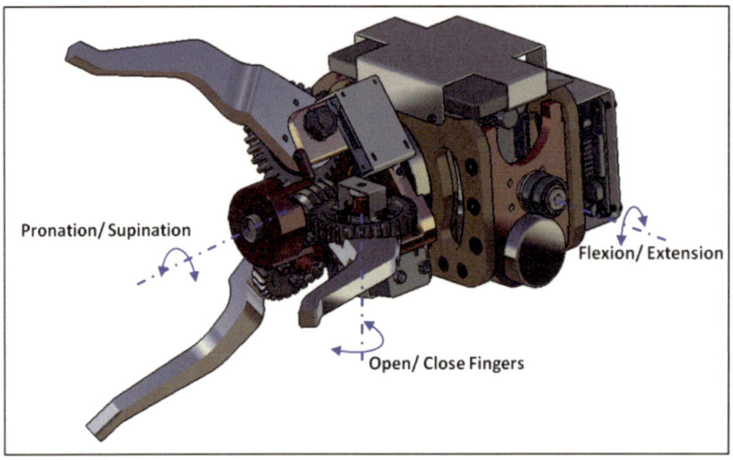

Fig. 8. GWU-Hand-I with integrated 2-DOF wrist motion and payload capability of 50 Kg

3.2.3 Actuation of highly-dexterous and self-contained end-effectors

The discussion introduced in the previous two sections identifies three major structural characteristics attributed to robotic hands: dexterity, size and payload. An ideal robotic hand encompasses all three aspects in a single structure, thereby providing a high level of dexterity within a small and self-contained structure that can handle large payloads. In reality, the size of the actuators employed to develop robotic hands prevents the accomplishment of this maximum performance objective. Few robotic hands however manage to combine the high level of dexterity in a self-contained and small structure, at the expense of lowering the payload capabilities of the fingers.

The DLR hand shown in Figure 9 is an example of a highly dexterous and self-contained robotic hand. The hand is not anthropomorphic as it includes four fingers instead of five. Each finger contains four joints and three degrees of freedom. The thumb is identical to the remaining three fingers and therefore possesses similar kinematics. The unique feature of the DLR hand resides in the palm structure, where the metacarpal frame is divided into two sections connected together by an articulated joint. This improves the compliant curling aspect of the palm and achieves optimal grasping performance of objects of random shapes. The thirteen articulated joints are actuated by brushless motors integrated inside the fingers and the palm frame, and powered using an external battery source. Due to the miniature size of the actuators, the payload capability of the DLR hand is reduced to 3kg with an overall weight of 1.8kg (excluding the weight of the external batteries) (Borst et al., 2003).

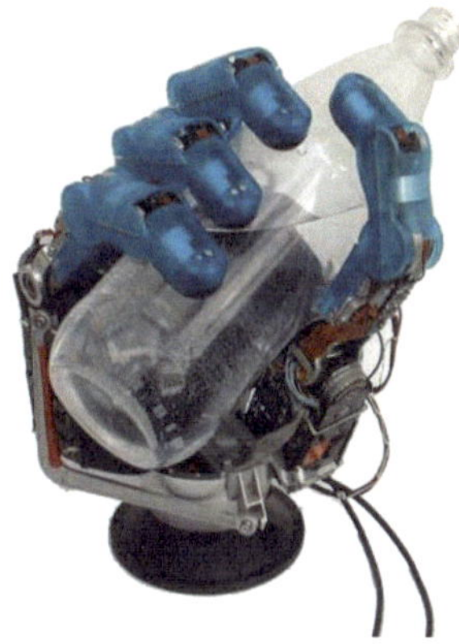

Fig. 9. The DLR hand with four dexterous fingers and a self-contained structure

4. Control methods

The interaction of a robotic arm and end-effector with the surrounding environment requires a high level of autonomy in operation. In general, the complicated kinematic nature of the manipulators and end-effectors makes tele-operation difficult to execute. This is merely due to the high number of articulated joints a robotic arm or hand contains, which makes it difficult to simultaneously actuate the degrees of freedom in order to accomplish a desired task. In most real-time applications, a robotic arm or hand is expected to possess a desirable level of autonomy that minimizes the amount of supervisory intervention from the operator. This, not only facilitates the process of human-machine interaction, but also ensures a consistent and robust operation for optimal performance.

The topic of autonomous manipulation in the broadest sense can be treated from two different perspectives. The first perspective relates to manipulator arms that operate inside a static environment where the tasks executed by the arm, or the dynamics of the assignments themselves are seldom modified. For example, a robotic arm on a static platform, loading microbial samples into a petri dish inside a laboratory environment, would always expect the target object to be located at stationary coordinates with respect to an inertial frame. This kind of operations can be preprogrammed in an exhaustive scheme and executed with extreme confidence provided no perturbations occur in the objective operation. The second perspective relates to robotic arms and end-effectors interacting with a dynamic environment where the trajectory of the arm depends on the target coordinates. In this case, an algorithm rather than a preprogrammed routine generates a control law - based on real-time sensor input - capable of defining an optimal trajectory that produces the desirable outcome.

In either case, the development of the control scheme requires a mathematical representation of the dynamics or the kinematics (or both) of the robotic arm and end-effector. If a dynamic model is available, the objective is to determine a control history that moves the arm along a trajectory from an initial point to a final point. For a kinematic model, the objective is to generate the trajectory that moves the articulated joints from an initial point to a final point in an optimal configuration, while minimizing a cost function subject to constraints. The discussion on robotic control methods can be reasonably lengthy given the significant amount of details in the literature. As such, a broad and abstract discussion on the topic is introduced and summarized under the following three major disciplines:

a. Kinematic Control
b. Dynamic Control
c. Supervisory Control

4.1 Kinematic control

In the kinematic analysis and control of autonomous robotic manipulation tasks, the first objective is to derive a global transformation matrix that maps the local kinematics (such as position or speed) of a point on a specific link into a global inertial frame 0. Every joint i of the manipulator and hand is assigned a local frame $i-1$. Thus, for a robot with n-links, there exist n-frames, where frame n is generally assigned to the end-link or the fingertip in the existence of an end-effector. A global transformation matrix ${}^{0}_{n}T$ that maps the tip coordinates to the global frame 0 can be established using the Denavit-Hartenberg parameters as follows:

$$ {}^{0}_{n}T = \prod_{i=0}^{n-1} {}^{i}_{i+1}T = {}^{0}_{1}T\,{}^{1}_{2}T\ldots\ldots\ldots\,{}^{n-1}_{n}T \tag{1} $$

where ${}^{i}_{i+1}T \in R^{4*4}$, representing the local transformation matrix between joint $i+1$ and joint i, is defined as

$$ {}^{i}_{i+1}T = \begin{bmatrix} {}^{i}_{i+1}R & {}^{i}_{i+1}d \\ 0 & 1 \end{bmatrix} \tag{2} $$

where ${}^{i}_{i+1}R$ expresses the orientation of frame $i+1$ relative to frame i and ${}^{i}_{i+1}d$ expresses the position of the origin of frame $i+1$ with respect to frame i.

Any kinematic property of any link of the arm and hand expressed in the local coordinates can be mapped into any other local or global frame using a variation of equation (1). For instance, for autonomous manipulation applications, the objective is to define the states x of the hand's fingers (frame n) with respect to the global frame of the manipulator according to

$$ X = {}^{0}_{n}T * x \tag{3} $$

where $X \in R^{n*1}$ defines the states of the fingers with respect to the global frame 0. The resulting kinematic equations for a robotic arm with more than 2 links are highly non-linear, which complicates the closed form analytical solution of the most common inverse kinematics problem. In the case of autonomous manipulation such as the pick-and-place operations, the states of the target object are known. These are either provided by the operator, or synthesized from real-time measurements performed by integrated sensors. The objective therefore is to solve the inverse of the problem stated by equation (3) to generate an optimal joint-configuration of the arm and hand in order to accomplish the desired task.

Optimality in robotic autonomous manipulation can only be derived in the existence of a cost function. Therefore, the purpose of the inverse kinematics problem is to minimize the cost function subject to the kinematics established in equation (3) or the dynamics or both. A variety of numerical algorithms have been investigated in the literature to solve the inverse kinematics problem, some of the most popular are the Newton descent and the Newton-

Raphson algorithms (Agirrebeitia et al., 2002). In most cases however, the solution depends on the dimensions and singularity properties of the Jacobian matrix J, which dictates the existence of J^{-1} and therefore the kinematic properties of the arm and hand. More versatile methods dealing with redundant manipulation are proposed to solve the inverse kinematics problem for robotic structures with more than 6 degrees of freedom (Klein & Huang, 1983; Seraji et al., 1993; Tarokh & Kim, 2007).

4.2 Dynamic control

A general model that represents the forward dynamic behavior of a robotic arm and hand with n-links can be illustrated in a conservative form in terms of the generalized coordinates q(t) as follows :

$$M(q,t)\ddot{q} + F(\dot{q},t)\dot{q} + V(q,\dot{q},t)\dot{q} + G(q,t) = \tau(t) \tag{4}$$

where $q(t)$, $\dot{q}(t)$ and $\ddot{q}(t) \in R^{n*1}$ represent the links position, velocity and acceleration of the arm and hand, respectively. $M(q,t) \in R^{n*n}$ represents the mass or inertia matrix, $F(\dot{q},t) \in R^{n*n}$ represents the dissipative terms such as Coulomb damping or friction, $V(q, \dot{q}, t) \in R^{n*n}$ represents the Coriolis matrix, and $G(q,t) \in R^{n*1}$ represents the non-dissipative components such as gravity. $\tau(t) \in R^{n*1}$ represents the torque input vector.
Mapping between the generalized coordinates q(t) (and their derivatives) and the work-space coordinates x(t) (and their derivatives) can be performed using the Jacobian matrix $J(q) \in R^{n*n}$ where:

$$\dot{x}(t) = J(q)\dot{q} \tag{5}$$

The objective of a dynamic control scheme is therefore to calculate a time history of the control law $\tau(t)$ that allows the links of the arm and hand to either follow a desired trajectory, or maintain a desired position (or speed) by overcoming the resistance from the environment. This control scheme includes methods that correlate the input vector to the position of the links, known as position control, or methods that correlate the input vector to the velocity of the links, known as speed control. In both cases, the time-dependant feedback of the work-space variables such as position or velocity needs to be integrated in the control loop in order to ensure the stability of the scheme.
In the case where the time history of the torque generated by the actuators is known, the objective of the control scheme is to derive a solution to the dynamic model (equation (4)) that defines the position $q(t)$ or the speed $\dot{q}(t)$ of the links in the generalized coordinate system in terms of the input vector $\tau(t)$. The solution can be mapped back into the work-space coordinates using the Jacobian matrix if an exact model is available or an approximate estimation of the Jacobian if the established model contains uncertainties (Cheah et al., 2003). In most cases however, the desired position history of the links or the desired velocity history is specified. In theory, direct substitution in equation (4) would generate the desired control law $\tau(t)$. However, the non-linear aspect of the model and the inherent uncertainties complicate the analytical solution. A method, known as inverse dynamics (Khalil & Guegan, 2004), exists in which the linearization of the model is possible. This involves seeking a control law

$$\tau(t) = f(q, \dot{q}, t) \tag{6}$$

expressed in terms of the generalized coordinates q(t), and substituting back into equation (4) to generate a Newton-Euler linear closed-loop representation of the non-linear and coupled model (Khalil et al., 2007). A critical requirement for the ideal operation of the inverse dynamics approach is an exact model of the arm and the hand. In reality, an uncertainty-free model in practical applications is seldom available.

4.3 Supervisory control

Supervisory control is the process of controlling robotic arms and hands in a closed-loop master-slave scheme where the human operator is the master initiating the orders, and the robot is the slave acting or reacting to these orders. State-of-the-art control methods presented in the previous two sections are generally task-driven or environment-driven in the sense that only very specific and tailored tasks can be performed autonomously - with confidence - by the robotic arm or hand. In reality however, the tasks assigned to robotic manipulators are so complicated that traditional dynamic tools fall short from being able to model their aspects accurately. To cope with this problem, the common practice is to place a human operator in the loop to supervise the process.

Supervisory control schemes are most desirable for applications requiring a high level of autonomy for anthropomorphic arms and hands with complex kinematic structures. Research in this field aims at minimizing the input required from the operator in order to strengthen the human-machine interaction. Most commonly, data gloves such as the one shown in Figure 10, are employed to control the joints of robotic hands. In the same fashion, similar sensors can be placed on the arm of the operator in order to control the joint motion of robotic manipulators.

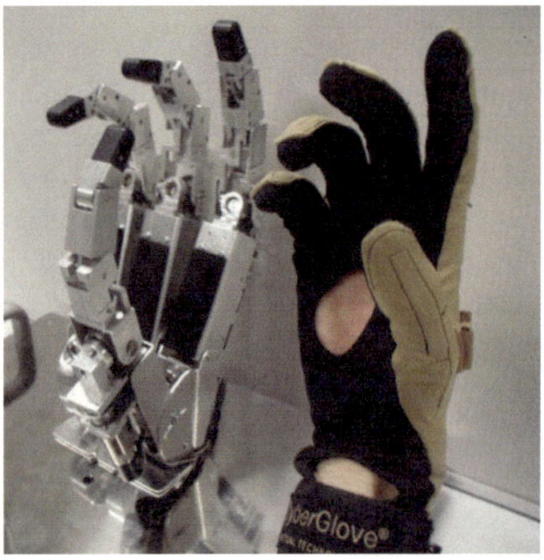

Fig. 10. A data glove controlling the joints motion of a robotic hand

Data gloves convert the motion of the operator's fingers into electrical signals. These are decoded and interpreted by a computer interface that allows the robotic hand to mimic the operator's gestures. Flex sensors (such as strain gages) mounted inside the gloves generate an electrical signal proportional to the bending amount of each phalange. A computer interface incorporated in the loop, converts these signals into angular measurements which are then communicated to the robotic hand to mimic the gestures. Other, more advanced data gloves employ acoustic, resistive or magnetic induction sensors to track the motion of the phalanx (Fahn & Sun, 2005).

Supervisory control is also pertinent to biomechatronic applications such as prosthetic limbs and hands. In this case however, more sophisticated algorithms are required to bridge the communication between the operator's mind and the prosthetic limb. Instead of data gloves, electrodes embedded in the operator's residual muscles are employed to measure the electromyographic (EMG) signals generated by the brain activity. These signals are recognized and interpreted using pattern-matching algorithms, and subsequent commands are initiated to the corresponding actuators in the prosthetic limb to perform the motion and restore the original biological functionality (Carrozza et al., 2002).

5. Challenges and opportunities

Despite the progress accomplished in the field of robotic arms and hands as outlined in the previous discussion, the objective of realizing man-like robotic structures with comparable dexterity, robustness and intelligence is far from being achieved. The problem in itself is significantly complicated owing its difficulty to the following aspects:

- With respect to sensing capabilities; the human skin anatomy possesses a near flat exo-layer that contains an infinite number of nerve endings, each powered individually and each providing more than one sensorial measurement to the brain. In comparison, the sensors employed in the robotic industry are dedicated measurement units that are in most cases sensitive to only one parameter. Moreover, the integration of such sensors in a skin-like morphology results in a discrete amalgamation of units that does not cover the whole surface, rather is restricted to some critical areas of interest on the robotic arm or hand.
- With respect to actuation capabilities; human muscles possess a very high fiber density that enables them to deliver a large amount of instantaneous power within a compact and linear morphology. In comparison, electrical motors possess a low power-to-weight ratio and often require additional inefficient amplification stages to deliver a large torque. Linear pneumatic actuators on the other hand, attempt to replicate the muscle's biological functionality; however, they lack the comparable compactness and require extra space to house the additional hardware.
- With respect to autonomy; the human upper limbs are capable of achieving a large number of highly dexterous tasks with extreme ease and extreme confidence. In comparison, the autonomy implemented on robotic arms and hands is task-driven and non-adaptive, where every specific task is modeled individually, and every task requires a dedicated mathematical representation in order to generate an optimal performance.

These challenges are well-known and understood in the research community, and opportunities to address their aspects are constantly considered. Through the development of novel materials and novel mathematical tools, the identified challenges can be addressed gradually, and new generations of sensors, actuators and control methods can be developed. For instance, novel materials such as nanowires, promise the synthesis of highly sensitive artificial skin that can be adapted to prosthetic arms in order to restore biological senses. (Takahashi et al., 2010). Polymers, flexinol and flexible magnetic actuators (FMA) also represent a future opportunity to advance the technology, and develop compact linear actuators with high power-to-weight ratios (Kim et al., 2010). Equal opportunities present themselves in the use of statistical and machine learning methods to promote adaptive robotic intelligence for robust manipulation. The subsequent integration of these new concepts is a promising scheme to bridge the gap between the limitations of robotic arms and hands, and the skillfulness of the human counterparts.

6. Conclusions

This book chapter introduced a comprehensive mechatronic perspective on the current stance of the technology related to robotic arms and end-effectors. Due to the multi-disciplinary nature of the topic, we presented our findings and the state-of-the-art contribution under three major categories: sensors, actuators and control methods.

In the context of sensor technology, proprioceptive, haptic, and exteroceptive sensors were discussed along with the physics adopted in each case to develop the sensing capabilities.

In the context of actuator technology, we distinguished between the actuation of robotic arms - often accomplished via motor actuation or cable-drive - and the actuation of robotic end-effectors. The latter encompassed three aspects of integration: high dexterity, self-containment and a combination of both. For all three categories, the discussion introduced the different techniques employed to actuate the joints of the wrist and fingers. In most cases, direct motor actuation or tendon-driven motor actuation is employed to articulate the joints. Some other techniques that use linear pneumatic air muscles are equally considered.

In the context of control methods and autonomy, dynamic control, kinematic control and supervisory control methods were introduced. For dynamic and kinematic control, a generic discussion on the topic was presented along with the most relevant numeric schemes employed to address the non-linear aspect of the governing equations, and their subsequent solutions. For the supervisory control, two examples of human-machine interaction were introduced, one accomplished through data gloves and the other through interpretation of EMG signals in prosthetic hands.

Future opportunities in the field lie in the development of novel material technology and novel mathematical tools that address the challenges associated with the current practice. Novel materials enable the development of sensor arrays that match the human skin in the anatomy and the versatility in function. Novel materials equally enable the development of compact actuators with high power-to-weight ratios. Mathematical tools on the other hand allow the integration of machine learning techniques and provide robotic arms and hands with adaptability levels comparable to human limbs.

This being said, the contribution of the technical content in this chapter lies in the synthesis of the multi-disciplinary nature of the field in a document that brings a comprehensive understanding of the current technology, identifies pertinent challenges and advocates for subsequent developmental opportunities.

7. References

Agirrebeitia, J., Aviles, R., de Bustos, I.F. & Ajuria, G. (2002). A method for the study of position in highly redundant multibody systems in environments with obstacles. *IEEE Transactions on Robotics*, Vol. 18, No. 2, (April 2002), pp. (257 - 262), ISSN: 1042-296X

Ben-Tzvi, P., Goldenberg, A.A. & Zu, J.W. (2008). Design and Analysis of a Hybrid Mobile Robot Mechanism with Compounded Locomotion and Manipulation Capability. *Journal of Mechanical Design*, Vol. 130, No. 7, (July 2008), pp. (1 - 13), ISSN: 1050 - 0472

Ben-Tzvi, P. (2010). Experimental Validation and Field Performance Metrics of a Hybrid Mobile Robot Mechanism. *Journal of Field Robotics*, Vol. 27, No. 3, (May 2010), pp. (250 - 267), ISSN: 1556 - 4959

Borst, C., Fischer, M., Haidacher, S., Liu, H. & Hirzinger, G. (2003). DLR Hand II: Experiments and Experiences with an Anthropomophic Hand. *Proceedings of the 2003 IEEE International Conference on Robotics and Automation, ICRA '03*, ISSN: 1050 - 4729, Taipei Taiwan, November 2003

Camarillo, D. B.,Milne, C. F., Carlson, C. R., Zin, M. R. & Salisbury, J. K. (2008). Mechanics Modeling of Tendon-Driven Continuum Manipulators. *IEEE Transactions on Robotics*, Vol. 24, No. 6, (December 2008), pp. (1262 - 1273), ISSN: 1552 - 3098

Carrozza, M.C., Micera, M.S., Zecca, L.M. & Dario, P. (2002). The Development of a Novel Prosthetic Hand - Ongoing Research and Preliminary Results. *IEEE/ASME Transactions on Mechatronics*, Vol. 7, No. 2, (June 2002), pp. (108 - 114), ISSN: 1083-4435

Castelli, F. (2002). An Integrated Tactile-Thermal Robot Sensor With Capacitive Tactile Array. *IEEE Transactions on Industry Applications*, Vol. 38, No. 1, (February 2002), pp. (85 - 90), ISSN: 0093-9994

Cheah, C.C., Hirano, M., Kawamura, S. & Arimoto, S. (2003). Approximate Jacobian Control for Robots with Uncertain Kinematics and Dynamics. *IEEE Transacations on Robotics and Automation*, Vol. 19, No. 4, (Augsut 2003), pp. (692 - 702), ISSN: 1042-296X

Chirikjian, G.S. (2001). Design and Analysis of Some Nonanthropomorphic, Biologically Inspired Robots: An Overview. *Journal of Robotic Systems*, Vol. 18, No. 12, (December 2001), pp. (710 - 713)

Cotton, D. P. J., Chappell, P.H., Cranny, A., White, N.M. & Beeby, S.P. (2007). A Novel Thick-Film Piezoelectric Slip Sensor for a Prosthetic Hand. *IEEE Sensors Journal*, Vol. 7, No. 5, (May 2007), pp. (752 - 761), ISSN: 1530-437X

Fahn, C-S. & Sun, H. (2005). Development of a Data Glove With Reducing Sensors Based on Magnetic Induction. *IEEE Transactions on Industrial Electronics*, Vol. 52, No. 2, (April 2005), pp. (585 - 594), ISSN: 0278-0046

Hwang, E-S., Seo, J-H. & Kim, Y-J. (2007). A Polymer-Based Flexible Tactile Sensor for Both Normal and Shear Load Detections and Its Application for Robotics. *Journal of Microelectromechanical Systems*, Vol. 16, No. 3, (June 2007), pp. (556 - 563), ISSN: 1057 - 7157

Khalil, W. & Guegan, S. (2004). Inverse and Direct Dynamic Modeling of Gough-Stewart Robots. *IEEE Transactions on Robotics*, Vol. 20, No. 4, (August 2004), pp. (754 - 761), ISSN: 1552-3098

Khalil, W., Gallot, G. & Boyer, F. (2007). Dynamic Modeling and Simulation of a 3-D Serial Eel-Like Robot. *IEEE Transactions on Systems, Man and Cybernetics*, Vol. 37, No. 6, (November 2007), pp. (1259 - 1268), ISSN : 1094-6977

Kim, S. H., Hashi, S. & Ishiyama, K. (2010). Methodology of Dynamic Actuation for Flexible Magnetic Actuator and Biomimetic Robotics Application. *IEEE Transactions on Magnetics*, Vol. 46, No. 6, (June 2010) pp. (1366 - 1369), ISSN: 0018-9464

Klein, C.A. & Huang, C.H. (1983). Review of pseudo-inverse control for use with kinematically redundant manipulators. IEEE *Transactions on Systems, Man and Cybernetics*, Vol. 13, No. 3, (March 1983), pp. (245 - 250)

Lee, H-K., Chang, S-I. & Yoon, E. (2009). Dual-Mode Capacitive Proximity Sensor for Robot Application: Implementation of Tactile and Proximity Sensing Capability on a Single Polymer Platform Using Shared Electrodes. *IEEE Sensors Journal*, Vol. 9, No. 12, (December 2009), pp. (1748 - 1755), ISSN: 1530 - 437X

Lotti, F., Tiezzi, P., Vassura, G., Biagiotti, L., Palli, G. & Melchiorri, C. (2005). Development of UB Hand 3: Early Results. *Proceedings of the IEEE International Conference on Robotics and Automation, ICRA '05*, ISBN: 0-7803-8914-X, Spain, April 2005

Moubarak, P., Ben-Tzvi, P. & Ma, Z. (2010). A Generic Configuration of a Compact Dexterous and Self-Contained End-Effector for Mobile Robotic Platforms. *Proceedings of the IEEE International Workshop on Robotic and Sensors Environments (ROSE 2010)*, pp. (116 - 121), Phoenix, Arizona, October 2010

Moubarak, P., Ben-Tzvi, P., Ma, Z., Sutherland, K.M. & Dumas, M. (2010). A Mobile Robotic Platform for Autonomous Navigation and Dexterous Manipulation in Unstructured environments. *Proceedings of the ASME International Mechanical Engineering Congress and Exposition (IMECE 2010)*, Vancouver Canada, November 2010

Mustafa, S. K., Yang, G., Yeo, S. H., Lin, W. & I-M, C. (2008). Self-Calibration of a Biologically Inspired 7 DOF Cable-Driven Robotic Arm. *IEEE/ASME Transactions on Mechatronics*, Vol. 13, No. 1, (February 2008), pp. (66 - 75), ISSN: 1083 - 4435

Nakano, K., Takahashi, T. & Kawahito, S. (2005). A CMOS Rotary Encoder Using Magnetic Sensor Arrays. *IEEE Sensors Journal*, Vol. 5, No. 5, (October 2005), pp. (889 - 894), ISSN: 1530-437X

Sato, K., Kamiyama, K. & Tachi, S. (2010). Finger-Shaped GelForce: Sensor for Measuring Surface Traction Fields for Robotic Hand. *IEEE Transactions on Haptics*, Vol. 3, No. 1, (January - March 2010), pp. (37 - 47), ISSN: 1939 - 1412

Seraji, H., Ling, M.K. & Lee, T.S. (1993). Motion control of 7-DOF arms: The configuration control approach. *IEEE Transactions on Robotic Automation*, Vol. 9, No. 2, (April 1993), pp. (125 - 139), ISSN: 1042-296X

Takahashi, T., Takei, K., Adabi, E., Fan, Z., Niknejad, A. M. & Javey, A. (2010). Parallel Array InAs Nanowire Transistors for Mechanically Bendable, Ultra High Frequency Electronics, *Journal of American Chemical Society ACS Nano*, in press

Tarokh, M. & Kim, M. (2007). Inverse Kinematics of 7-DOF Robots and Limbs by Decomposition and Approximation. *IEEE Transactions on Robotics*, Vol. 23, No. 3, (June 2007), pp. (595 - 600), ISSN: 1552 - 3098

Tobita, K., Ohira, T., Kajitani, M., Kanamori, M.S. & Ming, A. (2005). A Rotary Encoder Based on Magneto-Optical Storage. *IEEE/ASME Transactions on Mechatronics*, Vol. 10, No. 1, (February 2005), pp. (87 - 97), ISSN: 1083-4435

Yang, G-H., Kyung, K-U., Srinivasan, M.A. & Kwon, D-S. (2006). Quantitative tactile display device with pin-array type tactile feedback and thermal feedback. *Proceedings of the IEEE International Conference on Robotics and Automation, ICRA '06*, ISSN: 1050-4729, Orlando FL., May 2006

2

A Torque Cancelling System for Quick-Motion Robots

Daigoro Isobe
University of Tsukuba
Japan

1. Introduction

A new torque cancelling system that stabilizes mechanical sway in quick-motion robots is discussed in this chapter. Mechanical sway generated by reaction moments occurring in the motion of, for example, space stations is an important issue that cannot be neglected. Motion controls using the conservative law of angular momentum are applied by driving control momentum gyros (CMG) (Ookami et al., 2008) or reaction wheels (Yoshida et al., 2002). Applications of the gyros and the wheels are generally limited to slow momentum transitions relying upon high-gain feedback control. Although the same strategy can be used in the motion control of robots when the momentum transitions are slower, the dynamics of the robots should be considered to effectively control quicker motions. The torque cancelling system (TCS) discussed in this chapter cancels the reaction moment generated by the motion of an object by considering the precise dynamics of the object and the body of the robot itself. The reaction moment can be obtained accurately using a newly proposed solution scheme of inverse dynamics that handles the dynamics of complex robotic architectures by modelling them with finite elements. Once the reaction moment is known, it can be cancelled by applying an anti-torque to a torque-generating device.

The newly proposed solution scheme of inverse dynamics mentioned above is called the parallel solution scheme (Isobe et al., 2000; Isobe, 2004), and it was developed on the basis of a finite element approach. In contrast to conventional and revised schemes that use dynamic equations (for example: Nakamura & Ghodoussi, 1989; Nakamura, & Yamane 2000; Sugimoto, 2003), the proposed scheme can handle different types of configurations, such as open-loop, closed-loop, or multibranch link systems, and can also consider the elasticity of constituted links or passive joints by only changing the input numerical model, without the need to revise any part of the scheme. Its validity in various feedforward control experiments of various kinds of link systems has been verified (Isobe et al., 2003; Isobe et al., 2006; Isobe & Kato, 2008).

In this chapter, an outline of the parallel solution scheme is first described along with some numerical examples of flexible and underactuated models. Then, some examples of torque cancelling resulting from accurate calculations of dynamics are demonstrated by showing the experimental results carried out on a simple rotor-TCS system. The systems used for the rotor part include rigid and flexible outboard and inboard rotors, where difficult assumptions are normally required to consider the dynamics accurately. A numerical example of a complex robotic architecture is also presented to verify the use of the system.

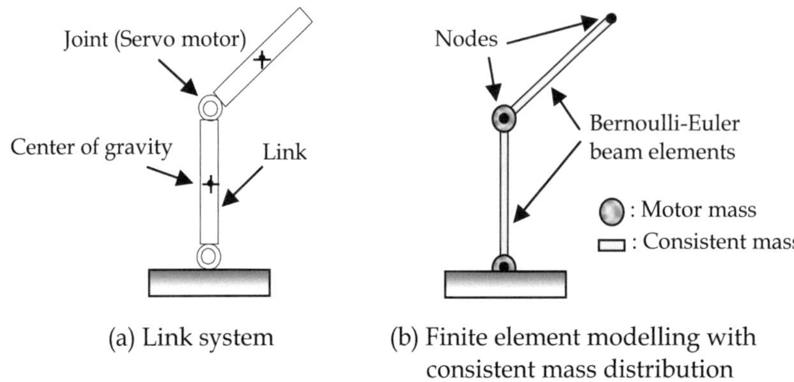

Fig. 1. Modelling of link systems using finite elements

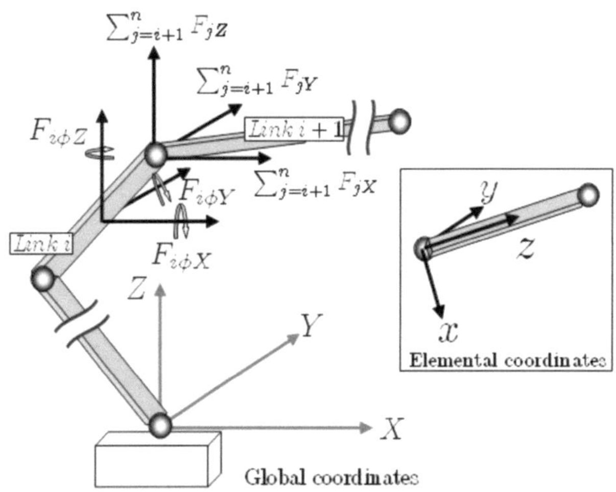

Fig. 2. Nodal forces acting on the i th link in an n-link system with consistent mass distribution

2. Parallel solution scheme of inverse dynamics

2.1 Calculation of joint torques

In the parallel solution scheme, a link system consisting of motor joints and links, as shown in Fig. 1(a), is modelled and subdivided using finite elements. A link is substituted with a single Bernoulli-Euler beam element in the revised version, as shown in Fig. 1(b), assuming a consistent mass distribution along the link (Isobe & Kato, 2008). The consistent mass matrix of a beam element is formulated in the same manner as the displacement function and does not require an expression of the centre of gravity. This type of modelling has the merits of reducing computational time without lowering accuracy, particularly in cases of occurrence of elastic deformations. It requires only one element subdivision per member for cases of infinitesimal deformation because the deformation of the element is defined using a high-order displacement function.

Figure 2 shows the nodal forces based on global coordinates acting on the i th link in a three-dimensional open-loop n-link system with a consistent mass distribution. The joint torque τ_{ix}

required around the x-elemental axis on the i th link is determined by adding the $i+1$ th joint torque $\tau_{(i+1)x}$ to the sum of the moments of inertia acting on this link and is expressed by the nodal forces based on elemental (or link) coordinates as

$$\tau_{ix} = l_i \left(\sum_{j=i+1}^{n} F_j \right)_y + F_{i\phi x} + \tau_{(i+1)x}, \ (i, j = 1 \sim n), \tag{1}$$

where l is the link length and F is the nodal force. $\sum F$ indicates the resultant force, the subscripts i, j the link number, x, y, z the elemental coordinate components, X, Y, Z the global coordinate components, and φ the angular components. The right hand side of Eq. (1) becomes different in a scheme with a lumped mass distributed at the centre of gravity (Isobe, 2004). By considering the other components around the y- and z-axes and arranging them into global coordinates (X, Y, Z) in matrix form, the joint torque vector is expressed as

$$\{\tau^n\} = [L^n][T^n]\{P^n\}, \tag{2}$$

where $\{P^n\}$ is a 6n×1 vector related to the nodal force and is defined as

$$\{P^n\} = \begin{Bmatrix} P_1 \\ P_2 \\ \bullet \\ \bullet \\ P_n \end{Bmatrix}, \ where \ \{P_i\} = \begin{Bmatrix} \sum_{j=i+1}^{n} F_{jX} \\ \sum_{j=i+1}^{n} F_{jY} \\ \sum_{j=i+1}^{n} F_{jZ} \\ F_{i\phi X} \\ F_{i\phi Y} \\ F_{i\phi Z} \end{Bmatrix}. \tag{3}$$

$[T^n]$ is a 6n×6n transformation matrix and is defined as

$$[T^n] = [h^n][T^n_{GE}], \tag{4}$$

where $[h^n]$ is a correction matrix between the x-y and z-x coordinate systems, which simply inverts the signs of the components in the y-axis direction. $[T^n_{GE}]$ is a transformation matrix between the global and elemental coordinates and is expressed as

$$[T^n_{GE}] = \begin{bmatrix} T_1 & & & & \\ & T_2 & & 0 & \\ & & \bullet & & \\ & 0 & & \bullet & \\ & & & & T_n \end{bmatrix}, \tag{5}$$

where

$$[T_i]=\begin{bmatrix} A_i & 0 \\ 0 & A_i \end{bmatrix} \tag{6a}$$

and

$$[A_i]=\begin{bmatrix} \cos\phi_{iXx} & \cos\phi_{iYx} & \cos\phi_{iZx} \\ \cos\phi_{iXy} & \cos\phi_{iYy} & \cos\phi_{iZy} \\ \cos\phi_{iZx} & \cos\phi_{iZy} & \cos\phi_{iZz} \end{bmatrix}, \tag{6b}$$

where φ_{iXx}, for example, represents the rotational angle between the X (global) and x (elemental) coordinates.

$[L^n]$ in Eq. (2) is a 3n×6n matrix related to link length and is expressed as

$$[L^n]=[T^n_\Lambda]\,[\Lambda^n], \tag{7}$$

where $[T^n{}_\Lambda]$ is a transformation matrix between each elemental coordinate and is expressed as

$$[T^n_\Lambda]=\begin{bmatrix} T_{11} & T_{12} & T_{13} & \bullet & \bullet & \bullet & T_{1n} \\ & T_{22} & T_{23} & \bullet & \bullet & \bullet & T_{2n} \\ & & T_{33} & \bullet & \bullet & \bullet & T_{3n} \\ & & & \bullet & \bullet & \bullet & \bullet \\ & & & & \bullet & \bullet & \bullet \\ & 0 & & & & \bullet & \bullet \\ & & & & & & T_{nn} \end{bmatrix}. \tag{8}$$

$[T_{ij}](i,j=1\sim n)$ is expressed using matrix $[A_i]$ shown in Eq. (6b) as

$$[T_{ij}]=[A_i]\,[A_j]^T. \tag{9}$$

$[\Lambda^n]$ is a matrix expressed as

$$[\Lambda^n]=\begin{bmatrix} \Lambda_1 & & & & \\ & \Lambda_2 & & 0 & \\ & & \Lambda_3 & & \\ & & & \bullet & \\ & 0 & & & \bullet \\ & & & & & \Lambda_n \end{bmatrix}, \tag{10}$$

where

$$[\Lambda_i]=\begin{bmatrix} 0 & l_i & 0 & 1 & 0 & 0 \\ l_i & 0 & 0 & 0 & 1 & 0 \\ 0 & 0 & 0 & 0 & 0 & 1 \end{bmatrix}. \tag{11}$$

Consequently, joint torques can be obtained, as shown in Eq. (2), by first converting the nodal forces in global coordinates to elemental coordinates and then by multiplying the matrix regarding link lengths.

2.2 Calculation of incremental nodal forces

The nodal forces used in Eq. (3) are obtained by solving the following equations of motion:

$$[M]\{\ddot{u}\}_{t+\Delta t} = \{F\}_{t+\Delta t}, \tag{12}$$

$$\{\ddot{u}\}_{t+\Delta t} = \{\ddot{u}\}_t + \{\Delta\ddot{u}\}, \tag{13}$$

where $[M]$ is the total mass matrix, $\{F\}$ is the external force vector, and $\{\ddot{u}\}$ is the acceleration vector. By adopting Newmark's β method (Newmark, 1959) as the time integration scheme and substituting the equation

$$\{F\}_{t+\Delta t} = \{F\}_t + \{\Delta f\} \tag{14}$$

into Eq. (12), one can derive the following equation for calculating the incremental nodal force vector:

$$\{\Delta f\} = -\{F\}_t + [M](\frac{1}{\beta\Delta t^2}\{\Delta u\} - \frac{1}{\beta\Delta t}\{\dot{u}\}_t - (\frac{1}{2\beta} - 1)\{\ddot{u}\}_t), \tag{15}$$

where $\{\Delta u\}$ is the incremental displacement vector and $\{\dot{u}\}$ is the velocity vector. The operation distance between each incremental step calculated from a target trajectory is used as an input for $\{\Delta u\}$. The velocity and acceleration vectors can also be directly substituted as input data; however, we used Newmark's β method (δ=1/2, β=1/4) for calculating the values in the actual program. Each corresponding term in the incremental nodal force vector is then substituted into Eq. (3) to constitute the vector $\{P^n\}$.

3. Calculation of target motions and estimation of inverse dynamics

3.1 Rigid body and flexible models

In the inverse dynamics calculation using the parallel solution scheme, accurate target motions that compensate for the inertial forces acting at the links and the stiffnesses of the links are required. Maintaining such accuracy is inevitable, especially for flexible models where elastic deformations occurring in link members greatly affect the torque values (Bayo, 1987; Asada et al., 1990; Shabana, 1990; Theodore & Ghosal, 1995). Moreover, the elastic vibration that occurs in the links becomes a serious issue for the control. Therefore, much effort has been made to model and describe robot dynamics with elastic links and to control them (Cetinkunt & Yu, 1991; Tzes & Yurkovich, 1991; Hui et al., 2002; Subudhi & Morris, 2002; Theodore & Ghosal, 2003; Wang & Mills, 2005; Feliu et al., 2006). In contrast, we decided to develop a solution scheme of kinematics using the finite element approach combined with the parallel solution scheme to handle analysed models comprehensively in a single calculation process (Isobe & Kato, 2008).

The following equation of motion can be used for rigid body models, where the occurrence of elastic deformations and internal forces can be neglected:

$$[M]\{\ddot{u}_m\}_{t+\Delta t} = \{F\}_{t+\Delta t}. \tag{16}$$

Here, $\{\ddot{u}_m\}$ is the acceleration vector for the overall system motion components at time $t+\Delta t$. For flexible models, the components for material deformations and internal forces should be considered and the following equation of motion at time $t+\Delta t$ is adopted (Bathe, 1996):

$$[M]\{\ddot{u}_m\}_{t+\Delta t} + [M]\{\ddot{u}_d\}_{t+\Delta t} + [C]\{\dot{u}_d\}_{t+\Delta t} + [K]\{\Delta u_d\} = \{F\}_{t+\Delta t} - \{R\}_t, \tag{17}$$

where $[C]$ is the total damping matrix, $[K]$ is the total stiffness matrix, and $\{R\}$ is the internal force vector. The vectors $\{\ddot{u}_d\}$, $\{\dot{u}_d\}$ and $\{\Delta u_d\}$ denote the acceleration vector, the velocity vector, and the incremental displacement vector for the material deformation components at time $t+\Delta t$, respectively. By applying Newmark's β method (δ=1/2) as a time integration scheme, the acceleration and velocity vectors are calculated as

$$\{\ddot{u}_m\}_{t+\Delta t} = \frac{1}{\beta\Delta t^2}\{\Delta u_m\} - \frac{1}{\beta\Delta t}\{\dot{u}_m\}_t - \left(\frac{1}{2\beta} - 1\right)\{\ddot{u}_m\}_t, \tag{18a}$$

$$\{\ddot{u}_d\}_{t+\Delta t} = \frac{1}{\beta\Delta t^2}\{\Delta u_d\} - \frac{1}{\beta\Delta t}\{\dot{u}_d\}_t - \left(\frac{1}{2\beta} - 1\right)\{\ddot{u}_d\}_t, \tag{18b}$$

$$\{\dot{u}_m\}_{t+\Delta t} = \frac{1}{2\beta\Delta t}\{\Delta u_m\} - \left(\frac{1}{2\beta} - 1\right)\{\dot{u}_m\}_t - \left(\frac{1-4\beta}{4\beta}\right)\{\ddot{u}_m\}_t\Delta t, \tag{18c}$$

$$\{\dot{u}_d\}_{t+\Delta t} = \frac{1}{2\beta\Delta t}\{\Delta u_d\} - \left(\frac{1}{2\beta} - 1\right)\{\dot{u}_d\}_t - \left(\frac{1-4\beta}{4\beta}\right)\{\ddot{u}_d\}_t\Delta t, \tag{18d}$$

where $\{\dot{u}_m\}$ is the velocity vector for the overall system motion components. Substituting Eqs. (18) into Eq. (17) yields

$$\begin{aligned}([K] + \frac{1}{\beta\Delta t^2}[M] + \frac{1}{2\beta\Delta t}[C])\{\Delta u_d\} &= \{F\}_{t+\Delta t} - \{R\}_t + [M](\frac{1}{\beta\Delta t}\{\dot{u}_d\}_t + (\frac{1}{2\beta} - 1)\{\ddot{u}_d\}_t) \\ -[M](\frac{1}{\beta\Delta t^2}\{\Delta u_m\} - \frac{1}{\beta\Delta t}\{\dot{u}_m\}_t &- (\frac{1}{2\beta} - 1)\{\ddot{u}_m\}_t) + [C]((\frac{1}{2\beta} - 1)\{\dot{u}_d\}_t + (\frac{1-4\beta}{4\beta})\{\ddot{u}_d\}_t\Delta t).\end{aligned} \tag{19}$$

By applying $\{\Delta u_m\}$ as the input in a time-integration loop of Eq. (19) and using the vectors of Eqs. (18) at time t, $\{\Delta u_d\}$ at each time step can be successively obtained. The displacement vector $\{u_m\}$ for the overall system motion components and the displacement vector $\{u_d\}$ for the material deformation components are calculated incrementally as

$$\{u_m\}_{t+\Delta t} = \{u_m\}_t + \{\Delta u_m\}, \tag{20a}$$

$$\{u_d\}_{t+\Delta t} = \{u_d\}_t + \{\Delta u_d\}. \tag{20b}$$

The total displacement $\{u\}$ is obtained by summing both the overall system motion and the material deformation components as

$$\{u\}_{t+\Delta t} = \{u_m\}_{t+\Delta t} + \{u_d\}_{t+\Delta t}. \tag{21}$$

The final target trajectories considering the effects of stiffness and damping are thus obtained using Eq. (21). The resultant forces acting on the elements can also be calculated using the obtained displacements.

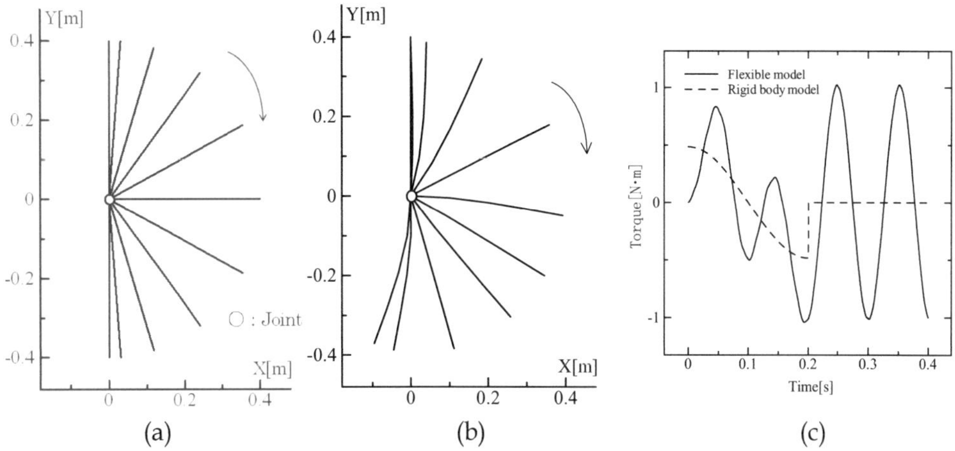

(a) Initial target motion (target motion for rigid-body model)
(b) Target motion for flexible model obtained by calculating the equation of motion
(c) Torque curves obtained for both models

Fig. 3. Inverse dynamics calculation of a one-joint rigid body/flexible link system

A numerical test was carried out on a one-joint flexible link system to verify the torque values obtained by the calculations. Damping is neglected in the calculations. Figure 3(a) shows a target trajectory of a 0.2-s motion in the horizontal plane for a rigid-body one-link system (link length: 40 cm; weight: 22.4 g; flexural stiffness: perfect rigidity). Using this target motion as input data for the algorithm described above, the motion considering the effect of stiffness is obtained, as shown in Fig. 3(b). Here, the flexural stiffness for the flexible model is given as 0.46 Nm^2. From the results, we can see the difference between the deformations of the two link systems. Figure 3(c) shows the torque curves for the two models obtained by the parallel solution scheme using the motions of Figs. 3(a) and 3(b) as input data. A smooth joint torque curve is obtained for the rigid-body model. In contrast, a vibration can be observed in the torque curve of the flexible model due to the flexural vibration that occurs in the model, particularly after the motion has stopped. The natural frequency of the model estimated from the torque curve agrees with the theoretical value, which is 10.02 Hz.

3.2 Underactuated models

The calculation process described in the previous section can be commonly used for rigid body, flexible, and underactuated models. However, some special treatments should be carried out for the underactuated model due to its large deformation. First, we use an elemental stiffness matrix equipped with only axial components for nonactuated links in order to exclude the transmission of bending moments at the nonactuated joints. However, the exclusion of the bending moment effect is not sufficient to realise the behaviour of passive joints because the accumulation of numerical errors due to large deformation occurs in the calculation. Therefore, an algorithm updating the coordinates of the system during time steps is also developed.

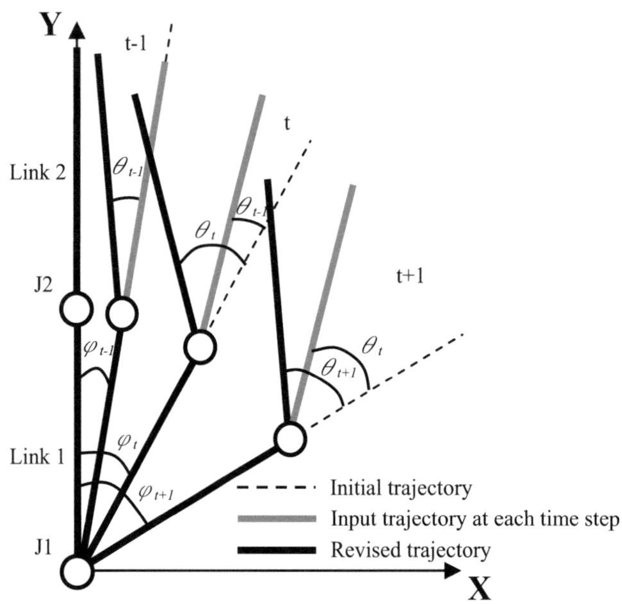

Fig. 4. Outline of updated transformation algorithm used in the calculation of target motions for underactuated models

The outline of the updated transformation algorithm is explained using a simple example of a two-joint underactuated link system as shown in Fig. 4. The broken lines in the figure are the initial trajectory, the gray lines are the input trajectory at each time step, and the dark lines are the revised trajectories at the end of each time step. φ is the rotational angle of the active joint (J1, in this case), which will not be revised by the algorithm, and θ is the rotational angle of the nonactuated joint (J2, in this case), which will be calculated using the deformation obtained from Eq. (20b). Suppose the posture at time step *t-1* is represented by the dark line shown in Fig. 4, with the rotational angle of J1 being φ_{t-1}. The input trajectory at the next time step *t* is calculated by only rotating the active joint to φ_t, whereas the overall posture is fixed as it is. The deformations of the link system at time step *t* are calculated starting from this posture (gray line at time step *t*). The calculated result at the end of time step *t* will become as the dark line (with the rotational angle of J1 being φ_t), and the revisions proceed successively to the next time step *t+1* in the same manner. In general *n*-link cases, the input trajectory can be revised by applying the above algorithm from the base joint to the end joint successively at each time step.
The kinematics calculation using the algorithm shown above is carried out for a two-joint underactuated link system. Let the initial target motion of a rigid model be a horizontal 2-s motion, as shown in Fig. 5(a). J1 is an active joint, whereas J2 is a nonactuated joint. The parameters of the links are as follows: material of the links, SUS430; flexural stiffness, 20.84 Nm^2; link lengths, 0.25 m; link mass, 9.75×10^{-2} kg; extra mass at the tip, 0.1 kg. The time

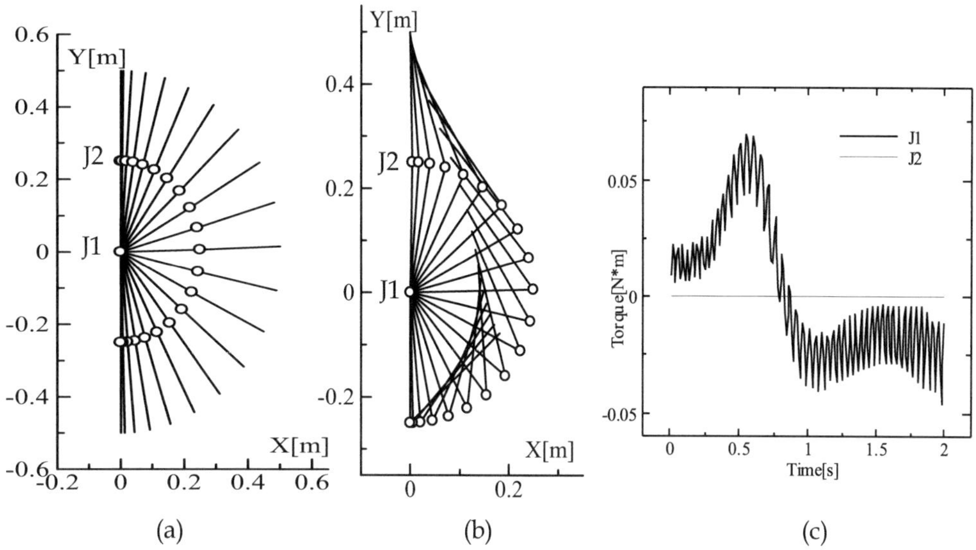

(a) Initial target motion
(b) Target motion calculated using the updated transformation algorithm
(c) Torque curves of a two-joint underactuated link system obtained using parallel solution scheme

Fig. 5. Inverse dynamics calculation of a two-joint underactuated link system

increment for the calculation is set to 10 ms. Figure 5(b) shows the calculated target motion of the two-joint underactuated model. Figure 5(c) shows the torque curves calculated using the data in Fig. 5(b) as input for the parallel solution scheme. The torque curve for J2 retains the value of 0 throughout the motion, which is proof that the joint is successfully expressed as a nonactuated joint. The vibration seen in the torque curve for J1 is proof that Link 1 has stiffness (or flexibility). It vibrates with a natural period in the range of 0.025 s - 0.04 s (note that the natural period changes with the posture), which matches the theoretical value for a two-link free vibration beam, which is 0.031 s. See the reference (Isobe, 2008) for further numerical examples and experimental results.

4. Torque Cancelling System (TCS)

4.1 General concept of TCS

A robot swinging around an arm in a quick motion, for example, normally needs a counterbalance motion to avoid any mechanical sway. The TCS proposed in this chapter is a system that generates anti-torque to suppress the mechanical sway by computing accurate inverse dynamics of the arm motion with an accurate consideration of the dynamics of the robot itself. The accurate consideration of the dynamics is carried out by implementing the parallel solution scheme described in the previous sections.

A simple explanation of the concept is provided in Fig. 6 for the uniaxial case. Suppose a rotor is rotated by a motor with a specific torque. A reaction moment will occur around the rotating axis that will make the whole body twist around the axis. To prevent the reaction moment from making the twist, a cancelling moment can be generated by another motor placed on the axis by supplying a specific, cancelling torque. The same concept is used in twin-rotator helicopters, where two rotators rotate in opposite directions to cancel the twist

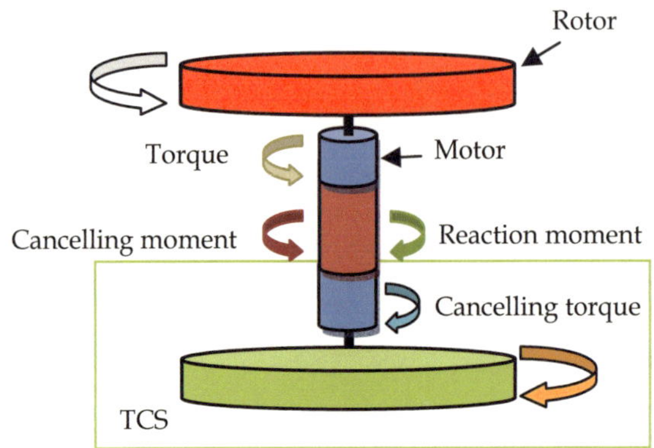

Fig. 6. General concept of uniaxial TCS

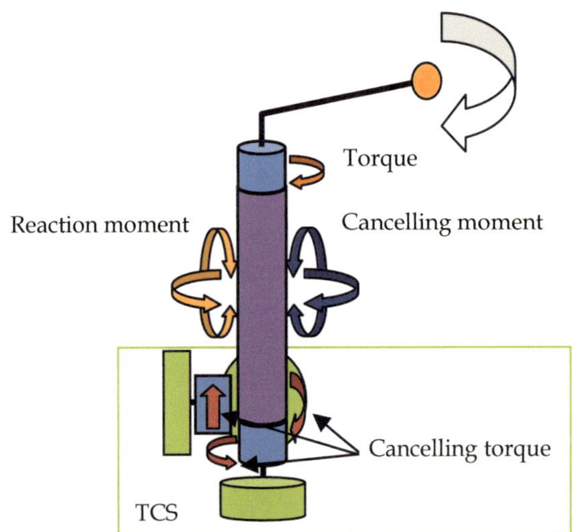

Fig. 7. General concept of triaxial TCS

motion. However, the difference between the twin-rotator helicopters and the proposed TCS is that the twin-rotators only rotate in opposite directions with the same rotational speed, whereas the TCS is driven using accurate torques by considering the dynamics related to the motion and the overall architecture.

A general concept of triaxial TCS is drawn in Fig. 7. If an object in motion follows the model of an outboard rotor (or a cantilever beam as shown in the figure), the reaction moments acting on the body are not as simple as those in the uniaxial case. The moments will act around three-dimensional axes because the centre of gravity of the overall body is at the offset position from the rotating axis. To cancel these moments, three TCSs, each set on each dimensional axis, should be placed as shown in Fig. 7. However, all of the TCSs do not have to be placed at one location or placed exactly on the rotating axis to maintain the function. Instead, a TCS can be mounted, literally, anywhere in the body. As shown later in this chapter, all a TCS has to do to suppress a mechanical sway is to cancel the moment generated at the precise location of the TCS.

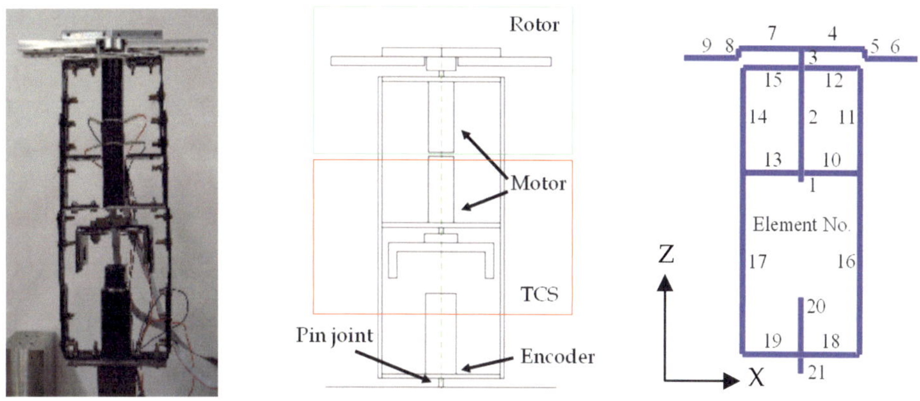

(a) (b) (c)

(a) Experimental setup
(b) Layout of a rotor and a TCS
(c) Finite element subdivision of the setup

Fig. 8. Experimental setup and numerical modelling of a rotor mechanism with uniaxial TCS mounted on the rotating axis

4.2 Uniaxial TCS mounted on rotating axis to cancel rigid-rotor motion

A simple rotor-TCS experimental setup, as shown in Figs. 8(a) and 8(b), was made to carry out a numerical estimation and some experiments for verification. In this setup, a rotor is driven by a gearless motor, and a prototype TCS is placed underneath, exactly on the rotating axis of the rotor. The frame body is connected to the ground with a rotation-free pin joint, where an encoder attached to a nonacutated motor is used to detect the rotational angle of the frame body. The rotor is made of aluminium, and the frame body is made of stainless steel, both with enough sectional area such that the whole architecture can be assumed to be rigid without any elastic deformation.

Figure 8(c) shows the finite element subdivision of the experimental setup, with each connection to connection modelled with a single Bernoulli-Euler beam element. The whole system is modelled with a total of 21 elements and 22 nodes. The capabilities of modelling this kind of complex architecture and computing the dynamics are the main features of the parallel solution scheme.

A rotational motion of 4π rad in 2 s is given to the rotor. The torque required for the rotor, which is actually computed at the upper node of element No. 3 in Fig. 8(c), is obtained by the parallel solution scheme as shown in Fig. 9(a). Reaction moments generated at the exact location of the TCS (lower node of element No. 1 in Fig. 8(c)) are calculated by the scheme as shown in Fig. 9(b). To cancel these moments, the torque shown in Fig. 9(c) should be supplied to the TCS. Note that the cancelling torque the TCS has to generate is a perfect duplicate of the reaction moment M_z (see Fig. 6).

Figure 10 shows the rotational angle of the frame body detected at the encoder attached on the pin joint when the motion torque is actually supplied to the rotor. It can be confirmed from the figure that the frame body is unstable during the motion when the TCS is not activated, whereas the rotation around the pin joint completely stops when the TCS is activated, which means that the reaction moments generated around the rotating axis are perfectly cancelled by supplying the accurately computed cancelling torques to the TCS.

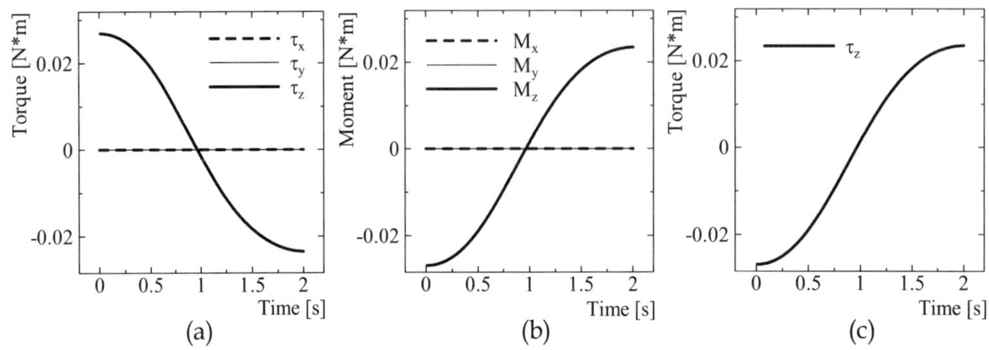

(a) Motion torques for driving a rotor
(b) Reaction moments generated at the exact location of TCS
(c) Input torque for TCS

Fig. 9. Motion torques, reaction moments, and input torque for uniaxial TCS mounted on the rotating axis

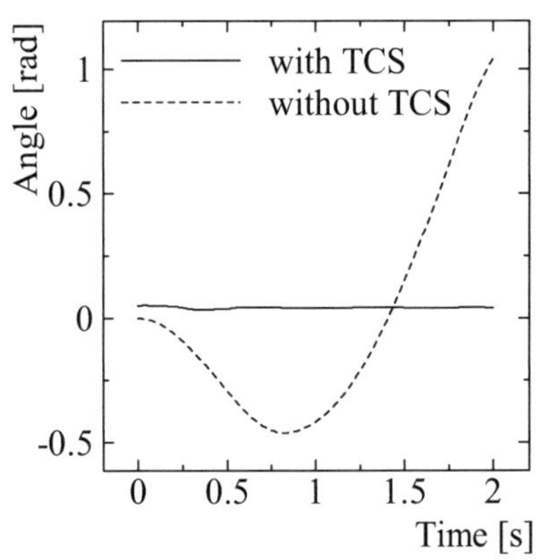

Fig. 10. Rotational angle of the frame body around the pin joint during the motion

4.3 Uniaxial TCS mounted in offset position to cancel rigid-rotor motion

Next, an experiment was carried out on a different configuration of the frame body, as shown in Figs. 11(a) and 11(b). In this case, a TCS is mounted in the offset position of the rotating axis to see if such a condition affects the capability of the system. The whole system is modelled with a total of 25 elements and 26 nodes as shown in Fig. 11(c). The same rotational motion is given to the rotor as in the previous experiment, and, consequently, there is no difference between Fig. 9(a) and Fig. 12(a). The motion torques for driving the rotor are the values calculated at the upper node of element No. 3, and the reaction moments are the values calculated at the upper node of element No. 25. The reaction moment and the input torque for the TCS around the z axis in Figs. 12(b) and 12(c) perfectly agree with those in Figs. 9(b) and 9(c). On the other hand, M_y in Fig. 12(b) is constantly shifted to the positive direction because a moment due to gravity subjected along the offset length is now generated. The moment generated by gravity, however, is supported at the pin joint and is not meant to be suppressed in this experiment.

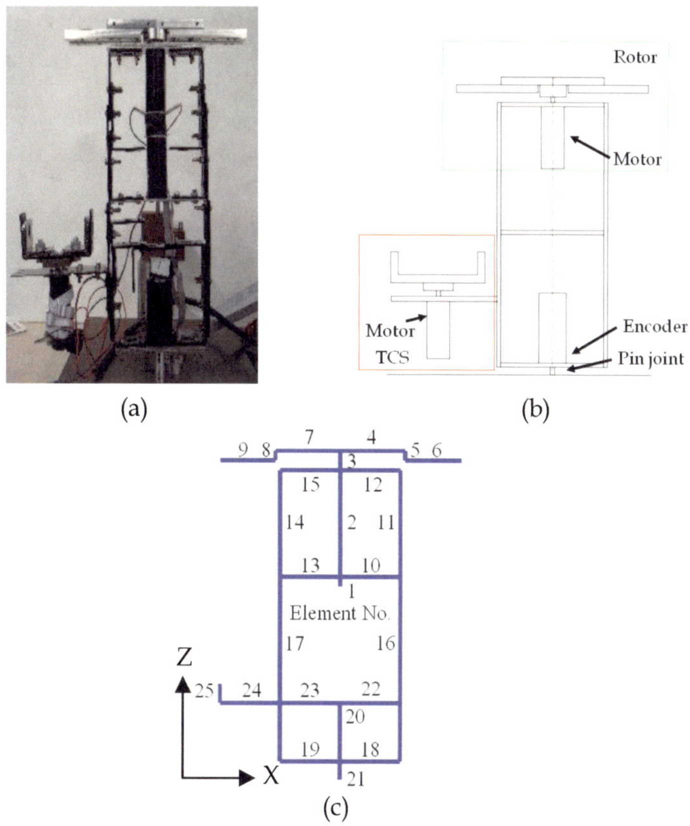

(a) Experimental setup
(b) Layout of a rotor and a TCS
(c) Finite element subdivision of the setup

Fig. 11. Experimental setup and numerical modelling of a rotor mechanism with uniaxial TCS mounted in the offset position of the rotating axis

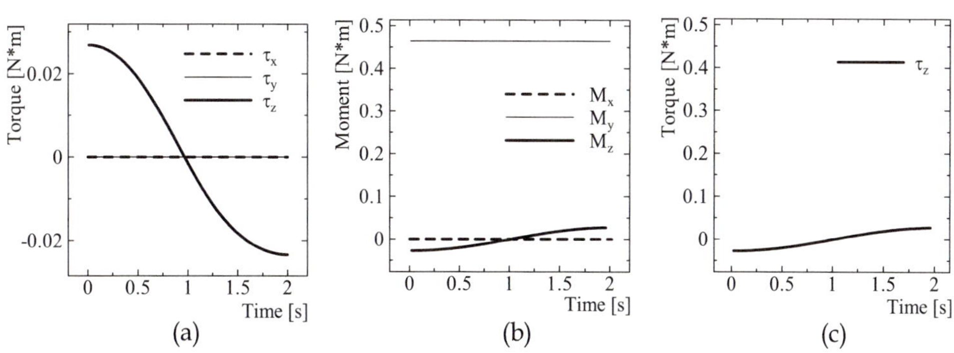

(a) Motion torques for driving a rotor
(b) Reaction moments generated at the exact location of TCS
(c) Input torque for TCS

Fig. 12 Motion torques, reaction moments, and input torque for uniaxial TCS mounted in the offset position

It can be confirmed from Fig. 13 that in this case, the mechanical sway around the pin joint is also successfully suppressed to a minimal amount by activating the TCS mounted in the offset position.

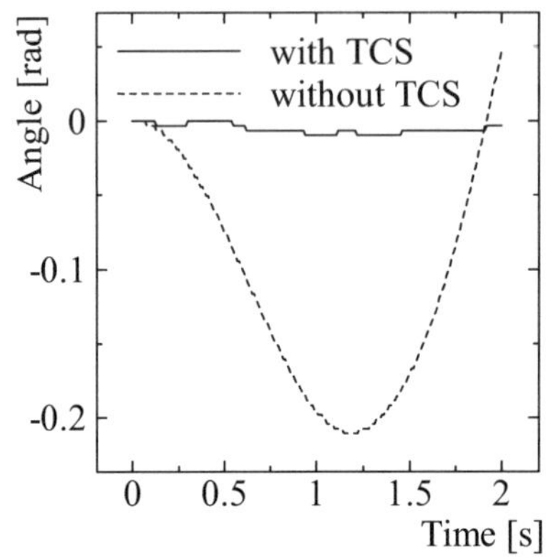

Fig. 13. Rotational angle of the frame body around the pin joint during the motion when a TCS is mounted in the offset position

4.4 Uniaxial TCS mounted in offset position to cancel flexible-link motion

Furthermore, a simple flexible link with TCS experimental setup, as shown in Figs. 14(a) and 14(b), was made to carry out a numerical estimation and some experiments for verification. In this setup, a flexible link is driven by a gearless motor, and a prototype TCS is placed at an offset position from the rotating axis of the active joint. The link system is connected to the ground with a rotation-free passive joint, where an encoder attached to a nonacutated motor is used to detect the rotational angle of the overall body. The flexible link is made of poly-carbonate, and the elastic deformation occurring in the link should not be neglected in the inverse dynamics calculation. The experimental setup is subdivided, with each connection to connection represented by a single Bernoulli-Euler beam element. The whole system is modelled with a total of 4 elements and 5 nodes as shown in Fig. 14(c). The capabilities of modelling this kind of architecture with flexible link and computing the dynamics are, again, the main features of the parallel solution scheme.

A rotational motion of π rad in 2 s is given to the active link. The torque required for the active joint is obtained by the parallel solution scheme as shown in Fig. 15(a). The reaction moment generated at the exact location of the TCS is calculated as shown in Fig. 15(b). To cancel this moment, the torque shown in Fig. 15(c) should be supplied to the TCS. Here again, note that the cancelling torque that the TCS must generate is a perfect duplicate of the reaction moment (see Fig. 6).

Figure 16 shows the rotational angle of the link system detected at the encoder attached on the passive joint when the motion torque is actually supplied to the active joint. It can be confirmed from the figure that the link system is unstable during the motion when the TCS is not activated, whereas the rotation around the passive joint completely stops when the TCS is activated, which means that the vibrating reaction moment generated around the rotating axis is perfectly cancelled by supplying the accurately computed cancelling torque to the TCS.

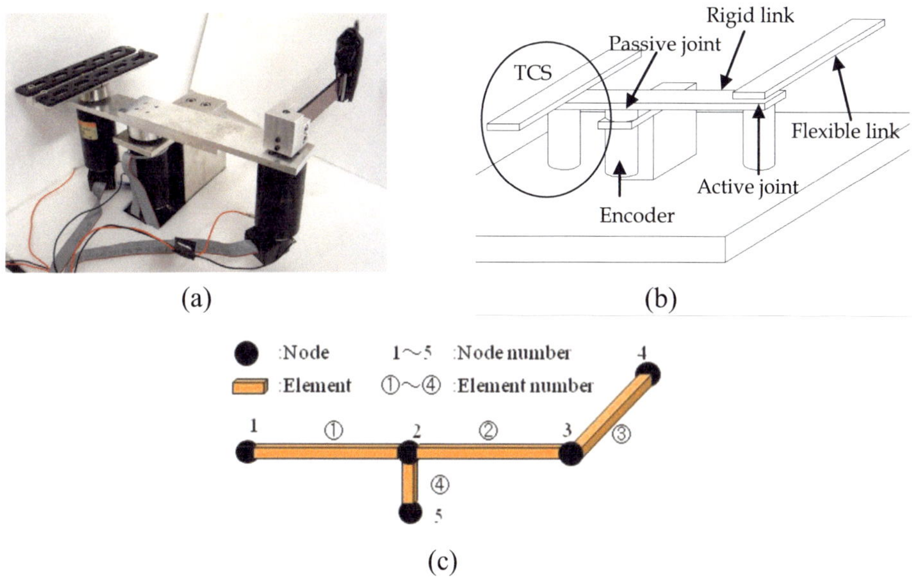

(a) Experimental setup
(b) Configuration of the setup and location of a TCS
(c) Finite element subdivision of the setup

Fig. 14. Experimental setup and numerical modelling of a flexible link mechanism with uniaxial TCS mounted in the offset position of the rotating axis

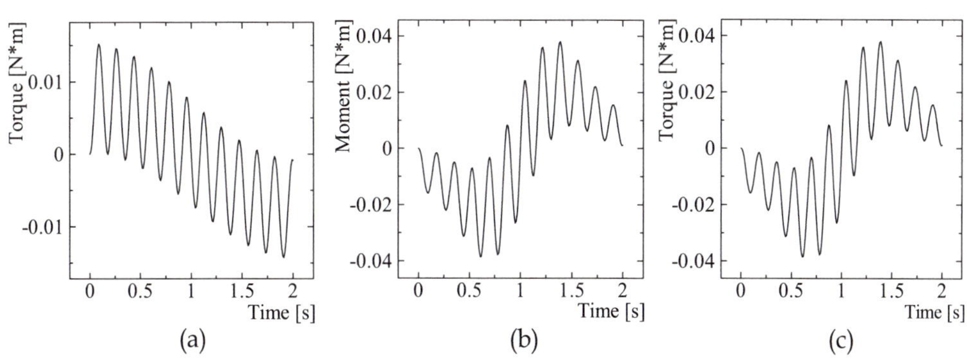

(a) Motion torque for driving the flexible link
(b) Reaction moment generated at the exact location of TCS
(c) Input torque for TCS

Fig. 15. Input torques and generated moments in the system

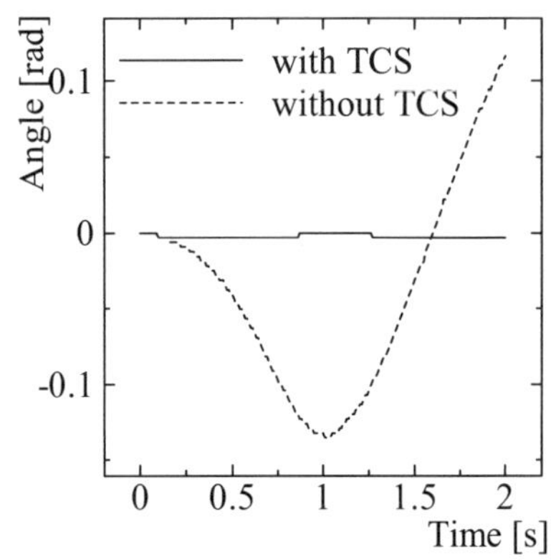

Fig. 16. Rotational angle of the link system around the passive joint during the motion

5. Application of TCS on a Robot

The TCS shown in this chapter can be applied to actual robotic architecture to suppress mechanical sway generated by rotational motions. A sample of a walking robot mounted with a TCS, along with its configuration and finite element subdivision, are shown in Fig. 17. A TCS shown in Fig. 17(a) is mounted at the central part of the robot, as shown in Fig. 17(b). As shown in Fig. 17(c), a caster at the central-bottom part supports the overall weight of the robot, while two electric magnet holders at the tip of two legs are used to get hold of a metallic base. Two motors (Motor 1 and Motor 2 in Fig. 17(c)) are activated to advance the steps, and the other two motors on the legs are used to lift up and lower the legs. Mechanical sway is often generated during the motions of Motor 1 and Motor 2 because the architecture possesses a comparatively large inertia. A TCS that is mounted at the central part of the body can be applied to suppress the sway during the gait motion. The robot is modelled numerically, as shown in Fig. 17(d), with finite elements and applied to the calculation of various required torques. A two-step gait for the walking robot is given as shown in Fig. 18. The motion torques for the two motors (Motor 1 and Motor 2) along with the input torque for the TCS during the motion is computed using the scheme described previously.

The calculated motion torques for the motors and input torque for the TCS are shown in Fig. 19. Large torques are required during the walking step motions at 1.0 s ~ 1.5 s and 3.5 s ~ 4.25 s (see Fig. 19(a) and 19(b)), and reaction moments are generated during those periods at various parts of the robotic architecture. The reaction moment generated at the exact location of the TCS is calculated by the parallel solution scheme as shown in Fig. 19(c), which is used directly as an input torque for the TCS. Some experiments using the setup are scheduled in the near future, and the latest results will be reported on our website (http://www.kz.tsukuba.ac.jp/~isobe/) and in technical papers.

(a)

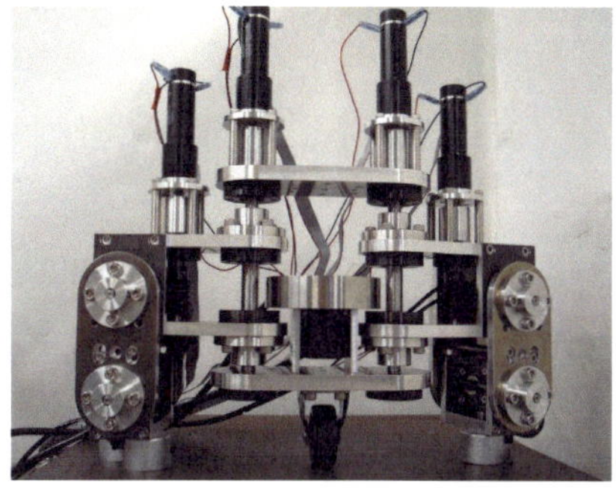

(b)

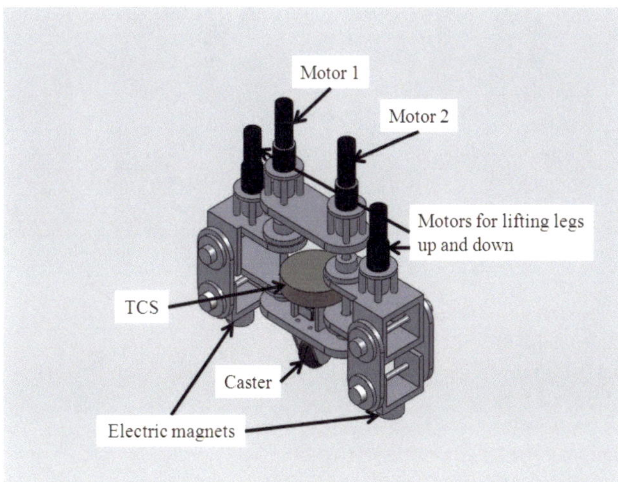

(c)

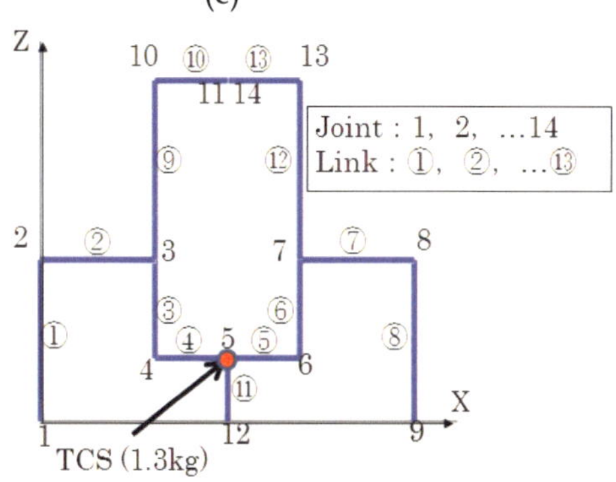

(d)

(a) View of a TCS mounted in a robot
(b) General view (TCS mounted at the central part)
(c) Configuration of the robot and location of the TCS
(d) Finite element subdivision of the robot

Fig. 17. A walking robot mounted with a TCS

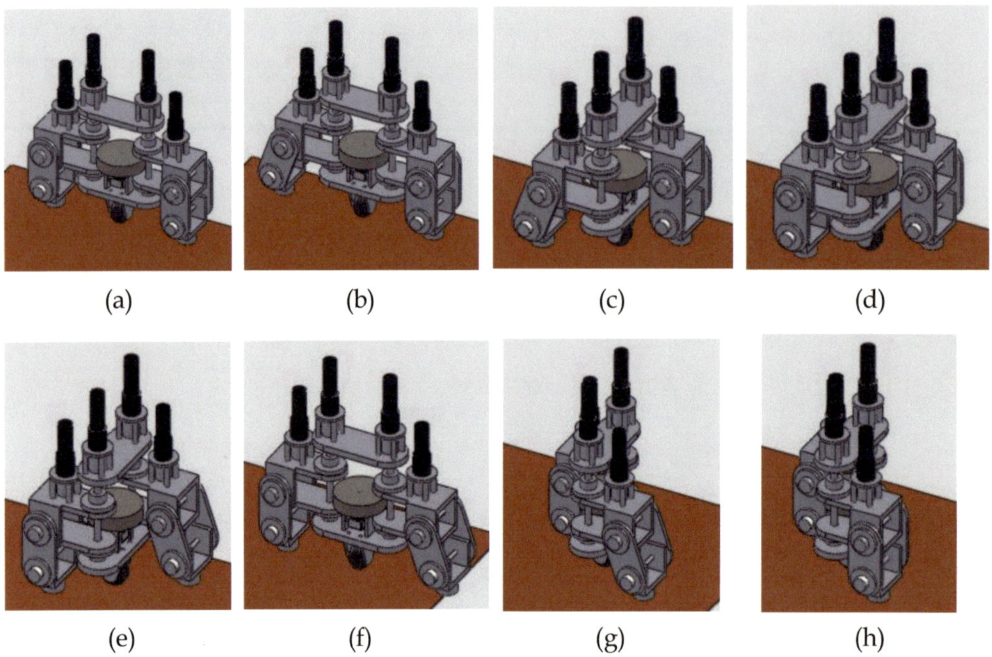

(a) Initial position (0.0 s)
(b) Lift up the right leg (0.0 s ~ 1.0 s)
(c) First step motion (1.0 s ~ 1.5 s)
(d) Lower the right leg (1.5 s ~ 2.5 s)
(e) Lift up the left leg (2.5 s ~ 3.5 s)
(f)(g) Second step motion (3.5 s ~ 4.25 s)
(h) Lower the left leg (4.25 s ~ 5.25 s)

Fig. 18. A two-step gait for the walking robot

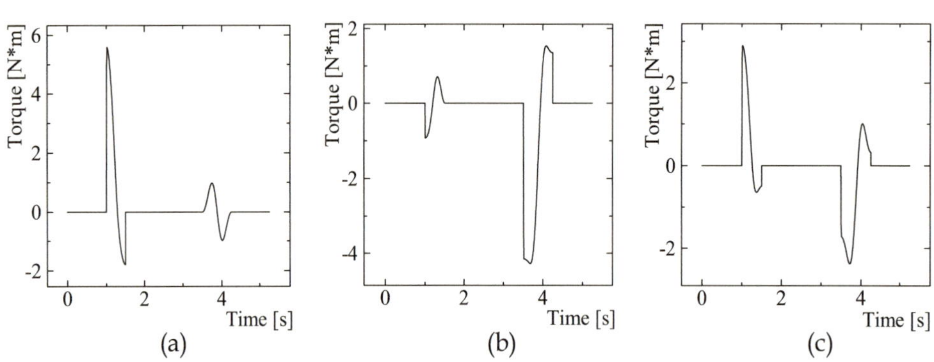

(a) Motion torque for Motor 1
(b) Motion torque for Motor 2
(c) Reaction moment generated at the exact location of TCS (or input torque for TCS)

Fig. 19. Motion torques for the motors and input torque for the TCS

6. Conclusion

Modelling the dynamics of a robotic architecture usually requires ambiguous assumptions and labour-intensive formulations at various stages. However, application of the finite element approach to inverse dynamics calculation enables us to consider the precise dynamics of robots using a very simple process. This approach leads us to simpler and more precise feedforward control of robots as well as the cancellation of mechanical sway during quick motions. We are now attempting to develop a compact TCS device that can be easily mounted on actual robots in both uniaxial and triaxial cases.

7. Acknowledgment

The author wishes to acknowledge the contributions of the following former and current graduate students of his laboratory: Takeo Ueda, Daisaku Imaizumi, Youichi Chikugo, Shunsuke Sato, Atsushi Yagi, Akihiro Kato, Koji Yamanaka, Hiromasa Ueda, Yuto Kitamura, Junya Hayakawa, Yasumasa Matsui, Kensuke Kondo, and Bumpei Nakayama.

8. References

Asada, H.; Ma, Z.D. & Tokumaru, H. (1990). Inverse Dynamics of Flexible Robot Arms: Modelling and Computation for Trajectory Control, *ASME J. Dyn. Sys. Measurement Control*, Vol. 112, pp. 177-185.

Bathe, K.J. (1996). *Finite Element Procedures,* Prentice-Hall.

Bayo, E. (1987). A Finite-Element Approach to Control the End-Point Motion of a Single-Link Flexible Robot, *J. Robotic Systems*, Vol. 4, No. 1, pp. 63-75.

Cetinkunt, S. & Yu, W.L. (1991). Closed Loop Behavior of a Feedback Controlled Fexible Arm: a Comparative Study, *Int. J. of Robotics Research*, Vol. 10, No. 3, pp. 263-275.

Feliu, V.; Pereira, E.; Díaz, I.M. & Roncero, P. (2006). Feedforward Control of Multimode Single-Link Flexible Manipulators based on an Optimal Mechanical Design, *Robotics and Autonomous Systems*, Vol. 54, No.8, pp. 651-666.

Hui, D.; Fuchun, S. & Zengqi, S. (2002). Observer-Based Adaptive Controller Design of Flexible Manipulators using Time-Delay Neurofuzzy Networks, *J. of Intelligent and Robotic Systems*, Vol. 34, No. 4, pp. 453-456.

Isobe, D.; Takeuchi, H. & Ueda, T. (2000). A Finite-Element Approach to Control Link Mechanisms: Its Concept and Basic Simulation, *Proc. Int. Conf. Computational Engineering Science 2000 -Advances in Computational Engineering and Sciences-*, pp. 1648-1653, Los Angeles, USA.

Isobe, D.; Imaizumi, D.; Chikugo, Y. & Sato, S. (2003). A Parallel Solution Scheme for Inverse Dynamics and Its Application in Feed-forward Control of Link Mechanisms, *J. of Robotics and Mechatronics*, Vol. 15, No. 1, pp. 1-7.

Isobe, D. (2004). A Unified Solution Scheme for Inverse Dynamics, *Advanced Robotics*, Vol. 18, No. 9, pp. 859-880.

Isobe, D.; Yagi, A. & Sato, S. (2006). General-Purpose Expression of Structural Connectivity in the Parallel Solution Scheme and Its Application, *JSME Int. J. Series C*, Vol. 49, No. 3, pp. 789-798.

Isobe, D. & Kato, A. (2008). Feedforward Control of Flexible Link Systems using Parallel Solution Scheme, *Int. J. Robotics and Automation*, Vol.23, No.1, pp.31-39.

Isobe, D. (2008). Unified Solution Scheme for Inverse Dynamics of Robotic Systems, In: *Progress in Autonomous Robot Research*, Mortensen, F.N. (Ed.), pp.125-173, Nova Science Publishers, ISBN:978-1-60456-281-1.

Nakamura, Y. & Ghodoussi, M. (1989). Dynamics Computation of Closed-Link Robot Mechanisms with Nonredundant and Redundant Actuators, *IEEE Trans. Robotics and Automation*, Vol. 5, No. 3, pp. 294-302.

Nakamura, Y. & Yamane, K. (2000). Dynamics Computation of Structure-Varying Kinematic Chains and Its Application to Human Figures, *IEEE Trans. Robotics and Automation*, Vol. 16, No. 2, pp. 124-134.

Newmark, N.M. (1959). A Method of Computation for Structural Dynamics, *A.S.C.E. J. Engineering Mechanics*, Vol. 85, pp. 67-94.

Ookami, Y.; Tomita, N.; Nakasu, S. & Matsunaga, S. (2008). *Introduction to the Space Station*, University of Tokyo Press, pp. 204-208, in Japanese.

Shabana, A.A. (1990). Dynamics of Flexible Bodies Using Generalized Newton-Euler Equations, *ASME J. Dyn. Sys. Measurement Control*, Vol. 112, pp. 496-503.

Subudhi, B. & Morris, A.S. (2002). Dynamic Modelling, Simulation and Control of a Manipulator with Flexible Links and Joints, *Robotics and Autonomous Systems*, Vol. 41, No. 4, pp. 257-270.

Sugimoto, K. (2003). Dynamic Analysis of Closed Loop Mechanisms on the Basis Vectors of Passive Joint Axes, *J. Robotic Systems*, Vol. 20, No. 8, pp. 501-508.

Theodore, R.J. & Ghosal, A. (1995). Comparison of the Assumed Modes and Finite Element Models for Flexible Multilink Manipulators, *Int. J. Robotics Research*, Vol. 14, No. 2, pp. 91-111.

Theodore, R.J. & Ghosal, A. (2003). Robust Control of Multilink Flexible Manipulators, *Mechanism and Machine Theory*, Vol. 38, No. 4, pp. 367-377.

Tzes, A.P. & Yurkovich, S. (1991). Application and Comparison of On-Line Identification Methods for Flexible Manipulator Control, *Int. J. of Robotics Research*, Vol. 10, No. 5, pp. 515-527.

Wang, X. & Mills, J.K. (2005). FEM Dynamic Model for Active Vibration Control of Flexible Linkages and Its Application to a Planar Parallel Manipulator, *Applied Acoustics*, Vol. 66, No. 10, pp. 1151-1161.

Yoshida, K.; Nenchev, D.N. & Hashizume, K. (2002). Flight Experiments of ETS-VII for Advanced Space Robot Control, *J. Society of Japanese Aerospace Companies*, Vol. 50, No. 584, pp. 351–359, in Japanese.

3

Locomotion Control for Legged Robot by Virtual Contact Impedance Method

Fumiaki Takemori
Tottori University
Japan

1. Introduction

In the current infrastructure, the accessible environment is not necessarily maintained, so that it seems to be hardest for user of wheelchair (invalid chair) such as the aged and the leg-handicapped to travel in the inaccessible environment. When the user of standard wheelchair travels up/down the stairs, accompany supporter is usually needed. Therefore a single action is resultingly restricted within narrow limits. From these reasons, it is desirable to develop a mobile robot carrying the human that enables to independently travel inaccessible environment.

In the mobile hardware with the locomotive function for the stairs, the wheeled and crawler type are major (iBOT , 2010), (SUNWA , 2010). Although the wheeled type is simple and high efficient for actuating method, because of not so redundancy, degree of freedom on mobility is not so high. In addition, wheel type tends to be large size due to its multi-wheel and crawler, so that it is difficult to operate in universal inhabited space. In terms of more useful and intelligence mobility, it is necessary for the robot itself to take avoidance action automatically in order to not damage the robot itself and circumferential obstacles. So, to realize an appropriate method for avoiding the obstacle and stairs is very important issue.

In this paper, a legged mobile robot carrying the human is developed considering the adaptiveness for inaccessible environment at existing infrastructure. Although the legged robot is complex as mobile-morphology, its redundancy with multi-links is advantage to high mobility in universal space. Though earlier researches (Wu, Y. & Higuchi, M. , 2004), (Sugahara et al. , 2004), have already been reported, these are not so advanced instance that stair detection and its avoiding method are integrated. So, as a method for detecting and avoiding the stairs/obstacles, the virtual impedance control method (Tsuji & Kaneko , 1999) is proposed. This method can avoid an obstacle without contacting by receiving virtual force from virtual spring and damper installed in virtual surface region. Finally, the mobility of the legged robot by non-contacting impedance control is presented through some experiments.

2. Legged robot with linear actuators

2.1 Outline of three-legged robot

In this paper, a legged mobile robot which has the ability to carry the human is developed. The photograph of prototype model is shown in Fig.1. This robot has three legs. Each leg consists of three linear actuators, so that total degree of freedom is 9. As shown in Fig.1 and Fig.2, to maintain the rigidity, three actuators are assigned to triangular configuration and farther the part of so-called *shin* has twin screw axle mechanism that are synchronously actuated by single DC-motor. Main specification of proposed legged robot is presented in Table 1.

Width	600[mm]
Height	710∼910[mm]
Weight	35[kg]
DC motor	9ch (150[W]∼14[W])
Power supply	7.2[v] battery ×5
Sensors	3-axis accelerometer : 3ch potentiometer : 9ch sole load sensor : 12ch PSD : 2ch
Control computer	Intel(R) Pentium(R) M 1.6GHz
I/O	AD : 32ch / DA : 16ch

Table 1. Specification

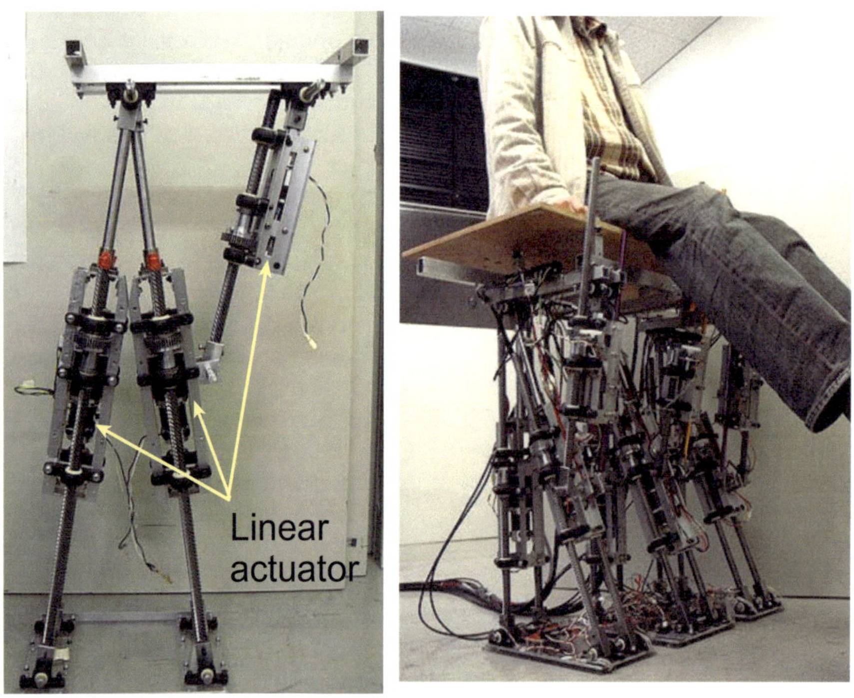

Fig. 1. Human carrying robot

2.2 Inverse kinematics

A local servo control in terms of linear actuator length is generally convenient to implement whole robot control algorithm. Then it is need to formulate the relationship between each link length and a specific position at free leg and/or at the seat. So, in this section, inverse kinematics for legged robot with linear actuated links is described.

As illustrated in Fig.3, define the heel point of grounded sole is the origin in sagittal plane and

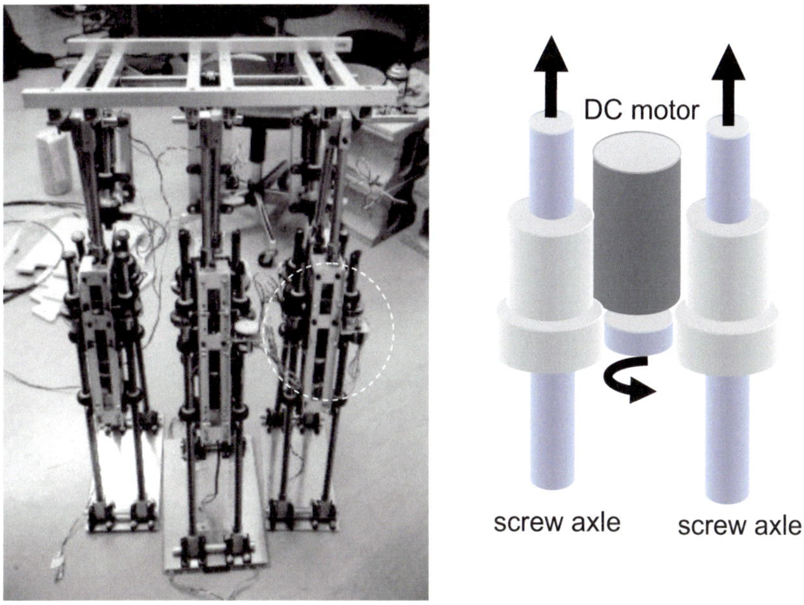

Fig. 2. Twin screw mechanism

also define $\boldsymbol{p}_t := [x_t, z_t, \delta_t]^T$ is the vector of free-leg's tip of toe and $\boldsymbol{p}_w := [x_w, z_w]^T$ is the vector of seat position. Farther, in Fig.3, F, W are sole and seat length, L is base length of each link and w_ε is short offset, which are all fixed.

At first, the approach to calculating link length $\boldsymbol{r}_g := [r_1^g, r_2^g, r_3^g]^T$ of the grounded leg is given. Based on geometrical structure,

$$r_1^g = \sqrt{x_w^2 + z_w^2} \tag{1}$$

$$r_2^g = \sqrt{(x_w - F)^2 + z_w^2} \tag{2}$$

can be immediately calculated. From the vector equation

$$\boldsymbol{L} + \boldsymbol{w}_\varepsilon + \boldsymbol{r}_3^g = \boldsymbol{W},$$

the following procedure in terms of r_3^g is given:

$$\begin{aligned} \|\boldsymbol{r}_3^g\|^2 &= (\boldsymbol{W} - \boldsymbol{L} - \boldsymbol{w}_\varepsilon)^T(\boldsymbol{W} - \boldsymbol{L} - \boldsymbol{w}_\varepsilon) \\ &= W^2 + L_0^2 - 2\boldsymbol{W}^T(\boldsymbol{L} + \boldsymbol{w}_\varepsilon) \\ &= W^2 + L_0^2 - 2W(L\cos\zeta - w_\varepsilon \sin\zeta) \end{aligned}$$

where

$$\zeta = \tan^{-1}\frac{-z_w}{F - x_w}. \tag{3}$$

As a result we obtain

$$\boldsymbol{r}_3^g = \sqrt{W^2 + L_0^2 - 2W(L\cos\zeta - w_\varepsilon \sin\zeta)}. \tag{4}$$

Next, link length $\boldsymbol{r}_f := [r_1^f, r_2^f, r_3^f]^T$ of free leg also can be given by similar approach. The followings are resultingly obtained:

$$\begin{aligned} r_1^f &= \sqrt{(\boldsymbol{p}_t - \boldsymbol{F} - \boldsymbol{p}_w)^T(\boldsymbol{p}_t - \boldsymbol{F} - \boldsymbol{p}_w)} \quad (5) \\ &= \sqrt{(x_t - F\cos\delta_t - x_w)^2 + (z_t - F\sin\delta_t - z_w)^2} \\ r_2^f &= \sqrt{(x_w - x_t)^2 + (z_w - z_t)^2} \quad (6) \\ r_3^f &= \sqrt{W^2 + L_0^2 - 2W(L\cos\phi - w_\varepsilon \sin\phi)} \quad (7) \end{aligned}$$

where

$$\phi = \tan^{-1}\frac{z_t - z_w}{x_t - x_w}. \quad (8)$$

According to above proceeding, if the kinematics in terms of positional vector is represented by

$$\begin{bmatrix} \boldsymbol{p}_w \\ \boldsymbol{p}_t \end{bmatrix} = \boldsymbol{f}(\boldsymbol{r}_g, \boldsymbol{r}_f) = \begin{bmatrix} \boldsymbol{f}^w(\boldsymbol{r}_g) \\ \boldsymbol{f}^t(\boldsymbol{r}_g, \boldsymbol{r}_f) \end{bmatrix},$$

then we define the inverse kinematics explicitly solving all links length as

$$\begin{bmatrix} \boldsymbol{r}_g \\ \boldsymbol{r}_f \end{bmatrix} = \boldsymbol{f}^{-1}(\boldsymbol{p}_w, \boldsymbol{p}_t) = \begin{bmatrix} \boldsymbol{f}_g^{-1}(\boldsymbol{p}_w) \\ \boldsymbol{f}_f^{-1}(\boldsymbol{p}_w, \boldsymbol{p}_t) \end{bmatrix}.$$

3. Control strategy for advanced mobility

3.1 Non-contact impedance method

In the case where mobile robot is used in public facility and the home environment, in order to not damage the robot itself and environmental objects, it is necessary to take into account for interactive force between end-effector and circumstance. Impedance control is often used for such control method that makes robot's end-effector have desired force and realizes mechanical compliance.

In a general way, the realization of impedance control necessitates a certain contact with external objects. For safety's sake, however, it is desirable for mobile robot to avoid obstacles without contacting external objects. From these reasons, this paper proposes a method of avoiding obstacles by non-contact impedance control.

The concept of non-contact impedance control is illustrated in Fig.4. Assume that virtual contact force occurs via virtual spring and damper, if an obstacle trespasses on virtual surface postulated at the tip of a toe. Assume $\boldsymbol{\xi} := [\xi_x, \xi_z]^T$ to be trespassing vector on virtual surface, virtual contact force is given by

$$\boldsymbol{u}_v = \begin{cases} \boldsymbol{J}^T(\boldsymbol{r}_f)(\boldsymbol{D}_v\dot{\boldsymbol{\xi}} + \boldsymbol{K}_v\boldsymbol{\xi}) & (\boldsymbol{\xi} < 0) \\ \boldsymbol{0} & (\boldsymbol{\xi} \geq 0) \end{cases} \quad (9)$$

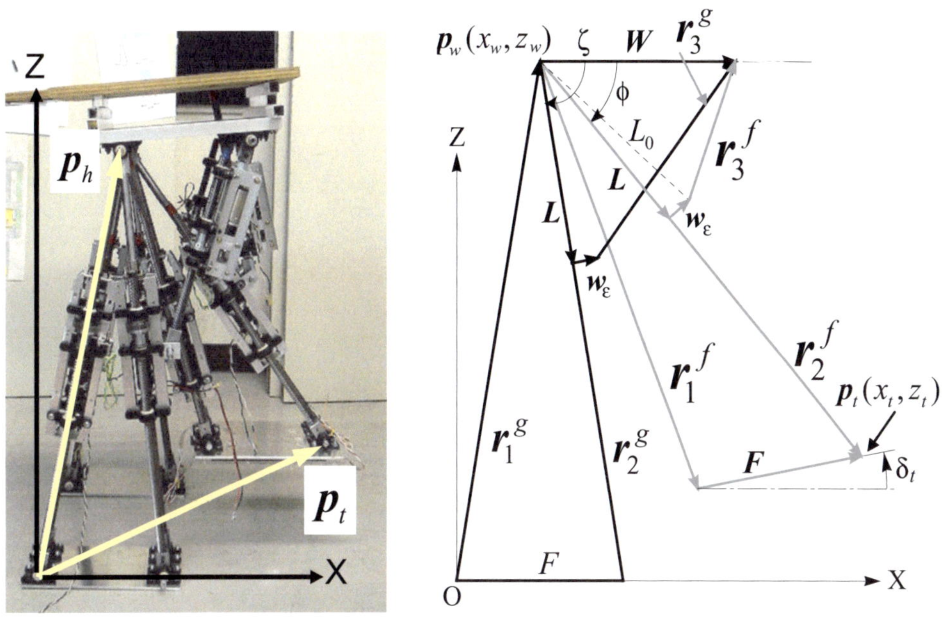

Fig. 3. Linear actuated legs configuration

here, K_v, D_v are virtual spring and viscosity coefficient matrix respectively. $J(r_f)$ is Jacobian-matrix defined by

$$J(r_f) = \begin{bmatrix} \frac{\partial f_1^t}{\partial r_1^f} & \frac{\partial f_1^t}{\partial r_2^f} & \frac{\partial f_1^t}{\partial r_3^f} \\ \frac{\partial f_2^t}{\partial r_1^f} & \frac{\partial f_2^t}{\partial r_2^f} & \frac{\partial f_2^t}{\partial r_3^f} \end{bmatrix}. \tag{10}$$

In actual robot control, virtual surface can be formed in realtime by measuring the distance between obstacle and end-effector using position sensitive detector (PSD).

3.2 Control method

The ideal task of mobile robot is to realize intelligent behavior robustly even if planed reference path is obstructed by irregular pitch steps or wall projection. As mentioned above section, our control strategy combines tracking control for planed reference path with force control by virtual contact impedance.

The proposed control method, as shown in Fig.5, consists of three control law Eq.(11), which are reference trajectory control u_t, virtual contact control u_v and sole counter force control u_s.

$$u = u_t + u_s + u_v \tag{11}$$

u_t is determined by servo control loop for reference trajectories in terms of links length which are precedently given to realize a walking pattern. u_s is used to compensate the gap between the sole and the floor if the robot does not poise owing to the transition of center of mass, external force, etc.

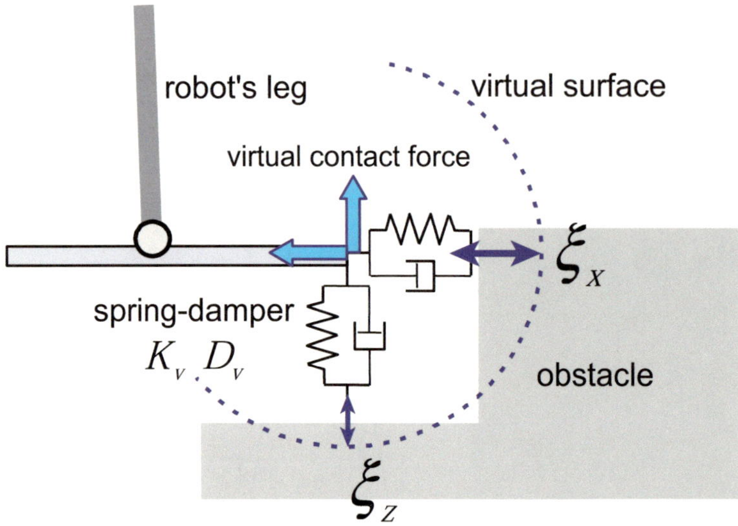

Fig. 4. Concept of non-contact impedance control

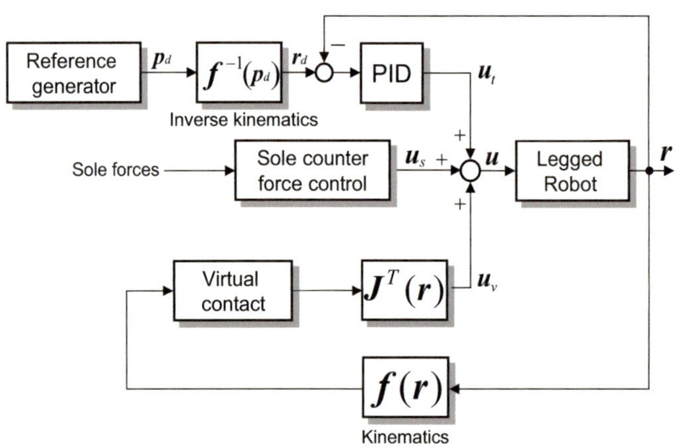

Fig. 5. Control system

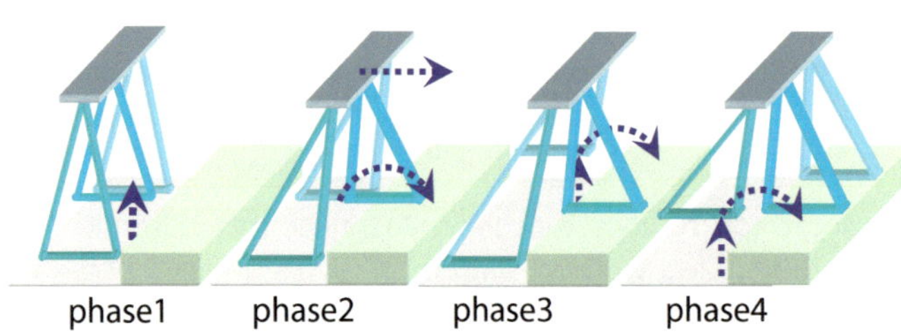

Fig. 6. Walking sequence

3.3 Walking pattern

In order to make the mobile robot proceed to a certain direction, some locomotion pattern is needed beforehand. Then it is useful to generate corresponding trajectories along the locomotion pattern in realtime and to realize feedback control for such trajectories.
A human carrying robot proposed in this paper must be able to move three legs individually in consideration of high mobility and stability. This paper considers the gait in which a single leg travels for free-leg and the other two legs keep on standing as fundamental walking pattern. Since such gait can posture robot's center of mass inside polygonal area formed by ground soles, static walk becomes to practicable. Each phase in proposed walking pattern, as illustrated in Fig. 6, is briefly explained as follows:

phase-1 : As swing up center leg, detect the edge of step based on the degree of change of PSD value. At the same time compute the height of step by downward PSD.

phase-2 : Give a reference elliptical orbit in order that for free sole to land on the step. Then give a horizontal orbit for seat position in keeping current height.

phase-3 : Swing left leg up to step height (which has already been given at phase-1. And give a reference elliptical orbit in order that for its sole to land on the step.

phase-4 : For right leg, give same operation as phase-3.

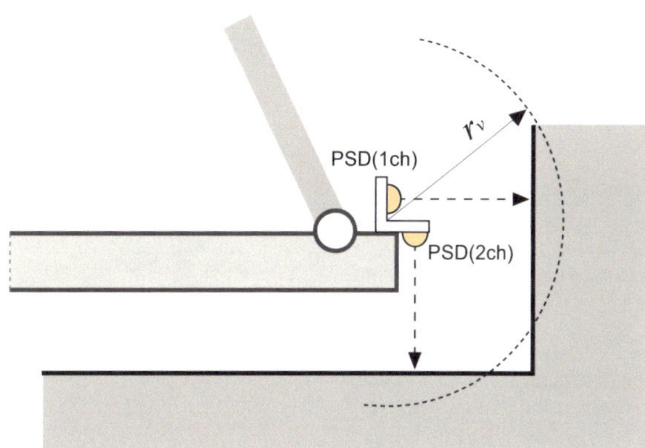

Fig. 7. Location of PSD

4. Experiment

4.1 Walking up stair

An experiment to avoid a stair installed in front of the robot is tried. In this experiment, while the distance between robot and stair can be measured by PSD in real time and the radius of virtual surface is given as $r_v = 5$ [cm], notice that the height of stair is unknown to human. Fig. 7 shows the location of 2ch PSD assigned for virtical and horizontal measures.
Fig. 8 presents sequential motion captures in walking up a stair, where a man with about 50[kg] weight is embarking on the seat. Indeed, because the robot travels in constantly standing by two legs, it is confirmed that the behavior of the robot loading a human is stable. Fig. 9(b) shows trajectories and virtual force at tip of toe, where dashed line means reference trajectory planned at phase-1, phase-2 in walking pattern. For comparison, the trajectory without virtual contact control is also shown in Fig. 9(a). As shown in Fig. 9, it is found

that as tip of toe approaches edge of stair closer, virtual force increases by virtual contact efficiency. Consequently it is confirmed that proposed virtual contact method can avoid the obstacle even if reference trajectory which approaches obstacle closely is applied.

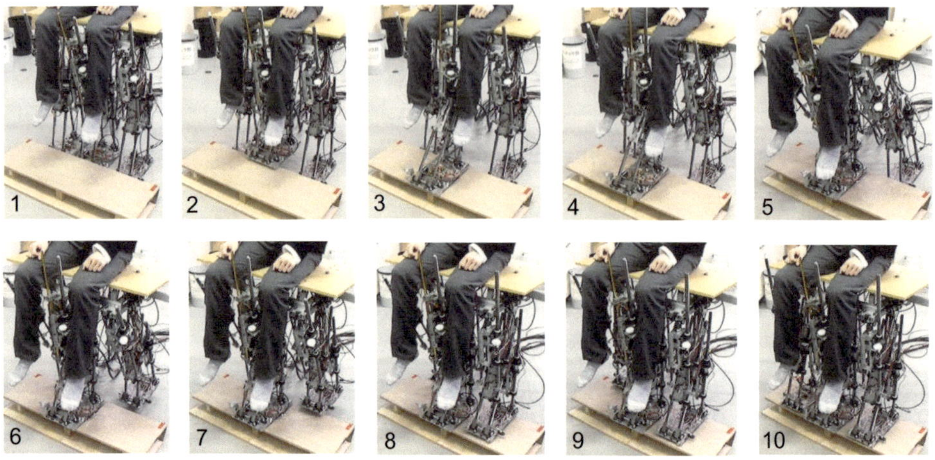

Fig. 8. Sequential motion captures

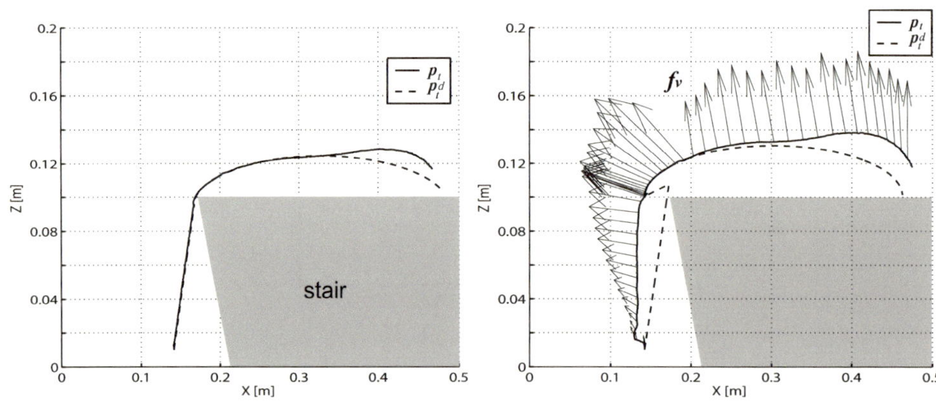

(a) without Virtual contact control (b) with Virtual contact control

Fig. 9. Trajectory and virtual force at tip of toe

4.2 Soft landing

The efficiency that alleviate the robot's impact in landing, i.e. soft-landing is expected by applying a variable gain $\boldsymbol{\alpha}$ to control law

$$\boldsymbol{u} = \boldsymbol{\alpha}\boldsymbol{u}_t + (I - \boldsymbol{\alpha})\boldsymbol{u}_v + \boldsymbol{u}_s \ \ (0 < \alpha_i < 1) \,. \tag{12}$$

Then, as second experiment, we verify the efficiency of soft-landing by measuring the degree of pitch in landing phase. Fig. 10(a) shows vertical acceleration and velocity at the seat when $\boldsymbol{\alpha} = I$, while Fig. 10(b) shows same datum when virtical contact control is given by

$$\boldsymbol{\alpha} = \mathrm{diag}\{0.15, \ \ 0.25\}$$

in trial and error. Obviously the degree of pitch controlled by virtual contact is more suppressed than the case not controlled. As a result, it is expected that mechanical landing impact and the fearfulness for the passenger are alleviated.

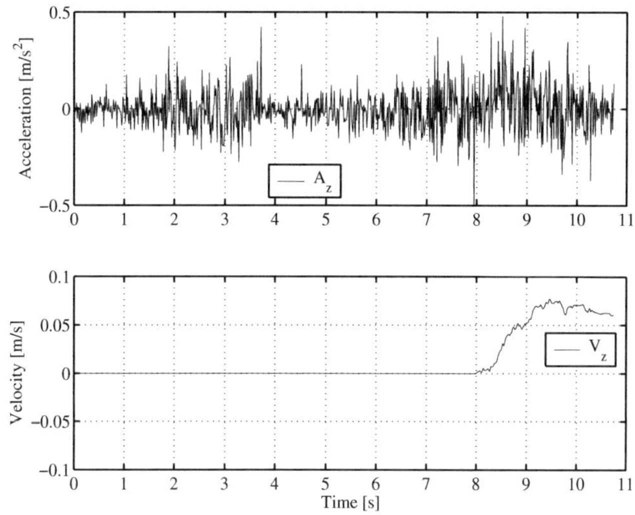

(a) without Virtual contact control

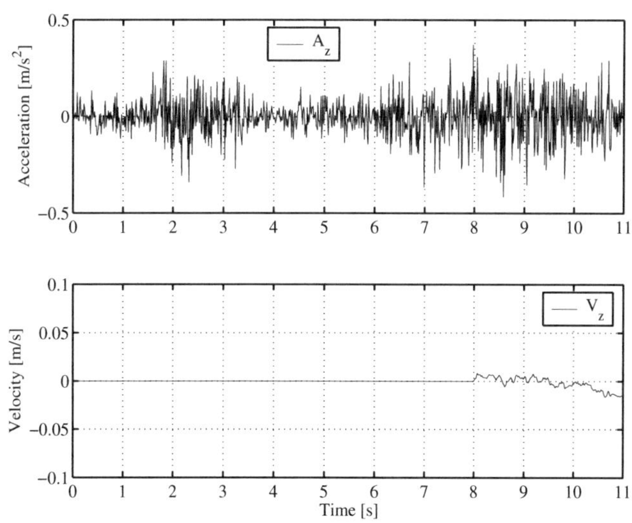

(b) with Virtual contact control

Fig. 10. Degree of pitch at the seat

5. Conclusion

In this paper, we have developed three-legged mobile robot which has the ability to carry the human by using has 9 linear actuated links. Each leg consists of three linear actuators assigned to triangular configuration in order to maintain the rigidity. For such linear structure, the inverse kinematics has been given. The inverse kinematics for local servo control is also given. As a method for avoiding the obstacles, we have proposed the virtual impedance control method which can avoid the stairs without contacting by receiving virtual force from virtual spring and damper installed in virtual surface region. Finally, the mobility for avoiding unknown height step and soft-landing motion have been confirmed through some experiments.

6. References

http://www.ibotnow.com

http://www.sunwa-jp.co.jp/en/

Sugahara, Y. et al. (2004). Realization of Dynamic Human-Carrying Walking by a Biped Locomotor, *Proceedings of the IEEE ICRA 2004*, pp.3055-3060.

Tsuji, T. & Kaneko, M. (1999). Noncontact Impedance Control for Redundant Manipulators, *IEEE Trans. on Systems, Man, and Cybernetics-Part A: Systems and Humans*, Vol.29, No.2, pp.184-193.

L., Villani et al. (2000). An Experimental Study of Adaptive Force/Position Control Algorithms for an Industrial Robot, *IEEE Trans. on Control Systems Technology*, Vol.8, No.5, pp777-786.

Wu, Y. & Higuchi, M. (2004). Development of a Power Assisting System of a Walking Chair, *Proceedings of 2004 IEEE/RSJ Int. Conf. on IROS*, pp.3207-3212.

4

Development of a Simulation Environment Applied to the Study of Fault-Tolerant Control Systems in Robotic Manipulators. Theoretical and Practical Comparisons

Claudio Urrea and John Kern
Departamento de Ingeniería Eléctrica, DIE
Universidad de Santiago de Chile, USACH, Santiago
Chile

1. Introduction

Industrial processes governed by automatic controllers can be subject to failures. Their incorrect operation can cause economic losses, workers' hazards, inconvenience for users, etc. Moreover, automation of processes by means of automatic control loops, although permitted freeing human operators from manual control and operation, has not immunized them from failures. A way to enhance their reliability is to provide them with tolerance mechanisms to face these failures (Blanke et al., 2001).

It is understood as 'failure' every change in the behavior of any component of the system (non-permitted deviation of any characteristic property or parameter), so the system cannot satisfy the function it has been designed for (Blanke et al., 2006). Control loops can hide failures preventing their detection to the point of causing irreparable damages forcing to stop the system or process.

Because of this, there is a growing need and concern for developing control systems capable of acceptable operation even in the case of failures, also being able to stop the process before irreparable damages arise. This kind of control systems are called 'failure tolerant': control systems that can, in the one hand, detect incipient failures in sensors and/or actuators; and, in the other, adapt control laws quickly to preserve the specified operation in terms of production quality, safety, etc. (Bonivento et al., 2004). Figure 1 shows the block diagram corresponding to a fault-tolerant control system.

(Blanke et al., 2006).

In this chapter the development of a simulation environment applied to the study of fault tolerant control systems for robotic manipulators by employing MatLab-Simulink programming tools is presented. By using this simulator is possible to represent the behavior of a manipulator: in the presence of a failure into one of its actuators and the status of correction of such failure, through an active fault tolerant control system. The end-effector Cartesian trajectories and the evolution of the joints of the manipulator are provided by means of animations and graphics in a comfortable and intuitive programming environment.

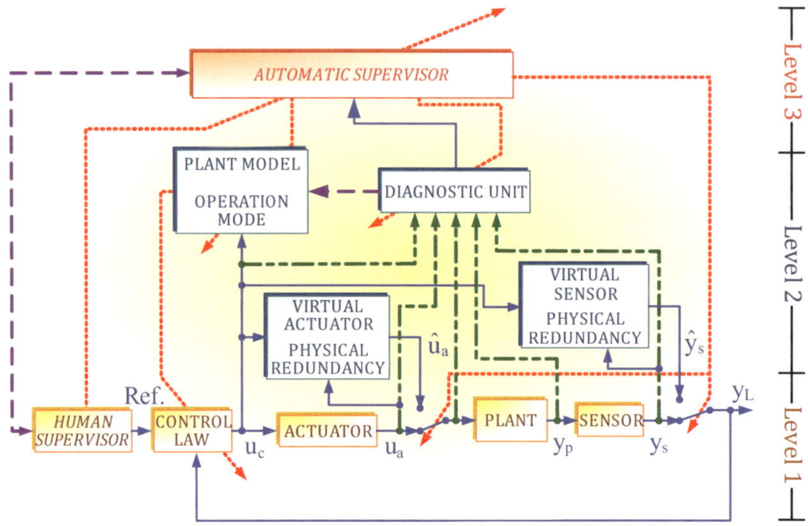

Fig. 1. Block diagram of a fault tolerant control system

2. Fault tolerant control

In simple terms, we can say that classic control is based in a control law ϑ that considers a set of objectives O and a set of constraints $\Re$, associated to the system's dynamic model. Such dynamic model frequently presents some inaccuracies, if we consider, for example (Arteaga et al., 2004).

- External perturbations.
- Time-varying parameters.
- Parameters with uncertainties.
- Noise in measurements.
- Unmodeled dynamics.

what brings as a consequence some limitations in the use of such classic controllers that need exact system parameters for a correct performance. A way to confront those uncertainties is to consider a fixed structure model that includes a set of unknown parameters to apply robust control and adaptive control techniques. Robust control and adaptive control for robots consider satisfactory performance in spite of uncertainties in inertial parameters of the joint model, external interference and unmodeled dynamics behaviour (Spong et al., 2005), (Kelly et al., 2005), (Craig, 1988). Specifically, robust control includes a control law that takes into account all the set of objectives and all the set of Θ parameters (known and unknown), and adaptive control is based on the estimation of on-line parameters, *i.e.* the value of unknown $\hat{\theta}$ parameters in the set of all possible system parameters.

When a failure in the system occurs, it is necessary to modify the set of objectives so to obtain an adequate control law to face the event, since both system constraints ($\hat{\Re}_f$) and unknown parameters ($\hat{\theta}_f$) can experience modifications. In order to overcome those requirements, fault tolerant control considers the modification of the set of objectives (O_f), creating the conditions of optimum operation and degraded operation for the system with no failures and in presence of failures, respectively, as shown in figure 2 (Blanke et al., 2001), (Puig et al., 2004), (Arteaga et al., 2004).

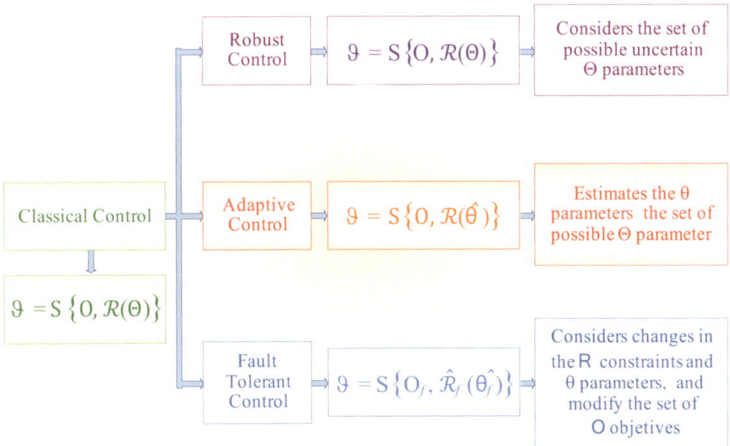

Fig. 2. Schematic diagram showing the types of control used in systems with parameter uncertainties and failures

The occurrence of failures in the system and the activation of tolerance mechanisms can be modeled as time discrete events, therefore fault tolerant control systems have and hybrid nature (Tsuda et al., 2001). When a sensor failure arises, this can be detected because the sensor reading will exceed some pre-determined threshold, in comparison with an expected value; under these conditions, fault tolerant control will not consider the information coming from the defective sensor, locking the associated joint. If there's a failure in a motor without a safety backup, the joint locks again, and the robot loses mobility of this joint. In any case, the robot loses a part of its movement range, and the ability to carry out its assigned tasks is limited (El-Salam et al., 2005).

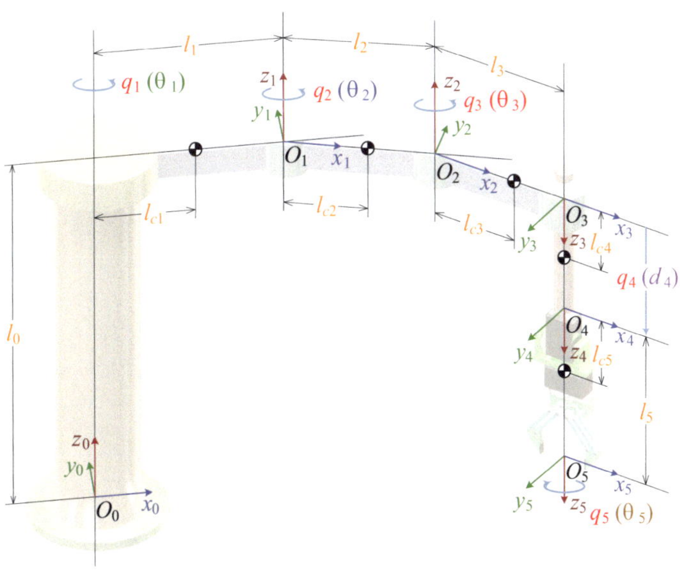

Fig. 3. Scheme of a redundant robotic manipulator of SCARA type

Having in mind the above mentioned failure behaviors, we will consider now the three first DOF of the robotic manipulator shown in figure 3, and the appearance of a failure in an actuator causing the blocking of the joint it commands, specifically joint number two, shown (in red) in figure 4. In such a situation, we consider a desired destiny point, given by coordinates (*x*,*y*), that under the action of a classic controller and in presence of the above mentioned failure, the manipulator is far from reaching.

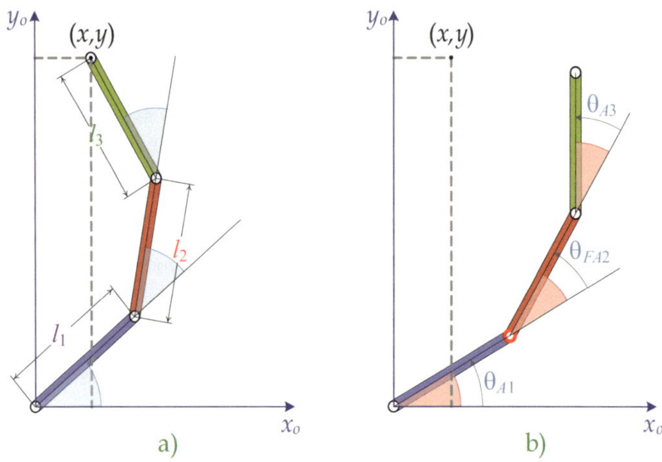

Fig. 4. Disposition of the three first DOF in the *x*-*y* plane of the redundant manipulator: a) Arriving to the destiny position (*x*,*y*); b) Not arriving to the destiny position (*x*,*y*): failure in actuator 2

For the design of fault tolerant control, it has been considered the generation of a virtual link, from the failing actuator that is blocked and its adjacent links, as seen in figure 5 (gray color):

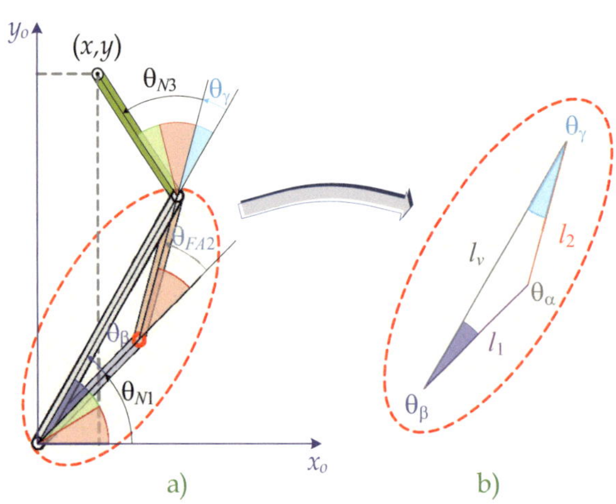

Fig. 5. Disposition of the three first DOF in the *x*-*y* plane of the redundant manipulator: a) Additional displacement of actuators 1 and 3, b) Equivalent diagram of displacement angles

3. Fault tolerant controller: Computed torque

Considering the hybrid nature of fault tolerant control, it is proposed an active fault tolerant control, including a different control law in function of the robotic manipulator status, *i.e.*, normal or failing, sensing possible failures on-line and, in accordance with that, reconfiguring the controller by selecting the most adequate control law (changing inputs and outputs) (figure 6). As classic controller it has been selected the computed torque controller. If we consider that the dynamic model of a manipulator with n joints can be expressed through the equation (1):

$$\tau = \mathbf{M}(q)\ddot{\mathbf{q}} + \mathbf{C}(q,\dot{q}) + \mathbf{G}(q) + \mathbf{F}(\dot{q}) \tag{1}$$

where:

τ : Vector of Generalized forces ($n\times1$ dimension).
$\mathbf{M}$: Inertia matrix ($n\times n$ dimension).
$\mathbf{C}$: Vector of centrifugal and Coriolis forces ($n\times1$ dimension).
q : Components of joint position vector.
$\dot{q}$: Components of joint speed vector.
$\mathbf{G}$: Vector of gravity force ($n\times1$ dimension).
$\ddot{q}$: Joint acceleration vector ($n\times1$ dimension).
$\mathbf{F}$: Friction forces vector ($n\times1$ dimension).

The control law by computed torque consists in the application of a computed torque in order to compensate centrifugal, Coriolis, gravitatory and friction effects, as can be seen in equation (2).

$$\tau = \hat{\mathbf{M}}(q)\left(\ddot{\mathbf{q}}_\mathbf{d} + \mathbf{K_v}\dot{\mathbf{q}}_\mathbf{e} + \mathbf{K_p}\mathbf{q}_\mathbf{e}\right) + \hat{\mathbf{C}}(\dot{q},\dot{q}) + \hat{\mathbf{G}}(q) + \hat{\mathbf{F}}(\dot{q}) \tag{2}$$

where:

$\hat{\mathbf{M}}$: Estimation of inertia matrix ($n\times n$ dimension).
$\hat{\mathbf{C}}$: Estimation of centrifugal and Coriolis forces vector ($n\times1$ dimension).
$\hat{\mathbf{G}}$: Estimation of the gravity force vector ($n\times1$ dimension).
$\hat{\mathbf{F}}$: Estimation of the friction forces vector ($n\times1$ dimension).

$$\mathbf{K_v} = \begin{bmatrix} \mathbf{K}_{v1} & & & \\ & \mathbf{K}_{v2} & & \\ & & \ddots & \\ & & & \mathbf{K}_{vn} \end{bmatrix} \tag{3}$$

Kv : Definite positive diagonal matrix ($n\times n$ dimension).

$$\mathbf{K_p} = \begin{bmatrix} \mathbf{K}_{p1} & & & \\ & \mathbf{K}_{p2} & & \\ & & \ddots & \\ & & & \mathbf{K}_{pn} \end{bmatrix} \tag{4}$$

Kp : Definite positive diagonal matrix ($n \times n$ dimension).
$\ddot{\mathbf{q}}_{\mathbf{d}}$: Joint desired acceleration vector ($n \times 1$ dimension).

$$\mathbf{q_e} = \mathbf{q_d} - \mathbf{q} \tag{5}$$

$\mathbf{q_e}$: Joint position error vector ($n \times 1$ dimension).

$$\dot{\mathbf{q}}_{\mathbf{e}} = \mathbf{q_d} - \dot{\mathbf{q}} \tag{6}$$

$\dot{\mathbf{q}}_{\mathbf{e}}$: Joint speed error vector ($n \times 1$ dimension).
Matching equations (1) and (2), we have:

$$\hat{\mathbf{M}}(q)\left(\ddot{\mathbf{q}}_{\mathbf{d}} + \mathbf{K_v}\dot{\mathbf{q}}_{\mathbf{e}} + \mathbf{K_p}\mathbf{q_e}\right) + \hat{\mathbf{C}}(q,\dot{q}) + \hat{\mathbf{G}}(q) + \hat{\mathbf{F}}(\dot{q}) = \mathbf{M}(q)\ddot{\mathbf{q}} + \mathbf{C}(q,\dot{q}) + \mathbf{G}(q) + \mathbf{F}(\dot{q}) \tag{7}$$

$$\ddot{\mathbf{q}}_{\mathbf{e}} + \mathbf{K_v}\dot{\mathbf{q}}_{\mathbf{e}} + \mathbf{K_p}\mathbf{q_e} = \hat{\mathbf{M}}(q)^{-1}\left(\left(\mathbf{M}(q) - \hat{\mathbf{M}}(q)\right)\ddot{\mathbf{q}} + \left(\mathbf{C}(q,\dot{q}) - \hat{\mathbf{C}}(q,\dot{q})\right) + \left(\mathbf{G}(q) - \hat{\mathbf{G}}(q)\right) + \left(\mathbf{F}(\dot{q}) - \hat{\mathbf{F}}(\dot{q})\right)\right) \tag{8}$$

If estimation errors are low, joint errors approach to a linear equation:

$$\ddot{\mathbf{q}}_{\mathbf{e}} + \mathbf{K_v}\dot{\mathbf{q}}_{\mathbf{e}} + \mathbf{K_p}\mathbf{q_e} \approx \mathbf{0} \tag{9}$$

The fault tolerant control law used for development of the simulator comprises two classic controllers per each computed torque that are "switched" to reconfigure the fault tolerant controller.

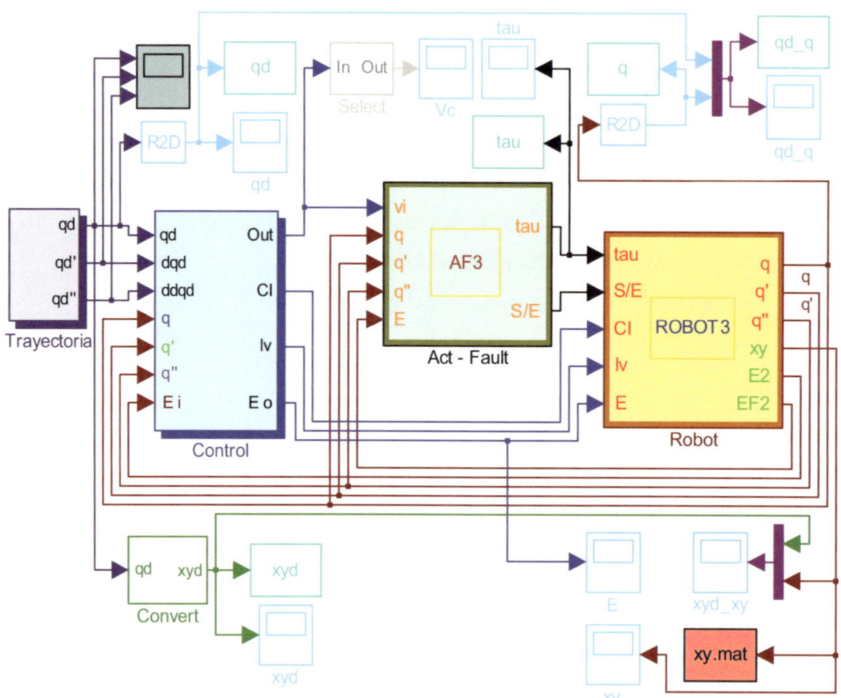

Fig. 6. Simulation diagram of the fault tolerant control system

4. Simulator development

In the development of the simulator, besides the control laws pointed in previous section and the respective manipulator dynamics, actuators dynamics is included corresponding to analogue servomotors with their characteristic position internal feedback. Each model has been carried out using MatLab embedded functions, grouped in distinctive blocks, as follows:

- Trajectory block.
- Control block.
- Actuator-failure block.
- Robot block.

as shown in figure 6.

Control block contains the control laws necessary to provide the manipulator with the ability to follow a desired trajectory, in normal operation and in presence of a failure, reconfiguring the controller in this last case to provide adequate signals to the actuators. In figure 7 we can see an inner view of the control block; this view has been divided in four stages for a better understanding.

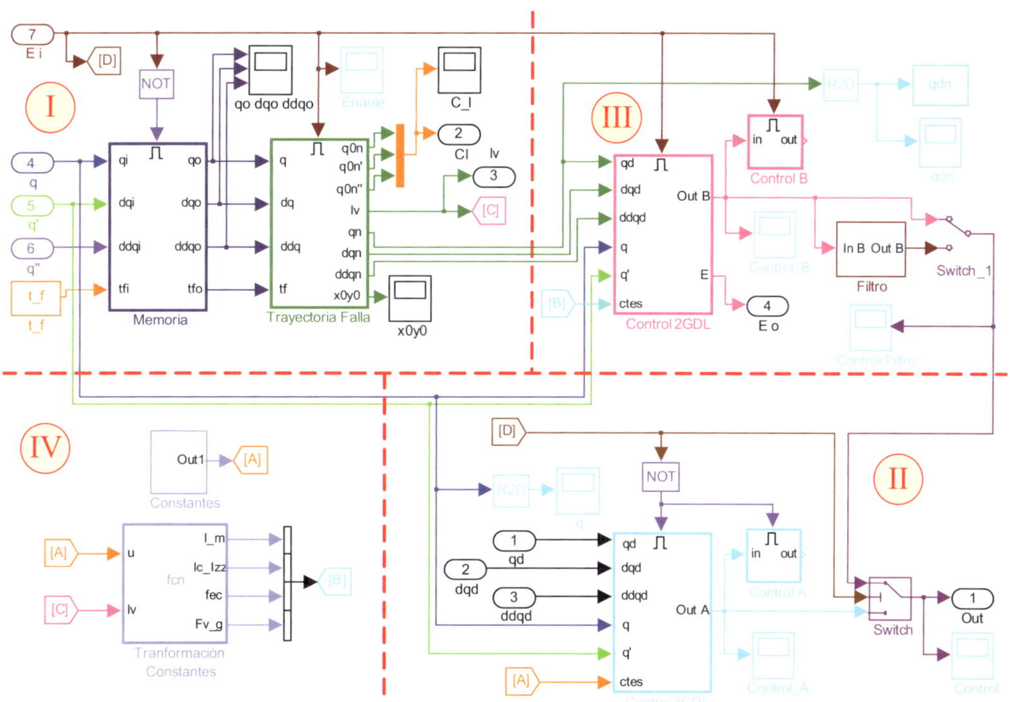

Fig. 7. Simulation diagram including the stages of the control block

Figure 8 shows stage I of the control block, dedicated to memorize the status of: position, speed and acceleration of the joints at the moment when a failure occurs, as well as the time when it happens. With this information, new joint trajectories can be generated.

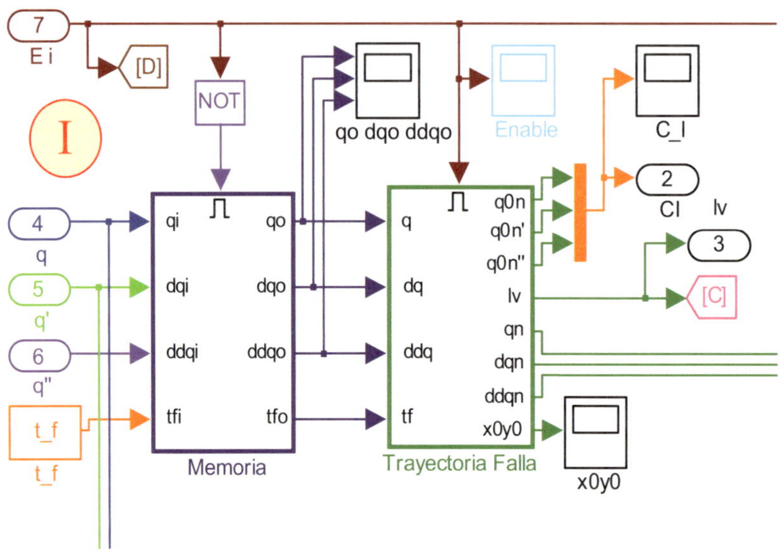

Fig. 8. Simulation diagram: stage I of the control block

Figure 9 shows stage II of the control block, used for processing the signals corresponding to the original desired trajectory.

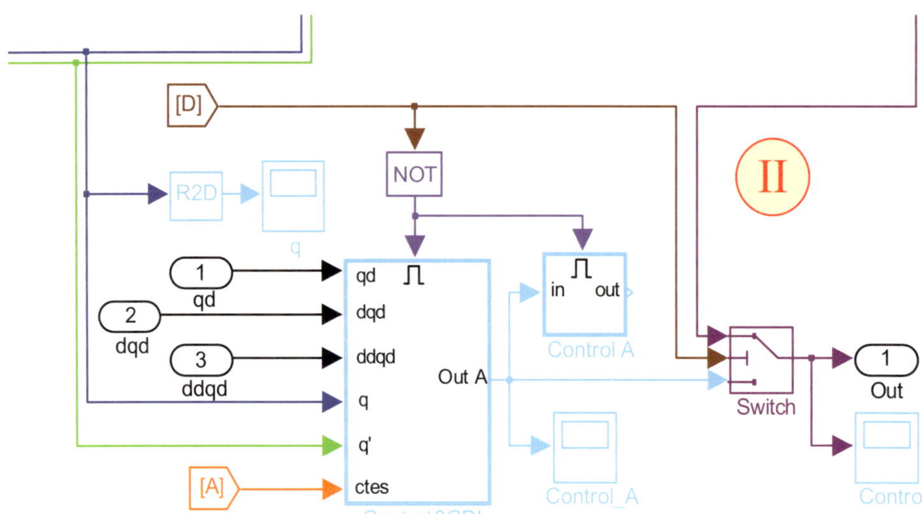

Fig. 9. Simulation diagram: stage II of the control block

The stage III of control block is shown in figure 10. This stage is dedicated to receive the signals from the new joint trajectories generated in stage I after occurring a failure, and to provide the new control signals into the actuators.

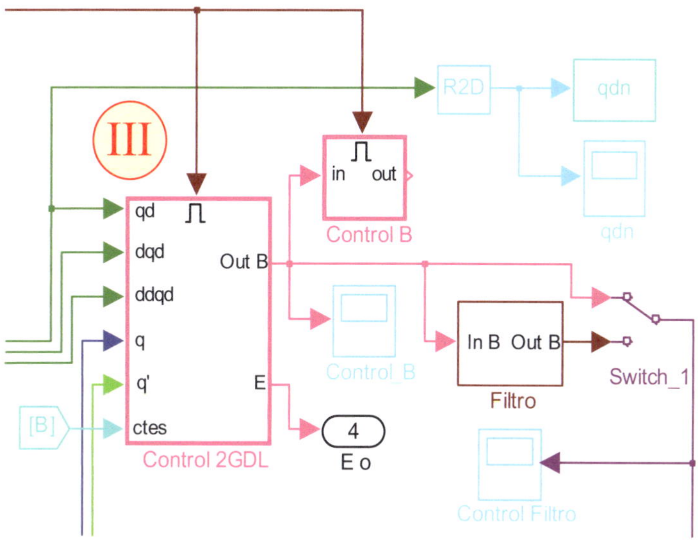

Fig. 10. Simulation diagram: stage III of the control block

When a failure arises in a real manipulator, the properties presented to actuators change, producing new inertias, longitudes, mass centers, etc. In the simulator, this function is carried out in stage IV of the control block, as pointed in figure 11.

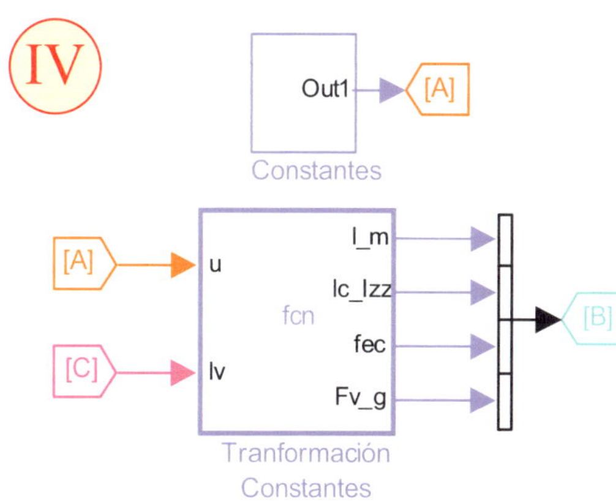

Fig. 11. Simulation diagram: stage IV of the control block

The necessary parameters used in functions developed at stages I to IV, corresponding to the control block, are entered via an interface generated through masking of MatLab embedded functions, this can be seen in figure 12. By means of this interface it is possible to enter the simulator with the controller, robot and failure time parameters.

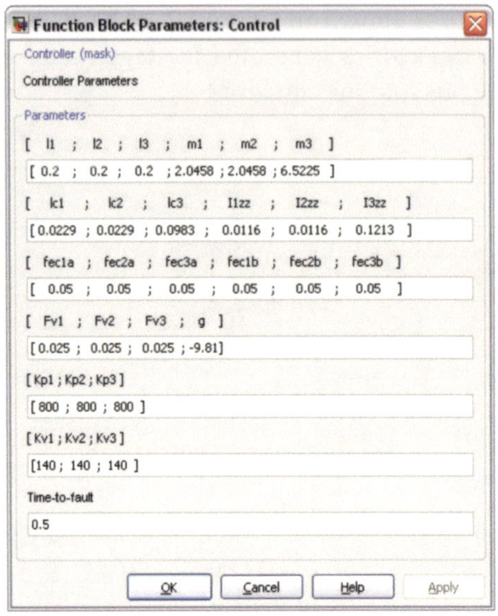

Fig. 12. Window for entering controller parameters

In figure 13 we have an inside view of the robot block, there we can see the blocks containing the dynamics representing manipulator's non-failure and failure behavior. It is shown; also, a block labeled *Switch-Out*, oriented to the switching function at the moment a failure arises.

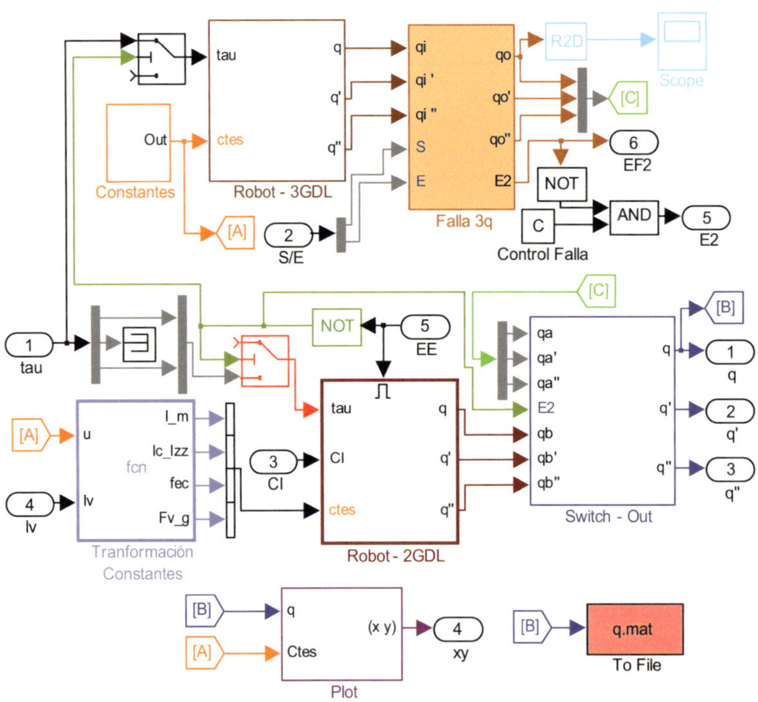

Fig. 13. Simulation diagram corresponding to failing and non-failing manipulator

The interface used for entering parameters of the robot manipulator is shown in figure 14.

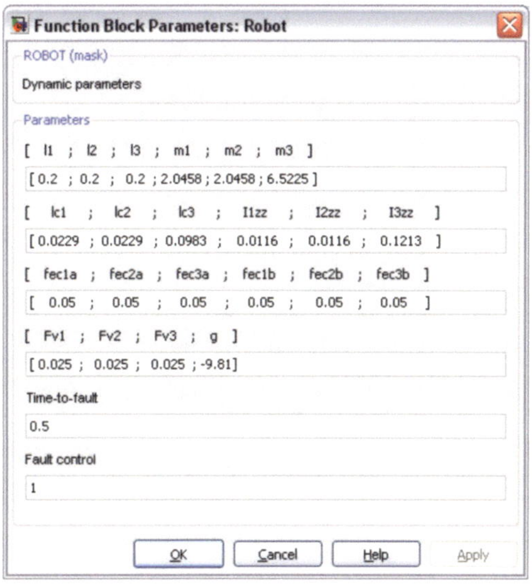

Fig. 14. Window for entering manipulator parameters

The data required for the generation of desired trajectory are charged through "mat" files in the trajectory block, as shown in figure 15.

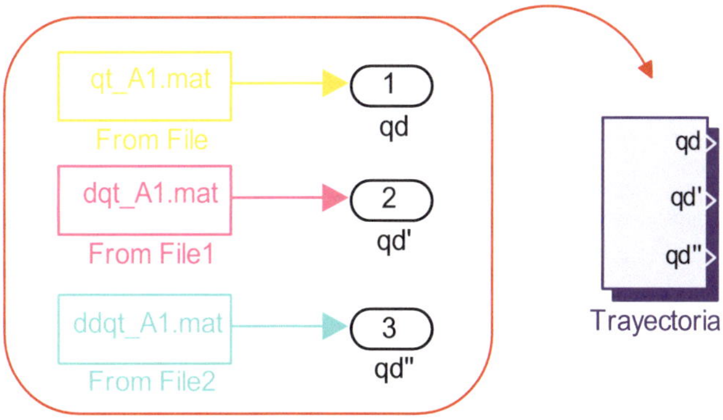

Fig. 15. Simulation diagram corresponding to trajectory generation

Figure 16 shows a view of the interface for analogue servomotor parameter entering.

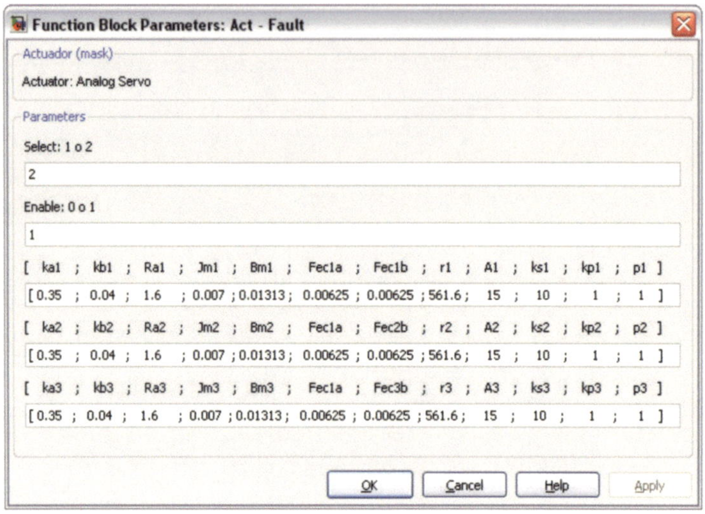

Fig. 16. Window for entering servomotor parameters

Finally, in figure 17 we can see the interface corresponding to the simulation environment for the fault tolerant control system applied on a manipulator. Through this interface, we can summarize the parameter entering windows showed in figures 14 and 16, and we can obtain the respective curves and animations of trajectories for the robot.

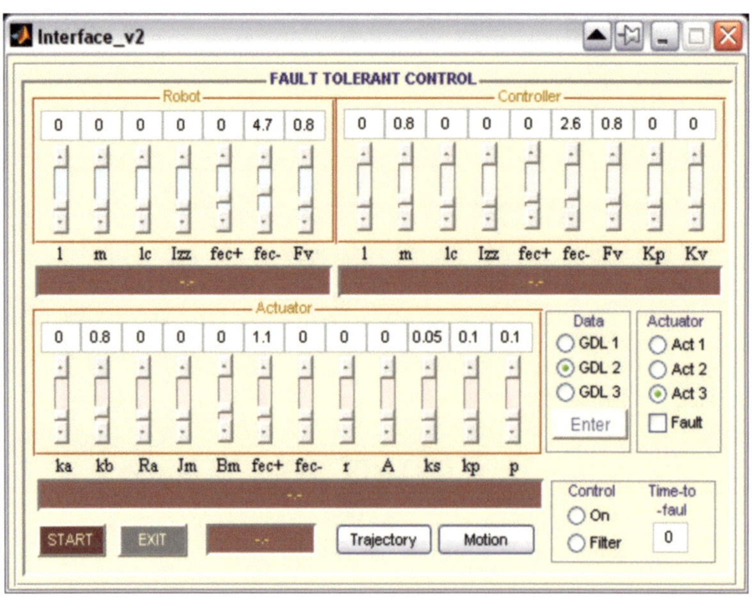

Fig. 17. Simulator user interface for parameter entering and generation of curves and animations

5. Results and conclusions

Here we show results obtained when subjecting the system to a trajectory composed by rectilinear sections. In figure 18 we can see the graphics corresponding to the simulation of trajectory tracking without failures, and in figure 19 a capture of the animation.

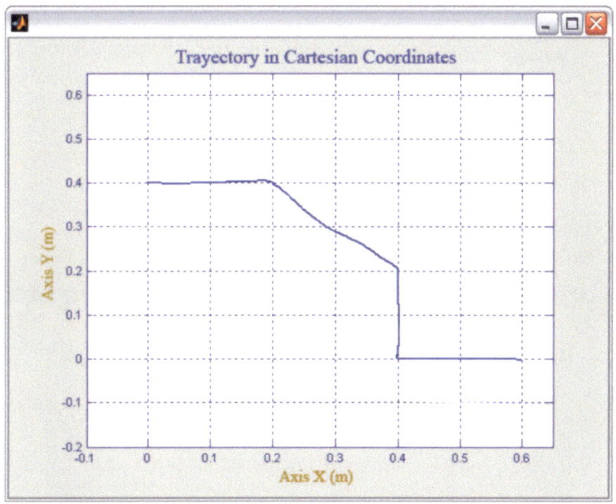

Fig. 18. Trajectory in Cartesian coordinates

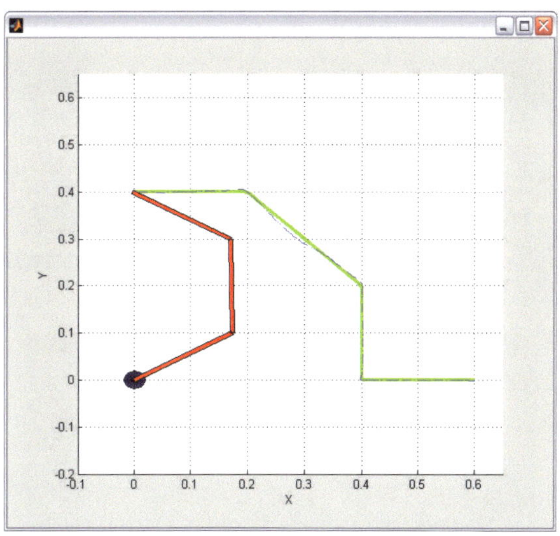

Fig. 19. Animation graphics of the manipulator following a desired trajectory

In figure 20 it is shown the curve corresponding to a simulation of trajectory tracking with a failure occurring in actuator n° 2 at 0.5 sec from initiating trajectory, and in figure 21 a capture of the respective animation, all this under the rule of classic control.

When we subject the manipulator having a failure under the action of the fault tolerant control, the remarkable deviation in Cartesian trajectory produced by classic control (figures 20 and 21) is significantly reduced. This can be seen in the simulation graphics in figure 22, and also in figure 23, showing a capture of the animation.

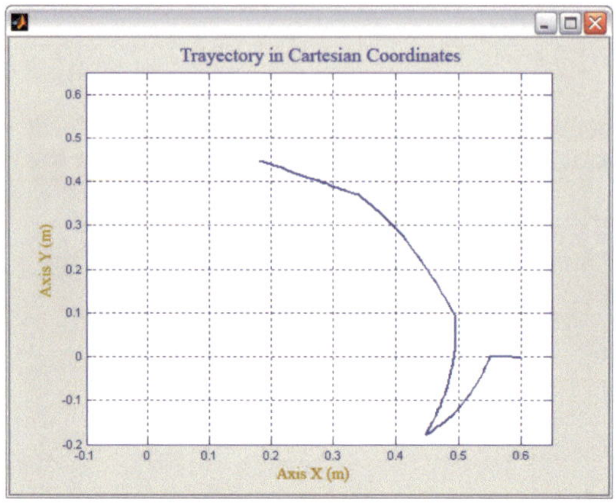

Fig. 20. Trajectory in Cartesian coordinates in presence of failure in an actuator

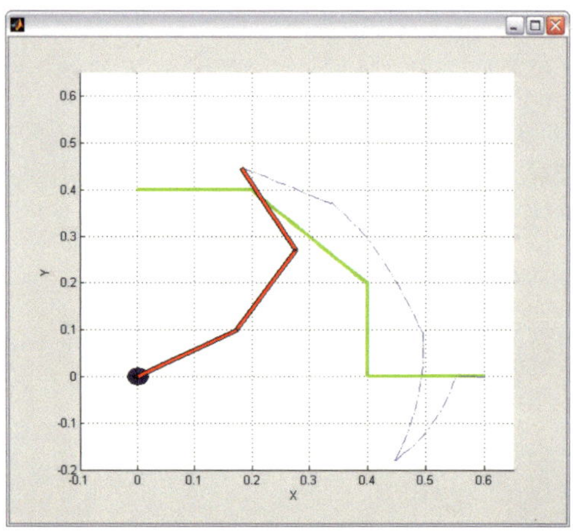

Fig. 21. Animation graphics of the manipulator following a trajectory in presence of failure in an actuator

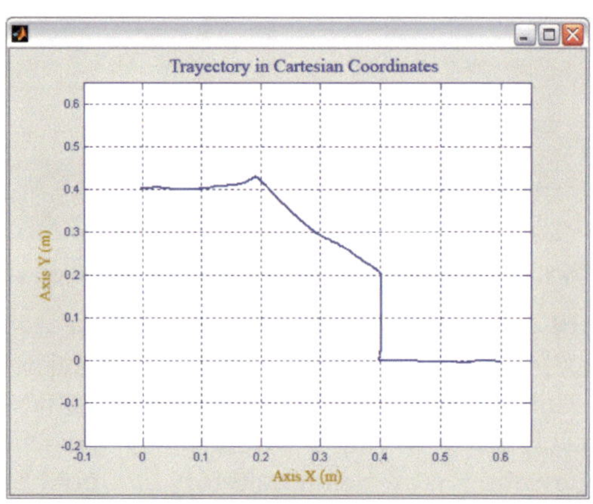

Fig. 22. Trajectory in Cartesian coordinates in presence of failure in an actuator, with failure correction

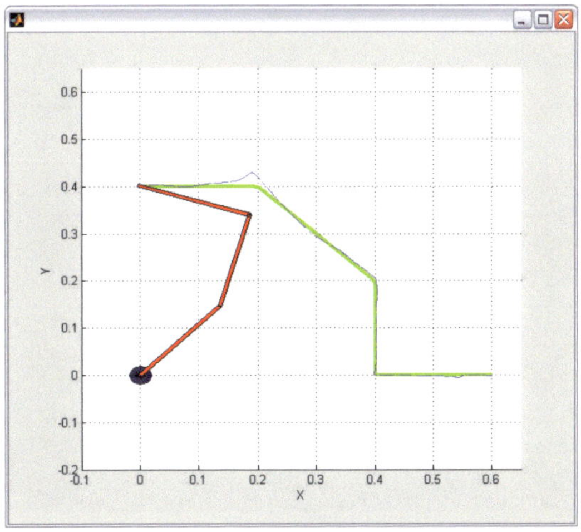

Fig. 23. Animation graphics following a trajectory in presence of failure in an actuator, with failure correction

6. Acknowledgements

This work had the support of the Department for Scientific and Technologic Research of the Universidad de Santiago de Chile, by means of Project 060713UO, Santiago, Chile.

7. Further developments

Thanks to the development of this work, from the implemented simulation tools and the obtained results, fault tolerant control systems essays are being currently carried out, in order to apply them to actual robotic systems, with and without link redundancy, like the SCARA-type robots shown in figure 24 and figure 25, respectively.

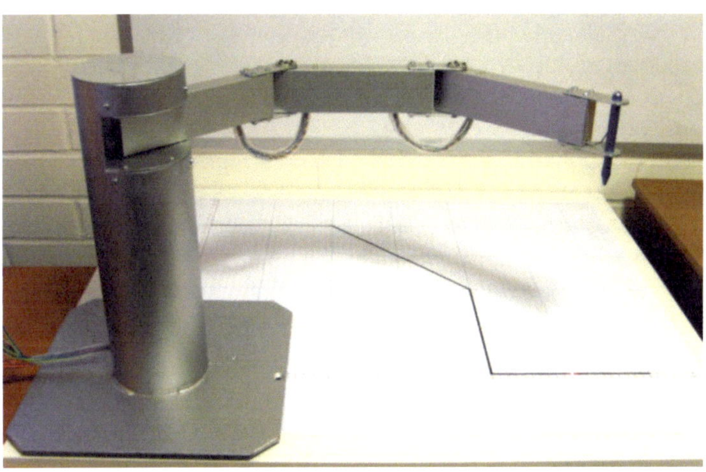

Fig. 24. SCARA-type redundant robot, DIE-USACH

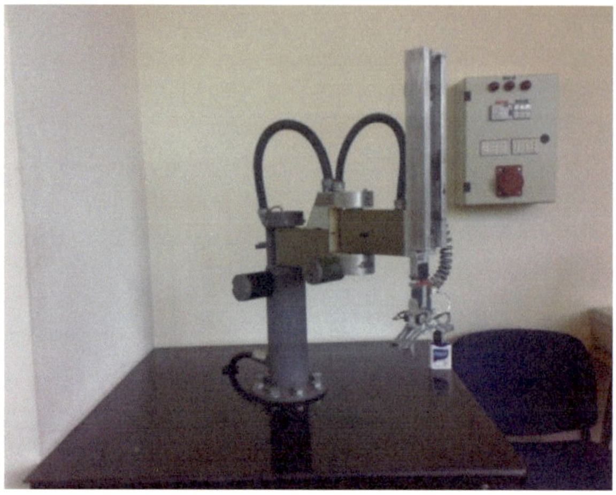

Fig. 25. SCARA-type robot, DIE-USACH

8. References

Blanke, M., Staroswiecki, M. & Wu, E., 2001. *Concepts and Methods in Fault-Tolerant Control*. In Arlington, VA, USA: American Control Conference. Proceedings of the 2001.

Arteaga, F. et al., 2004. Control Tolerante a Fallas Mediante Técnicas de Estabilización Simultanea con Múltiples Dominios de Estabilidad. *Revista Ingeniería UC*, 11(N° 3), pp.62-69.

Blanke, M., Kinnaert, M., Lunze, J. & Staroswiecki, M., 2006. *Diagnosis and Fault-Tolerant Control 2nd Edition*. New York: Springer-Verlag Berlin Heidelberg.

Blanke, M., Staroswiecki, M. & Wu, E., 2001. Concepts and Methods in Fault-Tolerant Control. In *Arlington, VA, USA: American Control Conference. Proceedings of the 2001.*, 2001. Pearson Education.

Bonivento, C., Isidori, A., Marconi, L. & Paoli, A., 2004. Implicit Fault-Tolerant Control: Application to Induction Motors. *Automatica, vol. 40, N° 3*, p.355–371.

Craig, J., 1988. *Adaptative Control of Mechanical Manipulators*. Reading, Massachusetts: Addison-Wesley Publishing Company, Inc.

Kelly, R., Santibáñez, V. & and Loría, A., 2005. *Control of Robot Manipulators in Join Space*. London: Springer-Verlag London Ltd.

Lewis, F., Dawson, D. & Abdallah, C., 2004. *Robot Manipulator Control Theory and Practice*. New York: Marcel Dekker, Inc.

Puig, V. et al., 2004. Control Tolerante a Fallos (Parte I): Fundamentos y Diagnóstico de Fallos. *Revista Iberoamericana de Automática e Informática Industrial*, 1(1), pp.15-31.

Spong, M., Hutchinson, S. & Vidyasagar, M., 2005. *Robot Modeling and Control, First Edition*. New York: John Wiley & Sons, Inc.

5

Kinematic Task Space Control Scheme for 3DOF Pneumatic Parallel Robot

Luis Hernández[1], Eduardo Izaguirre[2], Ernesto Rubio[3], Orlando Urquijo[4] and Jorge Guerra[5]

[1,2,3,4]*Automation, Robotic and Perception Group (GARP). Universidad Central de Las Villas (UCLV)*

[5]*SIMPRO*

Cuba

1. Introduction

Parallel robots have received special attention of the systems and control community based on its high force-to-weight ratio and widespread applications (Merlet, 2006). The control schemes for a parallel robot can be divided into two strategies: joint space control (Wang et al., 2009), (Gupta et al., 2008) and task-space control (Ting & Chen, 1999). The joint space control scheme is based on the information of each actuator length and can be implemented as a decoupled independent single-input single-output (SISO) control systems for each actuator, with in general, poor compensation of the uncertainties. On the other hand, a task space controller has a potential to provide a better control for the parallel robot under system uncertainties: inertia, modeling error, friction, etc. Nevertheless this type of control scheme needs the direct measurement of system task space state (Gao et al., 2008) or task space state estimation, normally cumbersome. Task space control schemes have been presented for direct inverse dynamics control, with joint space dynamic model compensation (Feng, 1995) or task space dynamics model compensation,(Li & Xu, 2008), including vision based computed torque in task space control (Paccot et al., 2008). Many of this schemes have been proved by simulation or in laboratory testbed, but not commonly on industrial motion platform.

The goal of this work is to improve the performance of the control system of an industrial 3DOF pneumatic parallel platform, used by SIMPRO company who develop driver simulator. The solution of the control problem in the joint space of the parallel structure with pneumatic actuators have been obtained by Rubio (Rubio et al., 2009), using a decoupled control based on actuator lineal model around the operational point. But the performance of this scheme is strongly dependent of the precision of the kinematic model and dynamic uncertainties. The solution proposed, the *kinematic task space control*, is based on the direct measurement of system task space state. The control system considers two loops in cascade, an internal loop solving the robot' joint control ($\mathbf{q}$), and an external loop implementing the task space control ($\mathbf{x}$).

The proposed control philosophy, based on robot kinematic model, have been used by Hernández (Hernández et al., 2008a),(Hernández et al., 2008b) in vision-based 2D and 3D control of robot manipulators. In this scheme type, for digital implementation, it is possible

to approximate the dynamic effect of the internal loop as an external loop time delay (Corke, 1996), (Hernández et al., 2010) and (Bonfe et al., 2002). In this conditions a stability analysis in discrete time is developed.
To illustrate the proposed controller, the control system stability and its performance, both co-simulation results via MATLAB/Simulink and ADAMS, as well as experimental results using the SIMPRO 3DOF pneumatic parallel robot are presented. Experimental results confirm the expected step response in the task space, with good time performance and zero steady-state error. The control scheme presented open a new research field in the task space control with algorithms for the solution of, trajectory control, feed-forward control, etc.

2. Parallel robot model

As shown in Fig. 1, the robotic system considered consist of a 3DOF parallel manipulator controlled by pneumatic actuators. The parallel robot is produced by SIMPRO for driving simulator purpose, the robot have sensors to measurement the joint displacements and task space states. The basic mathematical description of this system consists of the parallel robot kinematics and dynamic model and also the actuator models.

Fig. 1. SIMPRO 3DOF pneumatic driver simulator parallel platform

2.1 Robot kinematics model

The same manner that serial manipulators, kinematics relations of parallel robots gives the relationship between the joints variables **q** and the corresponding position (x, y, z) and angular orientation (θ, φ, ψ) of center of mass of mobile platform in cartesian space. For an n-axis parallel structure, the forward kinematic (FK) solution **T** could be numerically computed according with the number of joints of kinematic architecture (Merlet, 2006). Generic mathematical representation of forward kinematics could be:

$$\mathbf{x} = \begin{bmatrix} x\, y\, z\, \theta\, \varphi\, \psi \end{bmatrix}^T = f(q_1, q_2, \ldots, q_n) = \mathbf{T} \tag{1}$$

For parallel robots the complexity of FK equations increase noteworthy with the numbers of degree of freedom, the solution is non-unique and numerical methods are currently used to obtained the solutions, unfortunately there is no known algorithm that allows the

determination of the current pose of the platform among the set of solutions. Furthermore, the computation times involved in FK algorithms are still too large for use in a real time application (Merlet, 2006).
For robot path planning, the inverse kinematic (IK) expressions $\mathbf{T}^{-1}$ gives the joint coordinates $\mathbf{q}$ required to reach the specified pose of mobile platform. The mathematical expression of inverse kinematics can be written as:

$$\mathbf{q} = \begin{bmatrix} q_1 \, q_2 \, \ldots \, q_n \end{bmatrix}^T = g(x, y, z, \theta, \varphi, \psi) = \mathbf{T}^{-1} \tag{2}$$

Schematic of the 3-DOF parallel robot under study is shown in Fig. 2. The system consists of a fixed base connected to moving platform by three actuated kinematics chains, following the RPSU-2SPS architecture. A base coordinate frame designated as $Oxyz$ frame is fixed at the center of the base with its z-axis pointing vertically upward and the x-axis pointing backwards of the platform. Similarly a moving coordinate frame $Px'y'z'$ is assigned to the center mass of the moving platform, with the z'-axis normal to the mobile platform. By simplicity the directions of both z and z' axes are pointing in the same unit vector.

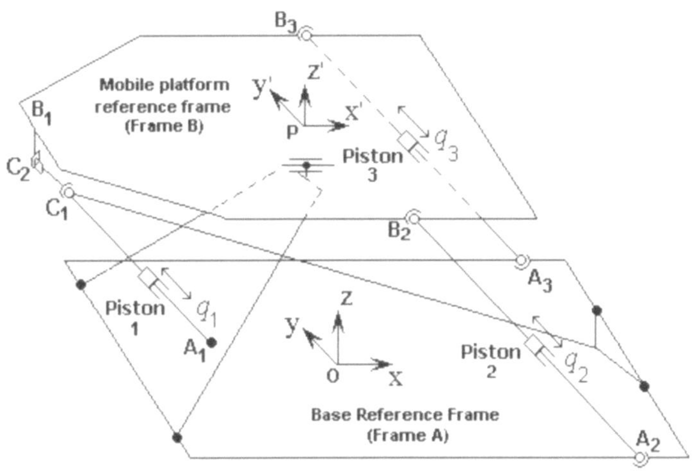

Fig. 2. The RPSU-2SPS kinematics structure of the 3-DOF motion platform.

The actuators are double effect electro-pneumatic cylinders whose lineal displacements produce the 3-DOF of robot, consisting in two rotations around the x' and y' axes, represented by roll (θ) and pitch (φ) angles respectively, and linear displacement along the z' axis (elevation), defined by the variable h. So, the moving platform can simulate different sceneries in correspondence with virtual reality world that is shown in LCD display located inside the cabin with is supported by mobile platform. This type of applications are developed by SIMPRO company.
The vectorial formulation allows us to simply construct a set of equations which contains the same number of equations as the unknown variables. According with this procedure a closed vector cycle is constituted between the points A_i and B_i in correspondence with the illustration represented in Fig. 3.
Establish the inverse kinematics model is essential for the position control of robot. Then, for each kinematics chain, a vectorial function can be formulating by expressing the actuated joint coordinates as a function of cartesian coordinates ($\mathbf{x}$) whose define the pose of the mobile

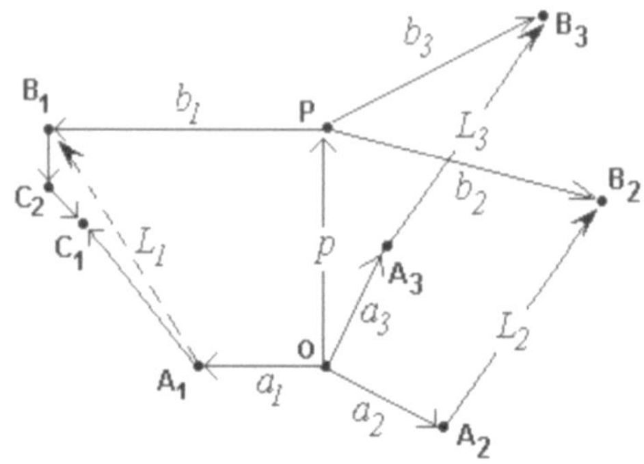

Fig. 3. Representation of closed loop vectors for each active legs.

platform. According to equation (2) is necessary to found the relation $\mathbf{A}_i\mathbf{B}_i = g(\mathbf{x})$ in order to calculate the inverse kinematics model of robot.
Considering the schematic presented in Fig. 3, follows that:

$$\mathbf{p} + \mathbf{b}_i = \mathbf{a}_i + \mathbf{L}_i \tag{3}$$

where:

$$\mathbf{p} = \mathbf{OP} = [P_x, P_y, P_z]^T - [O_x, O_y, O_z]^T \tag{4}$$

$$\mathbf{a}_i = ||\mathbf{OA}_i||_2 \tag{5}$$

$$\mathbf{b}_i = |\mathbf{PB}_i||_2 \tag{6}$$

$$\mathbf{L}_i = ||\mathbf{A}_i\mathbf{B}_i||_2 \tag{7}$$

The position vector of the mobile platform with reference to the fixed frame is defined by the vector $\mathbf{p} = \mathbf{OP}$, that is, platform position in cartesian space is defined by z' coordinates of point P. Consequently the orientation of mobile platform is determined by θ and φ angles.
Due to mechanical restrictions of robotic structure, the yaw angle is zero ($\psi = 0$), then, the rotation matrix is defined by only the roll and pitch angles. Following the ZYX convention for Euler angles, the rotation matrix can be computed by the expression:

$$\mathbf{R}^{xyz}_{x'y'z'} = \mathbf{R}^A_B = \begin{bmatrix} cos(\varphi) & sin(\varphi)s(\theta) & sin(\varphi)c(\theta) \\ 0 & cos(\theta) & -sin(\theta) \\ -sin(\varphi) & cos(\varphi)sin(\theta) & cos(\varphi)cos(\theta) \end{bmatrix} \tag{8}$$

It can be easily seen that the described architecture is characterized by 3-DOF's in cartesian space; in general case the appropriated linear displacements of each actuator are related with changes of orientation and elevation of the mobile platform. Then, the location of the moving platform can be described by a position vector $\mathbf{p}$ and a rotation matrix $\mathbf{R}^A_B$.
By construction the fixed reference frame coordinates of A_i's points are known, while the coordinates of B_i's may be determined from the position and orientation of mobile platform. Hence the vector $\mathbf{A}_i\mathbf{B}_i$ plays an important role in the solution of inverse and direct kinematics problems.

Considering equations (3) to (7) and (8), the general closed-loop equation for the SIMPRO motion simulator platform as the form:

$$\mathbf{A}_i\mathbf{B}_i = \mathbf{OP} + \mathbf{R}_B^A\mathbf{PB}_i - \mathbf{OA}_i \tag{9}$$

The central kinematics chain (corresponding with the actuator No. 1) have different architecture for the others, and then, new additional equation is necessary to be considered:

$$\mathbf{A}_1\mathbf{B}_1 = \mathbf{A}_1\mathbf{C}_1 + \mathbf{C}_2\mathbf{C}_1 - \mathbf{B}_1\mathbf{C}_2 \tag{10}$$

Substituting (10) in (9) and considering the all closed loop equations for each active leg, we have:

$$\mathbf{A}_1\mathbf{C}_1 = \mathbf{OP} + \mathbf{R}_B^A\mathbf{PB}_1 - \mathbf{OA}_1 + \mathbf{B}_1\mathbf{C}_2 + \mathbf{C}_2\mathbf{C}_1 \tag{11}$$

$$\mathbf{A}_2\mathbf{B}_2 = \mathbf{OP} + \mathbf{R}_B^A\mathbf{PB}_2 - \mathbf{OA}_2 \tag{12}$$

$$\mathbf{A}_3\mathbf{B}_3 = \mathbf{OP} + \mathbf{R}_B^A\mathbf{PB}_3 - \mathbf{OA}_3 \tag{13}$$

The design of robot included two additional rigid bodies interconnected between the base and the moving platform to provide mechanical stability and stiffness of robotic structure. This particular characteristic produce a curvilinear movement of mobile platform during elevation as is shown in Fig.4.

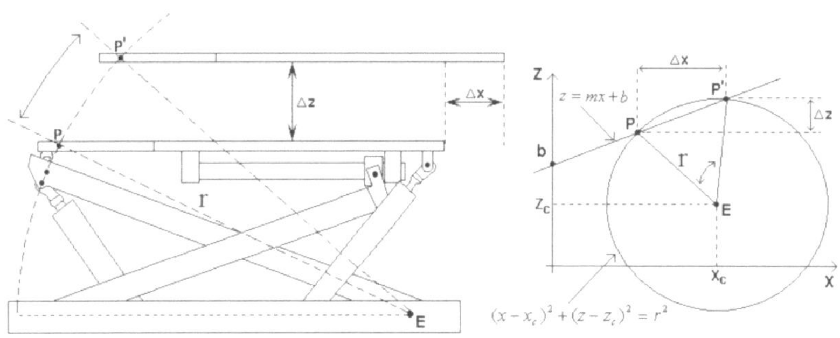

Fig. 4. Additional displacement $\triangle x$ of mobile platform during elevation.

Consider the straight line between points PP' is defined by equation $z = mx + b$, where m and b are the slope and intercept respectively, additionally the circle with center in $(x_c; z_c)$ and ratio $\mathbf{r}$ is described by the circumference equation $(x - x_c)^2 + (z - z_c)^2 = r^2$. Then, the movement along z-axis as a function of displacement of moving platform in the x-axis is calculated from:

$$\Delta z = m\left(\frac{2m(b - z_c) - 2x_c{}^2}{1 + m^2} - \frac{4k_1}{1 + m^2}\right) \tag{14}$$

where:

$$k_1 = x_c^2 + z_x^2 + b(b + 2z_c) - r^2$$

Because equation (14) compute the new coordinates of B_i's points of mobile platform during elevation, itt's incorporate to the solution of IK problem.

Given the initial position L_{oi} of limbs it is possible to calculate the displacements of lineal actuators $\mathbf{q_i}$ according to (15), considering the scheme of Fig. 5.

$$q_i = L_i - L_{oi} \tag{15}$$

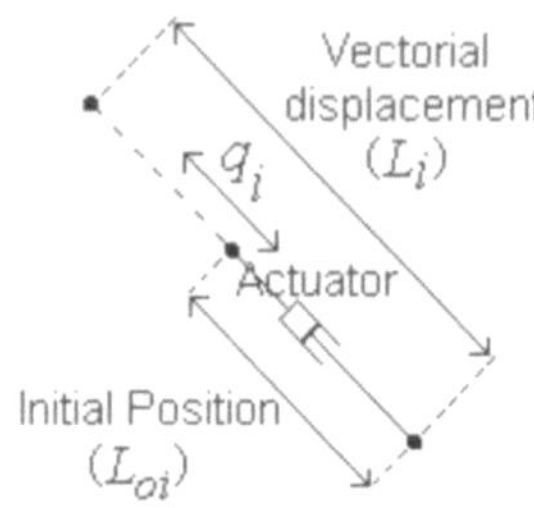

Fig. 5. Illustration of initial position of actuators in each robott's limbs.

2.2 Robot dynamics model

In the absence of friction or other disturbances, the dynamics of parallel robot can be written as (Merlet, 2006)

$$\mathbf{M}_x(\mathbf{x})\ddot{\mathbf{x}} + \mathbf{C}_x(\mathbf{x}, \dot{\mathbf{x}})\dot{\mathbf{x}} + \mathbf{G}_x(\mathbf{x}) = \mathbf{J}^T\mathbf{f} = \mathbf{f}_x \tag{16}$$

where:

$\mathbf{M}_x(\mathbf{x})$ $n \times n$ symmetric positive definite manipulator inertia matrix
$\mathbf{x}$ $n \times 1$ vector of space displacements
$\mathbf{C}_x(\mathbf{x}, \dot{\mathbf{x}})$ $n \times n$ centripetal and Coriolis matrix
$\mathbf{G}_x(\mathbf{x})$ $n \times 1$ vector of gravitational term
$\mathbf{J}$ $n \times n$ robot Jacobian matrix
$\mathbf{f}$ $n \times 1$ vector of applied joint forces

In other hand the dynamic model in joint space can have, in some configuration, the form, (Li & Xu, 2008)

$$\mathbf{M}(\mathbf{q})\ddot{\mathbf{q}} + \mathbf{C}(\mathbf{q}, \dot{\mathbf{q}})\dot{\mathbf{q}} + \mathbf{g}(\mathbf{q}) = \mathbf{f} \tag{17}$$

where:

$\mathbf{M}(\mathbf{q})$ $n \times n$ symmetric positive definite manipulator inertia matrix
$\mathbf{q}$ $n \times 1$ vector of joint displacements
$\mathbf{C}(\mathbf{q}, \dot{\mathbf{q}})$ $n \times n$ centripetal and Coriolis matrix
$\mathbf{G}(\mathbf{q})$ $n \times 1$ vector of gravitational term
$\mathbf{f}$ $n \times 1$ vector of applied joint forces

2.3 Pneumatic Actuators Model

Our platform actuators are pneumatic cylinders. To obtain the dynamic model of a pneumatic cylinder, normally (Brun et al., 2000), the following consideration are taken: only viscous friction is present, the temperature is constant and equal in both cylinder chambers and the gas is ideal. Additionally, according Rubio (Rubio, 2009), the influence of the underlap characteristic of the valve is considered. Under this conditions, the transfer function of the system, position $Y(s)$ versus control action $U(s)$, is obtained with the form,

$$\frac{Y(s)}{U(s)} = \frac{b_1 s + b_0}{s\,(c_3 s^3 + c_2 s^2 + c_1 s + c_0)} \tag{18}$$

Once defined the operating point of the valve, the coefficients c_i and b_i, are variables in dependent of the position y_0.
According the experimental work developed in (Rubio, 2009), a good approximation of equation (18) can be,

$$Gyu(s) = \frac{b}{s\,(s^2 + a_1 s + a_0)} \tag{19}$$

where b is the system gain, $a_1 = 2\,\tilde{\zeta}\,\tilde{\omega}_n$ and $a_2 = \tilde{\omega}_n^2$; where $\tilde{\omega}_n$ and $\tilde{\zeta}$ are respectively the open loop undamped natural frequency and damping ratio of the system. These parameter are not constants, they vary in dependency of the position y_0.

3. Control problem

The control problem is formulated as the design of a controller which computes a control signal Δ corresponding to the movement of the robot's in such a way that the desired task space position be reaches following wanted performances index.

3.1 Control problem formulation

For control the desired state, $\begin{bmatrix} \theta_d & \varphi_d & h_d \end{bmatrix}^T$, is the position of center of mass platform moving part. The *task state error* being defined as:

$$\tilde{y} = y_d - y = \begin{bmatrix} \tilde{\theta} \\ \tilde{\varphi} \\ \tilde{h} \end{bmatrix} = \begin{bmatrix} \theta_d \\ \varphi_d \\ h_d \end{bmatrix} - \begin{bmatrix} \theta \\ \varphi \\ h \end{bmatrix}$$

which could be calculated at every measurement time and used to move the robot in a direction allowing its decrease. Therefore, the control aims at ensuring that

$$\lim_{t\to\infty} \tilde{y} = \lim_{t\to\infty} \begin{bmatrix} \tilde{\theta} & \tilde{\varphi} & \tilde{h} \end{bmatrix}^T = 0$$

We make the as assumptions for the control problem that, only the control problem is evaluated with initial error $\tilde{\xi}(0)$ is sufficiently small and there exists a robot joint configuration $\mathbf{q}_d$ for which $\xi_d = \xi(\mathbf{q}_d)$. This condition ensures that the control problem is solvable.

3.2 Kinematic task space control

For our control problem formulation, the task space vector of the robot can be measured by external sensors. As such, a direct knowledge of the desired joint position $\mathbf{q}_d$ is not available. Nevertheless, the desired joints position can be obtained as a result of the estimated control signal Δ and the solution of the kinematics problems.
The implemented closed-loop block diagram can be described as shown in Fig. 6. The control system has two loops in cascade, the internal loop solving the robots' joint control, and the external loop implementing a *kinematic task space control.*

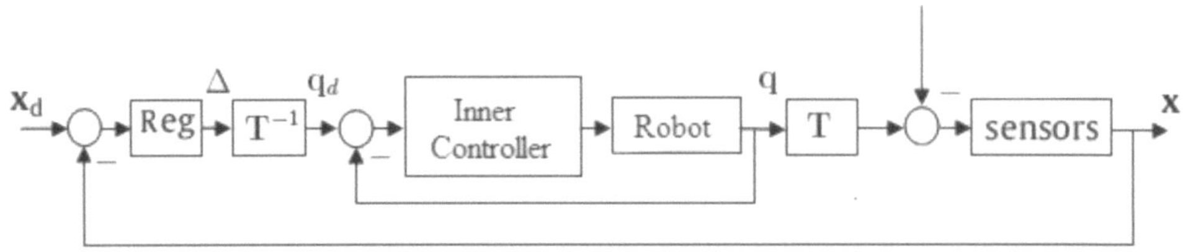

Fig. 6. Control scheme

The inner control loop has an open control architecture; in this architecture it is possible to implement any type of controller. One possibility is to use a non-linear controller in the state variables, called Model Based Computed Torque Control (Yang et al., 2008), having the following control equation, for a robot with n-DOF,

$$\tau = \mathbf{M}(\mathbf{q})[\ddot{\mathbf{q}}_d + \mathbf{K}_{vi}\dot{\tilde{\mathbf{q}}} + \mathbf{K}_{pi}\tilde{\mathbf{q}}] + \mathbf{C}(\mathbf{q}, \dot{\mathbf{q}})\dot{\mathbf{q}} + \mathbf{G}(\mathbf{q})$$

Where $\mathbf{K}_{pi} \in \Re^{n \times n}$ and $\mathbf{K}_{vi} \in \Re^{n \times n}$ are the symmetric positive-definite matrices and $\tilde{\mathbf{q}} = \mathbf{q} - \mathbf{q}_d$. It is possible to demonstrated that with this configuration the system behaves in a closed loop as a linear multivariable system, decoupled for each robot's joint, suggesting that the matrices could be specified as:

$$\mathbf{K}_{pi} = diag\{\omega_1^2, \ldots, \omega_n^2\}$$
$$\mathbf{K}_{vi} = diag\{\omega_1, \ldots, \omega_n\}$$

In this way each joint behaves as a critically damping second order linear system with bandwidth ω_i. The bandwidth ω_i determines the speed of response of each joint. In such way the dynamic effect of the internal loop could be independent with regard to the external loop, being under the conditions that:

$$\mathbf{q}(t) = \mathbf{q}_d(t) \quad \forall t > 0 \tag{20}$$

This control philosophy have been used by Hernández (Hernández et al., 2008b) in vision-based 3D control of robot manipulators.

Nevertheless, control systems is implemented fully as a sampled data systems. Åström (Åström K. J.and Wittenmark, 1990) established that the sampling rate of digital control systems should be between 10 and 30 time the desired closed loop bandwidth. For the case low cost driver simulators system the close loop bandwidth should be around 0.1 Hz. A sampling rate for the external loop between 100 to 30ms is excellent for the wanted close loop bandwidth of the system.

In the analysis of the control problem in the field of digital control systems, other dynamic representation can be made for the internal loop, of the robot control system could be as one or two delay units of the external loop (Corke, 1996) or for the robot (Bonfe et al., 2002). Using this consideration we modify Equation (20) as,

$$\mathbf{q}(k) = \mathbf{q}_d(k-1) \quad \forall k > 0 \tag{21}$$

4. Stability analysis

A simple ***I*** controller can be used in this control scheme (Hernández et al., 2008b), for that case the control law can be given by:

$$\mathbf{\Delta} = \mathbf{K}_I \int \tilde{\xi} \tag{22}$$

Where $\mathbf{K}_I \in \Re^{2\times 2}$ is the symmetric integral matrix:

$$\mathbf{K}_I = \begin{bmatrix} K_{I_1} & 0 & 0 \\ 0 & K_{I_2} & 0 \\ 0 & 0 & K_{I_3} \end{bmatrix} \tag{23}$$

Similarity as the visual control work (Hernández et al., 2010), Δ can be interpreted as the coordinates increment in the task space as a result of the direct mensuration of center of mass of moving platform position. Solving the inverse kinematics problem T^{-1} it is possible to obtain $\mathbf{q}_d$.
Considering that the task space is measured by linear sensors its can be represented by the gain matrix $\mathbf{K}$ as,

$$\mathbf{K} = \begin{bmatrix} K_\theta & 0 & 0 \\ 0 & K_\varphi & 0 \\ 0 & 0 & K_h \end{bmatrix}$$

Taking into account Fig. 6, obtaining the discrete equivalence of the controller of Equation (22), according to Equations (23) and (21) and taking a sampling period of $60ms$; a simplified diagram can be obtained as shown in Fig. 7.

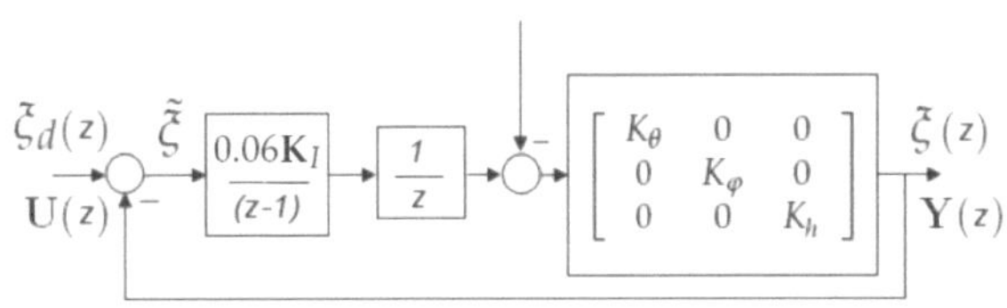

Fig. 7. Simplified control scheme

According to Fig. 7 the closed loop transfer function can be written as:

$$\frac{0.06\mathbf{K}_I\mathbf{K}}{(z^2 - z)}[\mathbf{Y}_d(z) - \mathbf{Y}(z)] = \mathbf{Y}(z) \tag{24}$$

Solving and taking the inverse Z transform we obtain:

$$\mathbf{y}(k+2) - \mathbf{y}(k+1) = -0.06\mathbf{K}_I\mathbf{K}\mathbf{y}(k) + 0.06\mathbf{K}_I\mathbf{K}\mathbf{y}_d(k) \tag{25}$$

The Equations (25) can be represented in the space state as,

$$\begin{bmatrix} \mathbf{y}(k+1) \\ \mathbf{y}(k+2) \end{bmatrix} = \begin{bmatrix} \mathbf{0} & \mathbf{I} \\ -0.06\mathbf{K}_I\mathbf{K} & \mathbf{I} \end{bmatrix} \begin{bmatrix} \mathbf{y}(k) \\ \mathbf{y}(k+1) \end{bmatrix} = \begin{bmatrix} \mathbf{0} \\ 0.06\mathbf{K}_I\mathbf{K} \end{bmatrix} \mathbf{y}_d(k) \tag{26}$$

where

$$\mathbf{G}=\begin{bmatrix}\mathbf{0} & \mathbf{I}\\ -0.06\mathbf{K}_I\mathbf{K} & \mathbf{I}\end{bmatrix}=\begin{bmatrix}0&0&0&1&0&0\\0&0&0&0&1&0\\0&0&0&0&0&1\\-0.06K_\theta K_{I_1}&0&0&1&0&0\\0&-0.06K_\varphi K_{I_2}&0&0&1&0\\0&0&-0.06K_hK_{I_3}&0&0&1\end{bmatrix}$$

and

$$\mathbf{H}=\begin{bmatrix}\mathbf{0}\\ 0.06\mathbf{K}_I\mathbf{K}\end{bmatrix}=\begin{bmatrix}0&0&0\\0&0&0\\0&0&0\\0.06K_\theta K_{I_1}&0&0\\0&0.06K_\varphi K_{I_2}&0\\0&0&0.06K_hK_{I_3}\end{bmatrix}$$

Following the the Equations (26)is easy to conclude that the control system is decoupled in each task space coordinate, with the form for the height,

$$\begin{bmatrix}h(k+1)\\h(k+2)\end{bmatrix}=\begin{bmatrix}0&1\\-0.06K_{I_3}K_h&1\end{bmatrix}\begin{bmatrix}h(k)\\h(k+1)\end{bmatrix}=\begin{bmatrix}0\\0.06K_{I_3}K_h\end{bmatrix}h_d(k) \qquad (27)$$

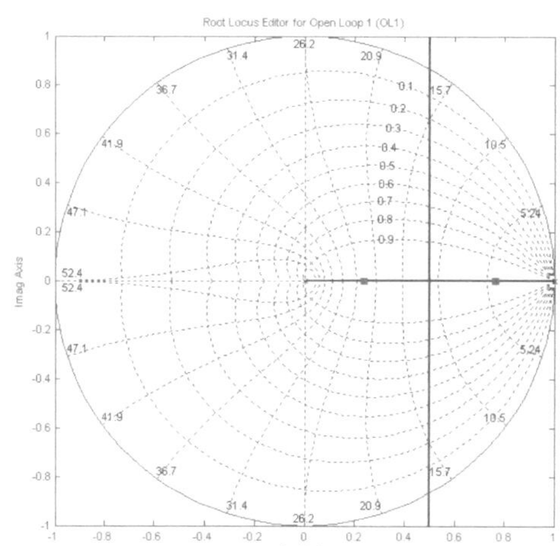

Fig. 8. Root-locus plot of height model for I controller with closed poles for gain $K_hK_{I_i}=3$.

According equation (27) the root loci of height control system is presented in Fig 8, in this case the closed loop poles are selected for the gain $K_hK_{I_i}=3$. An overdamped transient response to the step input is expected for this design. The system is stable with $K_hK_{I_i}<20$. A similar analysis can be done for Roll and Pitch task space coordinates.
A better system performance is obtained with a PI controller, for example with the controller,

$$C=\frac{0.06(z+0.2)}{(z-1)}$$

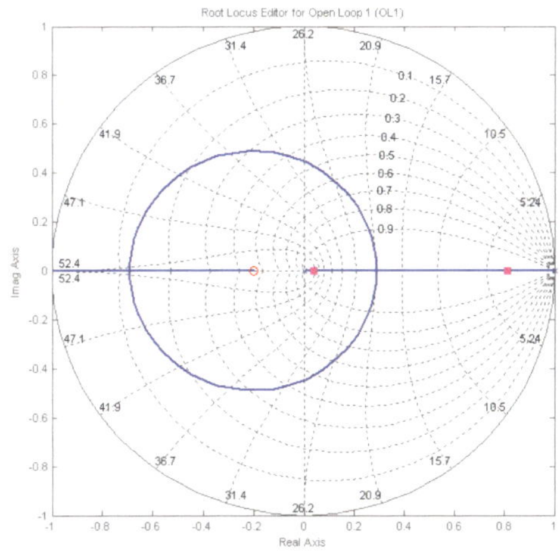

Fig. 9. Root-locus plot of height model for PI controller with closed poles for gain $K_h K_{I_i} = 20$.

The root locus plot is modified as is shown in the Fig. 9, in this case the system is stable with $K_h K_{I_i} < 50$.

5. Experimental study

A 3-DOF platform with open computer control architecture used by SIMPRO as movement simulator, is used as a study case, Fig. 1. The inner and the external loops are implemented in a Pentium-D 3.00-GHz connected to the robot through a Humusoft MF624 board which reads the potentiometer's joint position sensors, executes the control algorithm and gives the control signal to the electro-pneumatic valve with a sampling period of $1ms$. The task space position is acquired by the same card, reading a encoder for the height and Pitch and Roll via IMU. In this loop is solves the inverse kinematic problem and the values $\mathbf{q}_d$ are given to the internal loops with sampling period of $60ms$. The control algorithm has been implemented using MATLAB/Simulink with the *Real Time Workshop Toolbox* and *Real Time Windows Target*.

5.1 Practical design consideration

As shown in Fig. 6 the control system has two loops. The external loop execute the Kinematic task space control. For the case low cost driver simulators system the close loop bandwidth should be around $0.6rad/s$. According (Åström K. J.and Wittenmark, 1990) the external loop sampling period should be between 100 to $30ms$. The equation (21) is satisfied, if we choose the dynamic of the internal loop dominated by conjugated complex poles with $\zeta = 0.7$ and $\omega_n = 10rad/s$, with $60ms$ of sapling period for the external loop. A similar performance is obtained if the sampling period is $30ms$ and the condition (21) is modified as $\mathbf{q}(k) = \mathbf{q}_d(k-2) \quad \forall k > 0$.

5.1.1 Internal loop control

The parallel robot joint actuators are pneumatic cylinders. For this control a linear controller by pole placement is proposed.

The actuator dynamic have the form of the equation (19). According Rubio's work (Rubio, 2009), the controller have the form,

Table 1. Main specifications of the SIMPRO 3 DOF parallel robot

Description	Parameters
Actuator displacements	500 mm
Max. acceleration of actuators	980 mm/s^2
Initial elevation	1070 mm
Pitch and Roll angles	$\pm$ 18 $^\circ$
Elevation movement	$\pm$ 215 mm
Mass	1034 kg
Load/mass ratio	2.18

$$\frac{S(s)}{R(s)} = \frac{k_p\,(s+k_i)\,(s^2+a_1\,s+a_0)}{s\,(s+\omega_a)^2} \tag{28}$$

The transfer function of the internal closed loop is proposed as,

$$\frac{B_m}{A_m} = \frac{k_p\;b\,(s+k_i)\,(s^2+a_1 s+a_0)}{(s+p_1)(s+p_2)(s^2+2\zeta\omega_n\,s+\omega_n^2)(s^2+a_1 s+a_0)} \tag{29}$$

where p_1 and p_2 are two non dominant poles, selected by the designer joint to ω_n and ζ. With this parameters, based in the equation (29), the coefficients of controller (28) are found by the pole placement method.

5.2 Simulation

Following the scheme of Fig.6 a simulation process has been developed using MATLAB/Simulink-ADAMS. The principal parameters of the robot's geometric configuration Fig. 2 are specified in Table 1.

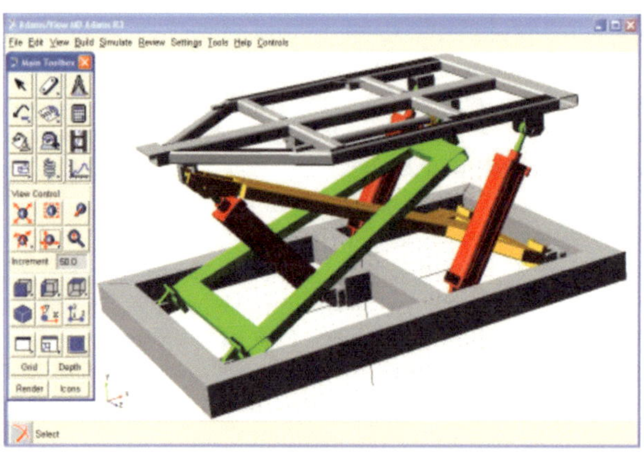

Fig. 10. Virual model of 3-DOF parallel robot development in ADAMS.

5.2.1 Kinematic model

Because kinematics studies the relations of the joint motions and the center of mass of moving platform movement, the interest of equations involving the kinematics relations are very important for control purposes of robots with parallel configurations.

The main specifications of motion platform under study is summarize in the Table 1, note the excellent load capacity and the relatively small workspace both typical characteristics of parallel robots.
Initial value of elevation of moving platform is h=1265 mm; and the corresponding coordinates locations (in mm) of points A_i, B_i and C_1, C_2 are given below.

$$\begin{aligned} A_1 &= [-180; 0; 0]^T \\ A_2 &= [180; -500; 0]^T \\ A_3 &= [180; 500; 0]^T \\ B_1 &= [-840; 0; h]^T \\ B_2 &= [896; -500; h]^T \\ B_3 &= [896; 500; h]^T \\ C_1 &= [-760; 0; h-210]^T \\ C_2 &= [-840; 0; h-150]^T \end{aligned}$$

Using the notation illustrated in Fig.3, the inverse kinematic problem can be solved by writing the loop closure equation for each actuated kinematic chain.
According with Table 1 and the equations (11) to (13) the of joints coordinates **q** are found, knowing the values of the elevation of moving platform (h), the roll and pitch angles (θ,φ), and the rotation matrix (8) as,

$$(L_1)^2 = [2076 - \lambda_0 - 940cos(\varphi)]^2 + [740 + h + 940sin(\varphi)]^2 \tag{30}$$

$$(L_2)^2 = [1397 - \lambda_0 + 720cos(\varphi) + \lambda_1]^2 + \lambda_2^2 + (\lambda_3 - \lambda_4)^2 \tag{31}$$

$$(L_3)^2 = [1397 - \lambda_0 + 720cos(\varphi) - \lambda_1]^2 + \lambda_2^2 + (-\lambda_3 - \lambda_4)^2 \tag{32}$$

where:

$$\begin{aligned} \lambda_0 &= \sqrt{1672^2 - 1720h - h^2} \\ \lambda_1 &= 500sin(\theta)sin(\varphi) \\ \lambda_2 &= 500cos(\varphi) - 500 \\ \lambda_3 &= 500sin(\theta)cos(\varphi) \\ \lambda_4 &= 720sin(\varphi) + 945 + h \end{aligned}$$

Substituting (30), (31) and (32) into (15) is possible to compute the vector of joints values $[q_1, q_2, q_3]^T$ of parallel robot from cartesian pose of center of mass of moving platform $[\theta, \varphi, h]^T$, establishing the inverse kinematic problem.

$$\begin{aligned} q_1 &= \sqrt{L_1} - L_{01} \\ q_2 &= \sqrt{L_2} - L_{02} \\ q_3 &= \sqrt{L_3} - L_{03} \end{aligned} \tag{33}$$

The parallel robot have been modeled in the software package MSC.Adams (ADAMS, 2005), excellent tool for the kinematic and dynamic simulation of complex mobile structures, (Li & Xu, 2008). So, the inverse kinematic equations have been validated using the robot's

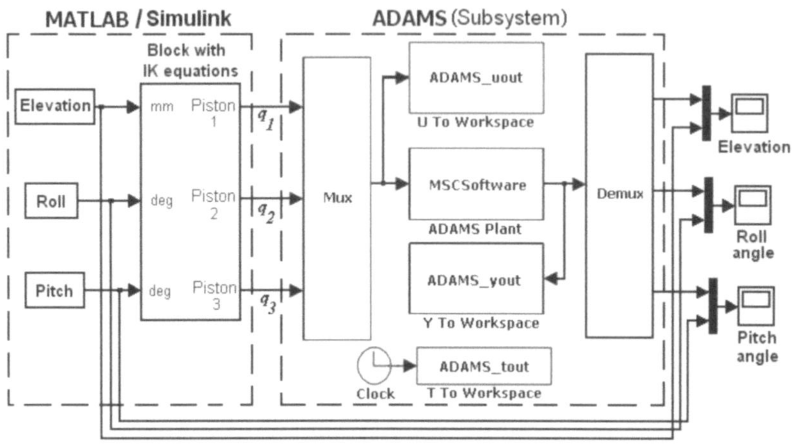

Fig. 11. Kinematic simulation Matlab/Simulink-ADAMS.

geometric virtual model in a co-simulation Matlab/Simulink-ADAMS, according with the block diagram of Fig. 11.
To obtain the desired values of actuated joints positions from the desired trajectory of mobile platform, the IK equations expressing in (33) are programming in Matlab/Simulink. The functions of $h(t)$, $\theta(t)$, and $\varphi(t)$ are the inputs of the IK block and the variables q_1, q_2, q_3 are the outputs; the block with the geometric model of robot (developed in ADAMS) received these variables as inputs which are then exported to ADAMS environment. The resulted outputs (position and orientation of the moving platform) of the virtual model in ADAMS are compared with the desired trajectory of center of mass of moving platform.

5.2.2 Joint control

The electro-pneumatic system consist of proportional flow valve type MPYE-5-3/8 connected to double-acting pneumatic cylinder FESTO DNC-125-500. To perform the dynamic identification of electro-pneumatic system the scheme of Fig. 12 was implemented. The bandwidth and magnitude of pseudo random binary signal (PRBS) in combination with the proportional gain K_p have been selected to obtain the adequate excitation of system.

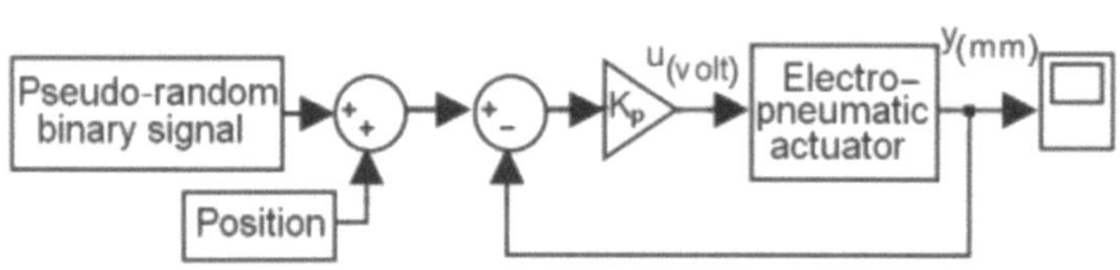

Fig. 12. Block diagram used in the dynamic identification of electro-pneumatic system.

The results of experimental dynamic identification, for the electro-pneumatic systems models (19) are,
Piston 1

$$\frac{Y_1(s)}{U_1(s)} = \frac{245.94}{s\,(s^2 + 27.72s + 253)} \tag{34}$$

Pistons 2 and 3

$$\frac{Y_2(s)}{U_2(s)} = \frac{Y_3(s)}{U_3(s)} = \frac{2008.3}{s\,(s^2 + 7.276 + 1349)} \tag{35}$$

The closed loop system is designed for two conjugated complex poles dominated with $\zeta = 0,7$ and $\omega_n = 10 rad/s$. According the equations (28) and (29), using pole placement method (Rubio et al., 2009), are obtained the transfers functions of controllers,
Controller for actuator 1

$$\frac{U_1(s)}{E_1(s)} = \frac{265(s^2 + 7.726\,s + 253)(s + 3.03)}{s(s^2 + 146.7s + 6267)} \tag{36}$$

Controllers for actuators 2 and 3

$$\frac{U_2(s)}{E_2(s)} = \frac{U_3(s)}{E_3(s)} = \frac{32(s^2 + 7.726\,s + 1349)(s + 3.03)}{s(s^2 + 146.7s + 6267)} \tag{37}$$

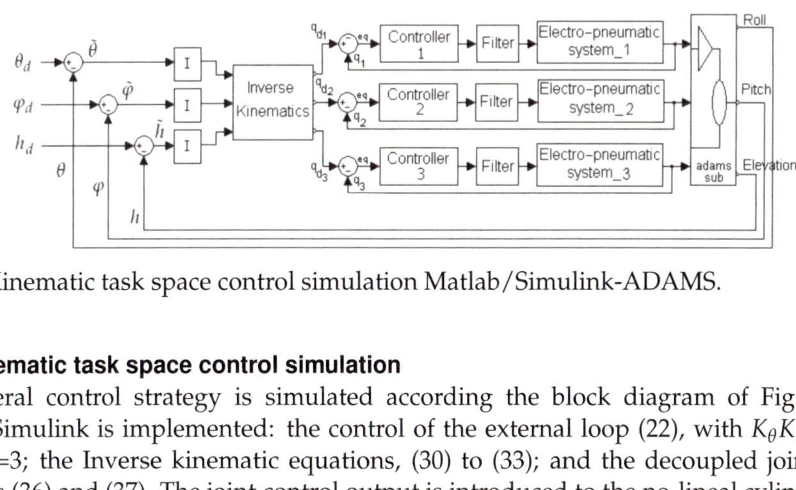

Fig. 13. Kinematic task space control simulation Matlab/Simulink-ADAMS.

5.2.3 Kinematic task space control simulation

The general control strategy is simulated according the block diagram of Fig. 13. In Matlab/Simulink is implemented: the control of the external loop (22), with $K_\theta K_{I_1} = K_\varphi K_{I_2} = K_h K_{I_3} = 3$; the Inverse kinematic equations, (30) to (33); and the decoupled joint control, equations (36) and (37). The joint control output is introduced to the no-lineal cylinder model of each joint (Rubio et al., 2009), which give the position input to the ADAMS geometric model. The ADAMS outputs are the task space position, used as feedback for the external loops.

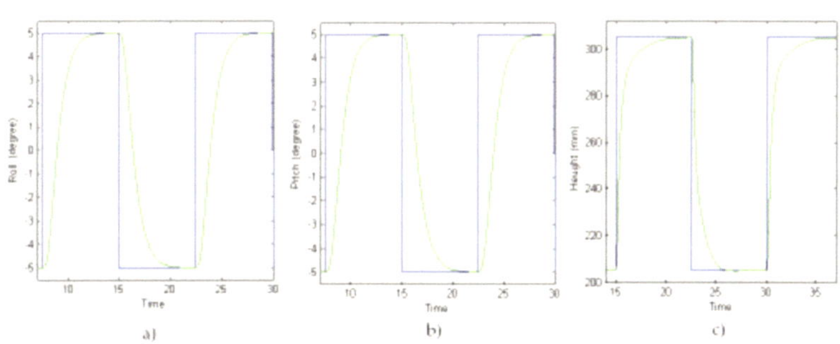

Fig. 14. ADAMS-Matlab/Simulink Simulation output, in blue the wanted value, in green the output. a) Roll, b) Pitch and c) Height.

The results of the simulation process is shown in the Fig. 14 where in blue are wanted value and in green the ADAMS output. In Fig. 14 a) Roll, in b) Pitch and in c) Height.

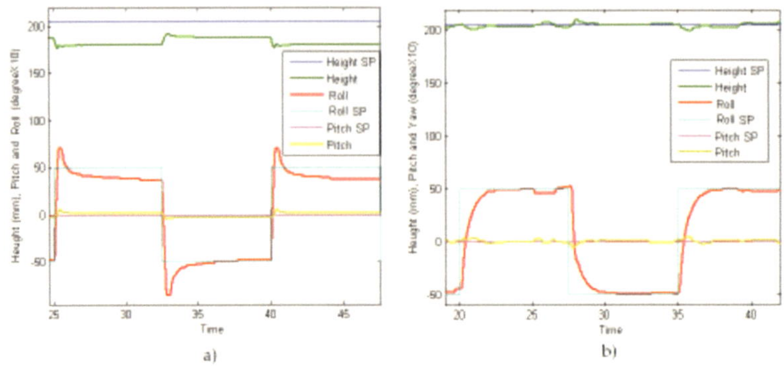

Fig. 15. Platform task space output to step variation in φ_d. a) With only joint control b) With joint control and kinematic task space control.

5.3 Experimental results

The control scheme proposed was implemented on the 3-DOF parallel robot used by SIMPRO as moving simulator. The control algorithm of the inner and the external loops have been implemented using MATLAB/Simulink with the *Real Time Workshop Toolbox* and *Real Time Windows Target*. The task space variables are measures with an encoder for the height and Pitch and Roll via IMU. Some experiments have been developed in real platform, initially the system receive step variation in φ_d, and the task space output is evaluated with only joint control, Fig. 15 a), and after the complete scheme proposed with joint control in the internal loops and kinematic task space control in the external loop, Fig. 15 b).

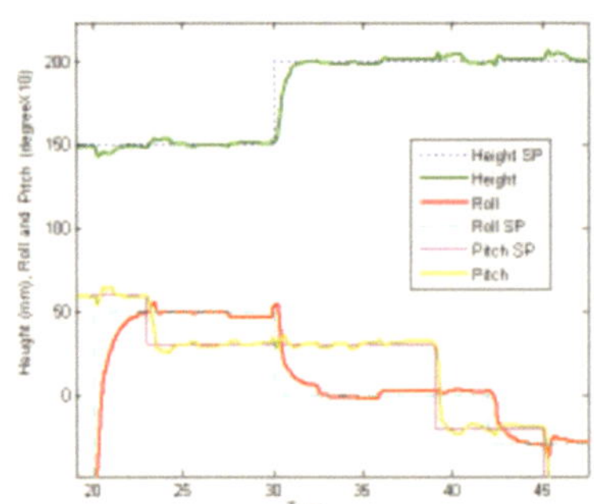

Fig. 16. Platform task space output with kinematic task space control with step variation in: θ_d, φ_d and h_d.

The better performance of the kinematic task space control is evident, the steady state error disappear, not only in φ, but also in h, which receive variation in the scheme with only joint control. The actual transient response is similar as the theoretically predict and similar as the simulation process with Matlab/Simulink-ADAMS, Fig.14. In the Fig. 16 is presented the platform task space output to simultaneous step variation in, θ_d, φ_d and h_d, with good performance in steady state and transient response.

6. Conclusion

In this work a *kinematic task space control* scheme is implemented in an 3DOF pneumatic parallel industrial robot, produced by SIMPRO company for motion simulator applications. The controller is structured in two loops in cascade, an internal loop solving the robot' joint control, and an external loop implementing the task space control. The dynamic effect of the internal loop is approximated as an external loop time delay, because the internal loop is designed faster than external loop. In these conditions the stability of the whole system in discrete time is balanced for I controllers in the external loop. The robott'inverse kinematics, the robott'model and simulation in ADAMS and the control of the pneumatic actuator of the internal loop are important solution in the control scheme implementation. In the chapter is demonstrated the improvement of the performance of the control system with the scheme *kinematic task space control*, in relation with the joint control scheme. In this direction co-simulation via MATLAB/Simulink and ADAMS, as well as experimental results using the SIMPRO 3DOF pneumatic parallel robot are presented. The experimental results confirm the expected performance of the system in the task space. As future works the control scheme presented can be improved with better controller in the external loop in order to be faster the control system, following trajectory, etc.

7. References

ADAMS (2005). *MSC.ADAMS Basic Full Simulation Package*, www.mscsoftware.com.

Åström K. J.and Wittenmark, B. (1990). *Compuer controlled systems: theory and design*, Englewood Cliffs (NJ).

Bonfe, M., Minardi, E. & Fantuzzi, C. (2002). Variable structure pid based visual servoing for robotic tracking and manipulation, *in* IEEE (ed.), *International Conference on Inteligent Robots and Systems*, Lausanne, Switzerland.

Brun, X., Belgharbi, M., Sesmat, S., Thomasset, D. & Scavarda, S. (2000). Control of an electropneumatic actuator, comparison between some linear and nonlinear control laws, *Journal of Systems and Control Engineering* (Control in Fluid Power Systems).

Corke, P. I. (1996). *Visual control of robots : high-performance visual servoing*, Robotics and mechatronics series ; 2, Research Studies Press ; Wiley, Taunton, Somerset, England New York.

Feng, G. (1995). A new adaptive control algorithm for robot manipulators in task space, *IEEE TRANSACTIONS ON ROBOTICS AND AUTOMATION*. Vol. 11.(NO. 3): 457–462.

Gao, J., Webb, P. & Gindy, N. (2008). Vevaluation of a low-cost inertial dynam. measurement system, *in* IEEE (ed.), *Proc. of the IEEE Conf. on Robotics, Automation and Mechatronics*, Chengdu, China, pp. 21–24.

Gupta, A., O'Malley, M. K., Patoglu, V. & Burgar, C. (2008). Design, control and performance of ricewrist: A force feedback wrist exoskeleton for rehabilitation and training, *The Internat. Journal of Robotics Research* Vol. 27(No. 2): 233–251.

Hernández, L., Sahli, H. & González, R. (2010). Vision-based 2d and 3d control of robot manipulators, *in* A. Jimenez (ed.), *Robot Manipulators Trends and Development*, IN-TECH, Vienna, Austria, pp. 441–462.

Hernández, L., Sahli, H., González, R. & González, J. (2008b). Simple solution for visual servoing of camera-in-hand robots in the 3d cartesian space., *Proceedings in the Tenth International Conference on Control, Automation, Robotics and Vision (ICARCV 2008).*, IEEE, Hanoi, Viet Nam, pp. 2020–2025.

Hernández, L., Sahli, H., González, R., Rubio, E. & Guerra, Y. (2008a). A decoupled control for visual servoing of camera-in-hand robot with 2d movement., *Proceedings in Electronics, Robotics and Automotive Mechanics Conference, 2009. CERMA '08.*, IEEE, Morelos, México, pp. 304–309.

Li, Y. & Xu, Q. (2008). Dynamic modeling and robust control of a 3-prc translational parallel kinematic machine, *Robotics and Computer-Integrated Manufacturing.* . www.elsevier.com/locate/rcim.

Merlet, J. P. (2006). *Parallel Robots, 2nd Edition*, Springer, France, ISBN: 1402041322.

Paccot, F., Lemoine, P., Andreff, N., Chablat, D. & Martinet, P. (2008). A vision-based computed torque control for parallel kinematic machines, *Proceedings of IEEE International Conference on Robotics and Automation.*, IEEE, Pasadena, USA.

Rubio, E. (2009). *Modelación, identificación y control de actuadores electro-neumáticos para aplicaciones industriales*, PhD thesis, Universidad Central "Marta Abreu" de Las Villas.

Rubio, E., Hernández, L., Aracil, R., Saltaren, R. & Guerra, J. (2009). Implementation of decoupled model-based controller in a 2-dof pneumatic platform used in low-cost driving simulators., *Proceedings in Electronics, Robotics and Automotive Mechanics Conference, 2009. CERMA '09.*, IEEE, Cuernavaca, Morelos, México, pp. 338 – 343.

Ting, Y. & Chen, Y. S.and Wang, S. M. (1999). Task-space control algorithm for stewart platform., *Proceedings of the 38th conference on decision and control*, IEEE, Phoenix, Arizona, USA, p. 3857Ű3862.

Wang, J., Wu, J., Wang, L. & You, Z. (2009). Dynamic feed-forward control of a parallel kinematic machine, *Mechatronics* Vol. 19(No. 3): 313–324.

Yang, Z., Wu, J., Mei, J., Gao, J. & Huang, T. (2008). Mechatronic model based computed torque control of a parallel manipulator, *International Journal of Advanced Robotic Systems* 5(1): 123–128. www.elsevier.com/locate/rcim.

Part 2

System Interfacing, Instrumentation and Control

6

Blind Source Separation Based Classification Scheme for Myoelectric Prosthesis Hand

Ganesh R. Naik and Dinesh Kumar
RMIT University, Melbourne
Australia

1. Introduction

For over three decades, researchers have been working on using surface electromyography (sEMG) as a means for amputees to use remaining muscles to control prosthetic limbs (Baker, Scheme, Englehart, Hutcinson, & Greger, 2010; Hamdi, Dweiri, Al-Abdallat, & Haneya, 2010; Kiguchi, Tanaka, & Fukuda, 2004). Most research in this domain has focused on using the muscles of the upper arms and shoulders to control the gross orientation and grasp of a low-degree-of-freedom prosthetic device for manipulating objects (Jacobsen & Jerard, 1974). Each measured upper arm muscle is typically mapped directly to one degree of freedom of the prosthetic. For example, tricep contraction could be used for rotation while bicep flexion might close or open the prosthetic. More recently, researchers have begun to look at the potential of using the forearm muscles in hand amputees to control a multi-fingered prosthetic hand. While we know of no fully functional hand prosthetic, this is clearly a promising new area of EMG research. One of the challenges for creating hand prosthetics is that there is not a trivial mapping of individual muscles to finger movements. Instead, many of the same muscles are used for several different fingers (Schieber, 1995).
To identify hand gestures and actions that are a result of multiple active muscles, relative muscle activity from the different muscles in the forearm has to be identified. For this purpose, the sEMG needs to be recorded using multiple electrodes. However due to the close proximity of the different active muscles, each of these electrodes record muscle activity from multiple muscles, referred to as cross talk. In case of the hand there are number of muscles in the close proximity and often crossing over each other and cross talk is a major cause of low reliability in identifying the action. This is further exaggerated when the muscle activity is weak such as during maintained isometric gestures due to the relative low signal to the background noise. Spectral and temporal overlap makes the use of conventional filtering quite useless. Differences between people make accurate modeling and generalization of the sEMG not possible. To overcome this, number of researchers have used generalization pattern recognition tools and some of these works have got good results (Cheron, Draye, Bourgeios, & Libert, 1996; Koike & Kawato, 1996; Meyer, 2000). However these require manual intervention and it is obvious that such techniques are not suitable for reliable operations that can be automated.
Numbers of researchers have attempted to use sEMG for controlling the prosthetic devices (Doerschuk, Gustafon, & Willsky, 1983;Zardoshti-Kermani, Wheeler, Badie, & Hashemi, 1995). While the current techniques to classify sEMG are suitable for identifying gross

actions that are the result of the contraction of a single muscle, these are unable to accurately identify actions that have multiple active muscles or where the strength of contraction is weak. Previous EMG-based input methods have classified gross movements such as wrist flexion or bicep activation, and hand prosthetics research has placed sensors on the forearm to detect finger movements. However, these approaches involve a restrictive setup procedure including fixing the hand to a board or placing sensors at many places on the arm in order to recognize only a few finger movements.

Recently, blind source separation (BSS) by independent component analysis (ICA) has received attention because of its potential in many signal processing fields. In this research, we present results demonstrating accurate gesture classification with an off-the-shelf electromyography (EMG) device. Here ICA is applied to the electromyography (sEMG) signal analysis. The experiment shows that ICA can decompose sEMG signal and separate source and noise effectively. A unique classification scheme is adopted using ICA and neural network to control the adjustment of myoelectric prosthetic hand's movement. Many experiments show that some steady independent components always appear when muscle does the same tasks. This result will provide us with a promising method in the classification of muscle pattern recognition and the research on the Human-Computer Interface (HCI) technology.

2. Related work

There are two broad approaches that have been used for classification of the sEMG to obtain command signals for controlling a prosthetic device. The first approach is to use the amplitude of the *steady state* electromyography (EMG) signal where the subject exerts constant force with a chosen muscle. A single recording site or channel on this muscle is then used to control in a manner proportional to the amplitude. Many commercially available prosthetics fall into this category, but generally afford the use of only a single degree of freedom such as a gripper for grasping and releasing objects due to the inability of such a system to identify small changes in muscle activity. The second approach uses the initiation of an action as a trigger, coding of which produces functional movements.

(Graupe & Cline, 1975) were one of the pioneers to classify sEMG signals for prosthetic control. They obtained 85% accuracy in classifying magnitude data from a single channel with autoregressive coefficients as features. While their work is impressive, it is neither suitable for automation nor for identifying number of complex actions. (Englehart & Hudgins, 2003) identified four discrete elbow and forearm movements using the transient structure in the signal by using time-frequency representations of the signal. They achieved accuracy up to 93.7% with four channels of EMG data from the biceps and triceps. Such a scheme is suitable when there are single prime movers but not for hand gestures where there are number of multiple active muscles. (Nishikawa, Yu, Yokoi, & Kakazu, 1999) classified ten discrete movements of the wrist and fingers using four electrodes placed on the forearm. They proposed an online learning scheme, and obtained an average accuracy of 85.1%. The shortcoming in this approach is the large inter-group variations with accuracy of identifying the correct action ranging from 75.2% to 91.7%. (Ju, et al., 2000) report finger action identification for consumer electronics applications and their technique achieved 85% accuracy in classifying four finger movements with the aid of two pair of electrodes placed close to the wrist. Such accuracy may be suitable for consumer electronics but is unsuitable

for rehabilitation and aged care applications. Further, this system requires the user to train the system for each experiment limiting the suitability of this approach.
Numerous approaches have been applied to solve the problem of visual interpretation of gestures. Many of these approaches have been chosen and implemented to focus on a particular aspect of gestures: Hand tracking, pose classification, or hand posture interpretations (Rehg & Kanade, 1993; Schlenzig, Hunter, & Jain, 1994). (Farry & Walker, 1993) presented myoelectric control of NASM Johnson Space Center's sixteen degree-of-freedom Utah/MIT Dextrous Hand for two grasping (key and chuck) options and three thumb motions (abduction, extension, and flexion). Their work discussed myoelectric signal processing approaches, data collection apparatus, and a real-time teleoperation implementation. They also demonstrated results in real-time discrimination of key and chuck grasps and offline discrimination of thumb motions. Their research outcome included a 90% correct grasp selection rate and an 87% correct thumb motion selection, both using the myoelectric spectrum. (Peleg, Braiman, Yom-Tov, & Inbar, 2002) used a combination of a K-nearest neighbour (KNN) classifier and a genetic algorithm (GA) for feature selection which results in an average error rate of approximately 2%, thereby making it feasible to operate a robotic replacement arm with relatively few errors using only two pairs of electrodes. Four subjects were participated in the experiment.
Most recent work includes the investigation of hand gestures for six distinct actions (Chan & Englehart, 2005) and a framework where myoelectric signals from natural hand and finger movements can be decoded with a high accuracy. (Chan & Englehart, 2005) presented the work related to prosthetic applications. Their work includes the investigation of twelve normally limbed subjects (eight males and four females) for six distinct limb motions: *wrist flexion*, *wrist extension*, *supination*, *pronation*, *hand open*, and *hand close*. Each subject underwent four 60-s sessions, producing continuous contractions. They used short time Fourier transform and a linear discriminant analysis (LDA) classifier to classify the data. (Tenore et al., 2007) presented a hand gesture identification device using 32 surface-EMG electrodes were placed on the forearm of an able-bodied subject while performing individual finger movements. Using time-domain feature extraction methods as inputs to a neural network classifier, they showed that 12 individuated flexion and extension movements of the fingers can be decoded with accuracy higher than 98%. While this approach has the merits of reliability, the difficulty with this approach is that it requires multiple actions, leading to an unnatural and slow system. There is a need for systems that can recognize a range of hand actions that are more subtle and that are a result of multiple active muscles. Such a technique will allow the user to give natural control commands and would not require learning to make a series of actions for a specific command. To reliably identify actions that are based on contraction of multiple muscles requires identifying the level of muscle activity of the different muscles. This chapter reports a paradigm that is based on identifying activity from the different muscles by separating sEMG and the results indicate that the system is successful in identifying different complex hand actions. It is a combination of model based approach with blind source separation technique.

3. Myoelectric signal

Surface electromyogram (sEMG) is a non-invasive recording of muscle activity. It requires relatively simple equipment, and is suitable for numerous control applications. The close relationship of sEMG with the force of contraction of the muscle is useful for number of applications such as sports training, urinary incontinence and for machine control. The relationship of sEMG spectrum with muscle fatigue is also very useful for occupational health and sports training. While there are numerous applications for sEMG, these are limited due to reliability issues arising due to the complexity of the signal. sEMG may be affected by various factors such as

- The muscle anatomy (number of active motor units, size of the motor units, the spatial distribution of motor units).
- Muscle physiology (trained or untrained, disorder, fatigue).
- Nerve factors (disorder, neuromuscular junction).
- Contraction (level of contraction, speed of contraction, isometric/non-isometric, force generated).
- Artefacts (crosstalk between muscles, ECG interference).
- Recording apparatus factors (recording-method, noise, electrode's properties and recording sites).

The anatomical/ physiological processes such as properties and dimensions of tissues, and force and duration of contraction of the muscle are known to influence the signal. Peripheral factors such as spacing, type and size of electrodes may also have an influence on the signal (Basmajian & Deluca, 1985), and to obtain reliable information, considering such factors is critical. Some of these factors may be handled through careful skin preparation, and by selecting proper anatomical landmarks for the placement of electrodes. While these factors influence Surface EMG in general, these factors are more apparent when the sEMG signal strength is very small, such as during static posture.

The complexity of the human body along with the high level of environmental noise results in the bioelectric signal recordings getting corrupted. Environmental noise and electrical activity from parts of the body other than the target muscles gets super-imposed on the recordings. There are number of techniques that are employed to reduce the environmental noise and artefacts from other organs such as by careful selection of the electrode location and with suitably designed acquiring equipment. But there are number of conditions where the artefacts may be very strong and prevent the useful interpretation of the desired signals. Examples of such situations include EMG recordings from the back and thorax region that have ECG and breathing artefacts, and the artefacts maybe of much greater magnitude than the signal. The other example is during the sEMG recordings of the digitas muscles to identify the hand gestures where the cross talk due to the different muscles can result in unreliable recordings. One property of the sEMG is that the signal originating from one muscle can generally be considered to be independent of other bioelectric signals such as electrocardiogram (ECG), electro-oculargram (EOG), and signals from neighbouring muscles. This opens an opportunity of the use of blind source separation (BSS) for this application.

4. Basic principles of BSS techniques

Signals from different sources can get mixed during recording. Often it is required to separate the original signals, and there is little information available of the original signals. An example is the cocktail party problem. Even if there is no (limited) information available of the original signals or the mixing matrix, it is possible to separate the original signals using independent component analysis (ICA) under certain conditions. ICA is an iterative technique that estimates the statistically independent source signals from a given set of their

linear combinations. The process involves determining the mixing matrix. The independent sources could be audio signals such as speech, voice, music, or signals such as bioelectric signals (Bell & Sejnowski, 1995; Aapo Hyvärinen, Karhunen, & Oja, 2001).
Independent Component Analysis (ICA) is a new statistical technique that aims at transforming an input vector into a signal space in which the signals are statistically independent. The drawback of ICA, namely the need of high order statistics in order to determine ICA expansion, is counterbalanced by its performances, which are more meaningful compared with other methods like PCA -Principal Component Analysis
ICA assumes the mixing process as linear, so it can be expressed as:

$$x = As \tag{1}$$

Where $x = [x_1(t), x_2(t), \ldots, x_n(t)]$ are the recordings, $s = [s_1(t), s_2(t), \ldots, s_n(t)]$ the original signals and A is the mixing matrix of real numbers. This mixing matrix and each of the original signals are unknown. To separate the recordings to the original signals (estimated original signals u), the task is to estimate an un-mixing matrix W so that:

$$s = Wx = WAs \tag{2}$$

For this purpose, ICA relies strongly on the statistical independence of the sources s. The block diagram approach of ICA for source separation is shown in figure 1.

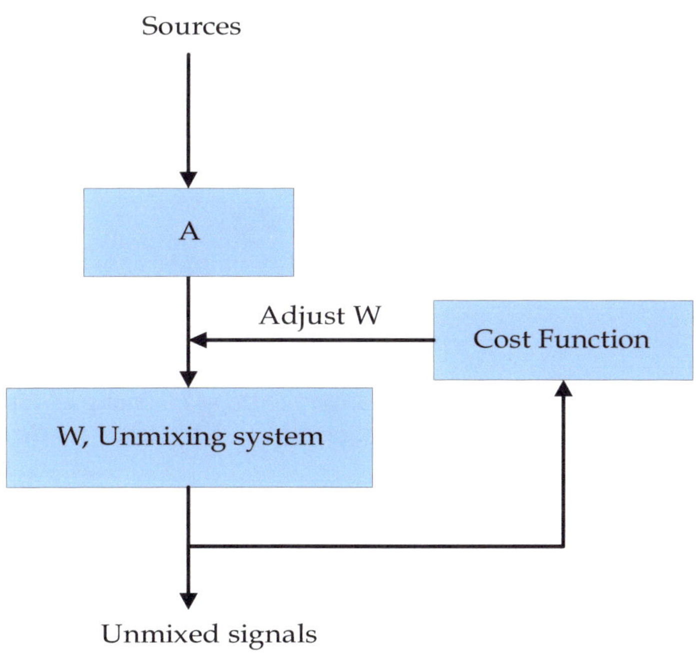

Fig. 1. An example of ICA source separation system

The ICA technique iteratively estimates the un-mixing matrix using the maximisation of independence of the unmixed signals as the cost function. Signals are statistically independent if the joint probability density of those components can be expressed as a multiplication of their marginal probability density. It is important to observe the distinction between independence and uncorrelatedness, since decorrelation can always be performed by transforming the signals with a whitening matrix to get the identity covariance matrix I. Independent signals are always uncorrelated but uncorrelated signals are not always independent. But in case of Gaussian signals, uncorrelatedness implies independence. Transforming of a Gaussian signal with any orthogonal un-mixing matrix or transform results in another Gaussian signal, and thus the original signals cannot be separated. Hence Gaussian signals are forbidden for ICA. Thus the key of independent component estimation is measuring the non-Gaussianity of the signals (Bell & Sejnowski, 1995; Comon, 1994; Aapo Hyvärinen, et al., 2001; Lee, 1998).

To summarise from the above, the signals that can be separated need to be non-Gaussian and independent. For the purpose of applying ICA to sEMG recordings, there is a need to determine the conditions under which these signals can be considered as independent and non-Gaussian, and the mixing matrix can be considered to be stationary and linear. This chapter analyses and tests these conditions using a stationary mixing matrix.

The ICA technique iteratively estimates the unmixing matrix using the maximisation of independence of the unmixed signals as the cost function. Substantial research has been conducted on algorithms using higher order statistics for estimation of ICA. One of the widely used techniques among these is FastICA. FastICA is a fixed point algorithm that employs higher order statistics for the recovery of independent sources (A. Hyvärinen, 2000; Aapo Hyvärinen, et al., 2001; Aapo Hyvärinen & Oja, 1997a, 1997b). Separation is performed to obtain uncorrelated and independent sources whose amplitude distributions are as non-Gaussian as possible are obtained. The non Gaussianity is measured with the differential entropy J, called negentropy, which is defined as the difference between the entropy of a Gaussian random variable y_{Gauss} (having the same mean and variance of the observed random variable y) and the entropy y (Aapo Hyvärinen, et al., 2001; Aapo Hyvärinen & Oja, 1997b).

$$J(y) = H(y_{Gausss}) - H(y) \tag{3}$$

Where the entropy H is given by

$$H(y) = -\int f(y)\log(f(y))dy \tag{4}$$

Since Gaussian random variables have the largest entropy H among all random variables having equal variance, maximizing the negantropy, $J(y)$ leads to the separation of independent source signals.

FastICA can estimate independent components (ICs) one by one (deflation approach) or simultaneously (symmetric approach), and the extracted number of ICs can be lower than the number of mixtures so that the unmixing matrix $\boldsymbol{W}$ can be rectangular. FastICA uses simple estimates of negentropy based on the maximum entropy principle (Aapo Hyvärinen, et al., 2001; Aapo Hyvärinen & Oja, 1997b).

4.1 Relevance of ICA for sEMG

This section establishes the relevance of the application of ICA for sEMG. The first step is test sEMG against the assumptions that underpin the theory of ICA. ICA is suitable for source separation when;

- The sources are statistically independent

- Independent components have non-Gaussian distribution
- The mixing matrix is invertible
- Sources and sensors are fixed.
- Signal transmission delays are negligible.

These assumptions are satisfied by sEMG because; (i) the MUAPs are statistically independent, (ii) have non-Gaussian distributions, (iii) if the number of recordings is same as the number of sources, the mixing matrix will be square and invertible, (iv) the sources and electrodes are fixed and (v) volume conduction in the tissue is essentially instantaneous (T. P. Jung et al., 1998; T. P. Jung et al., 2000; Makeig, Bell, Jung, & Sejnowski, 1996). Based on the above, ICA is suitable for separating sEMG recordings to obtain muscle activity if the number of channels is same as the number of active muscles.

One measure to test the quality of separation is to determine the dominant values and sparseness in the global matrix. Global matrix for ICA are generated by multiplying un-mixing matrix of signals from one time window, *Wp*, with the inverse un-mixing matrix of the other time window, *Wq*. If the separated signals are independent, each of the un-mixing matrices should be the inverse of the mixing matrix but for the ambiguity due to the order and arbitrary scaling (Cichocki & Amari, 2002; Aapo Hyvärinen, et al., 2001; T.-P. Jung, et al., 1998; T. P. Jung, et al., 2000; Makeig, et al., 1996), and the product of these, the global matrix should be sparse with typically one dominant cell in each row and column while the other cells should be close to zero (Meyer, 2000).

To test the efficacy of the use of ICA for sEMG, preliminary experiments were conducted where the global matrix was computed for the recordings. Four channels of sEMG that had been recorded during finger and wrist flexion experiments were separated using Fast ICA and the unmixing matrices were estimated for a series of time windows. Below are two matrices corresponding to two set of time frames. From these, it is observed that in each of these matrices, there is only one dominant cell in each row and column and the others are close to zero, and thus the matrix is sparse. This indicates that the quality of separation of the signals is good.

Trial 1:

$$G_{hand} = \begin{bmatrix} -0.0126 & -0.0519 & -0.0670 & 1.2294 \\ 0.0701 & -1.5252 & -0.0699 & 0.0263 \\ -1.4221 & -0.0233 & -0.0144 & -0.1761 \\ -0.0266 & -0.0045 & 0.6180 & 0.0100 \end{bmatrix}$$

$$\text{Det}\left(G_{semi\_blind\_ICA\_Hand}\right) = 1.6533$$

Trial 2:

$$G_{hand} = \begin{bmatrix} 0.0800 & -1.0094 & 0.0271 & 0.0927 \\ 0.0670 & -0.0046 & 0.0307 & -1.2610 \\ 0.0143 & 0.0295 & 0.8062 & 0.0273 \\ 2.1595 & 0.3787 & -0.0729 & 0.0686 \end{bmatrix}$$

$$\text{Det}\left(G_{semi\_blind\_ICA\_Hand}\right) = 2.2588$$

From these trials, it is also evident that one shortcoming with the use of ICA is the ambiguity related to the order of the outputs. The other ambiguity is due to the arbitrary scaling making the absolute value of the output arbitrary. To overcome these issues, this research has proposed to estimate the unmixing matrix only once for an individual and with the help of this, to train a supervised neural network with the targets being the known actions. It is proposed that this unmixing matrix and the trained network be used for identifying the hand actions. Such a scheme would be extremely fast and would not need supervision because during operations the computations would only require the pre-estimated unmixing matrix and pre-trained neural network weight matrix.

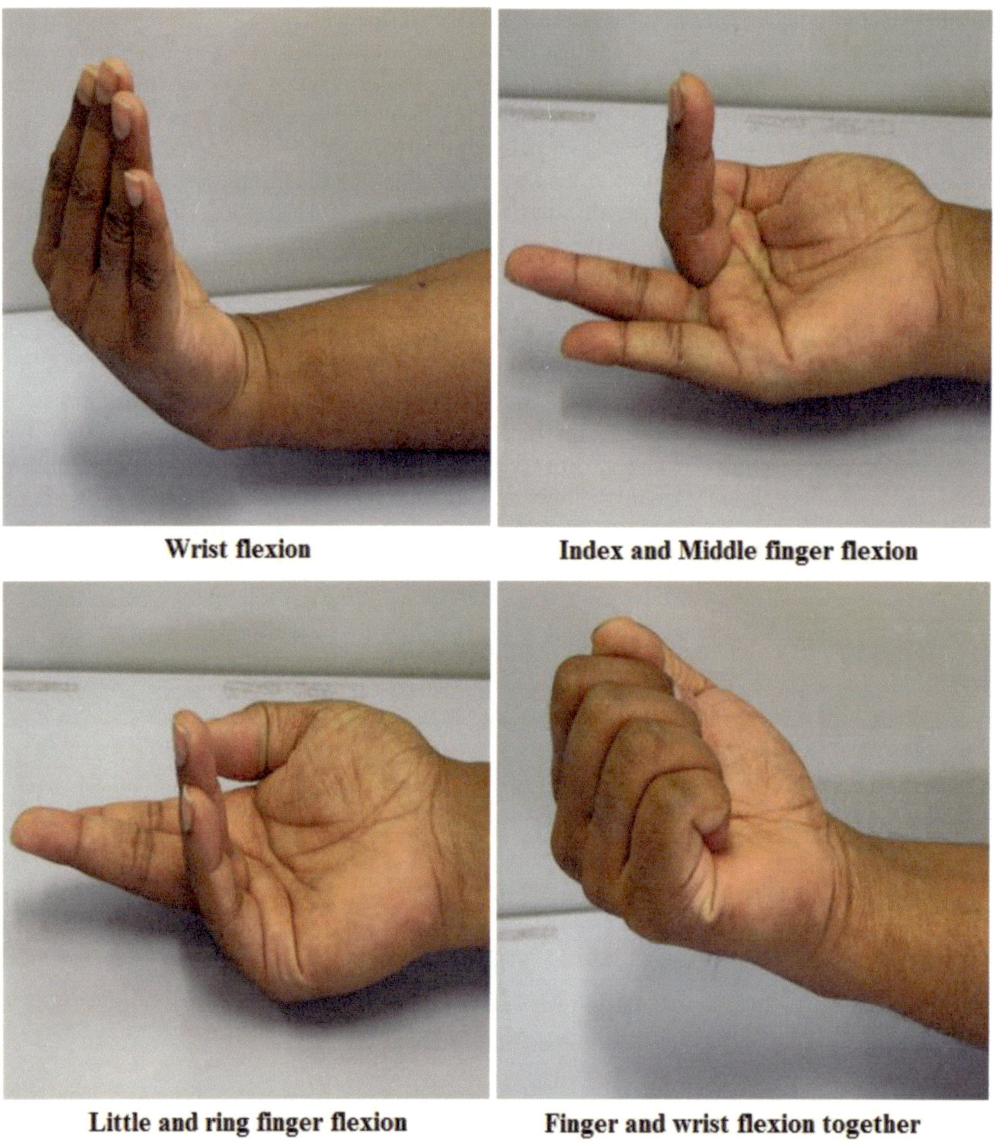

Wrist flexion — **Index and Middle finger flexion**

Little and ring finger flexion — **Finger and wrist flexion together**

Fig. 2. Four Isometric Hand gestures performed during the experiment.

Experiments were conducted to evaluate the performance of the proposed ICA based technique. sEMG was recorded while the participant maintained four isometric finger flexions (fig. 2). The recordings were statistically tested to evaluate the reliability of separation and then tested using a backpropogation neural network. Overall methodology approach is shown in fig. 3.

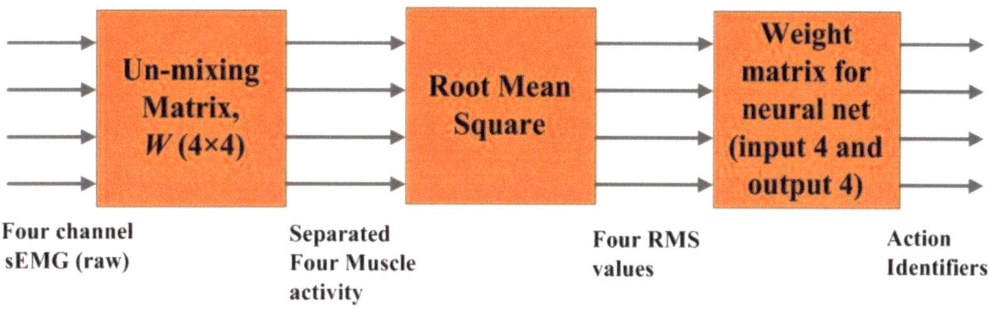

Fig. 3. Flowchart of the system.

4.2 sEMG recording

Experiments were conducted after obtaining approval from RMIT University human experiments ethics committee. Five able subjects, ages ranging from 21 to 32 years (four males and one female) and one amputee subject are volunteered for the experiments. Surface EMG was recorded using a Delysis eight channel sEMG acquisition system (Boston, MA, USA). Each channel has a pair of electrodes mounted together with a fixed inter-electrode distance of 10mm and a gain of 1000. Four electrode channels were placed over four different muscles. A reference electrode was placed at Epicondylus Medialis. Before placing the electrodes, the subject's skin was prepared by lightly abrading with skin exfoliate to remove dead skin. This was done to reduce the skin impedance to less than 60 kOhm. Skin was also cleaned with 70% v/v alcohol swab to remove any oil or dust on the skin surface.

The experiments were repeated on two different days. The forearm was resting on the table with elbow at an angle of approximately 90 degree and in a comfortable position. Four isometric wrist and finger flexions were performed and each was repeated for a total of 24 times for each action over the two sessions. The signal was sampled at 1024 samples/second. The actions were selected because these required four multiple muscles to be contracting at the same time and thus could test the ability of the system and this ensured that the estimated unmixing matrix was square. Markers were used to obtain the isometric contraction signals during recording. A suitable resting time was given between each experiment. There was no external load.

4.3 Data analysis

As a first step, sEMG recordings were segmented to remove the start and end of each recording. This was done based on the temporal location of the markers. FastICA was then used to separate the four channels of sEMG using 4×4 matrix structures for the first day experiments. The estimated unmixing matrix, **W**, was saved and corresponded to the

participant. Root mean square (RMS) was computed for the four estimated separated signals to obtain one number corresponding to each muscle for each action. The above was repeated for each of the five participants.
The unmixing matrix, ***W***, was then multiplied with the recordings of the experiments of the test data corresponding to the balance twenty experiments (not used for training). RMS was computed for each of the separated signals and this resulted in a set of four RMS values for each of experiment.

4.4 Classification of data

A neural network with four inputs, four outputs, and twenty hidden neurones was used to classify the data. Four RMS values of the muscles were the inputs and numbers identifying the four actions were the target. For each participant and for each action, there were 24 recordings. Data from four randomly selected recordings were used to train the network. The weight matrix obtained at the end of the training and the unmixing matrix generated by ICA was saved to correspond to the participant. The training was repeated for each participant.
The system was tested using the balance twenty experiments data that had not been used for training. The input to the neural network was the set of four RMS values of the separated sEMG signal using the unmixing matrix corresponding to each participant. The weight matrix for each participant was used for the testing of the data of that participant. The output of the network was recorded and compared with the known corresponding actions and accuracy of identifying the action was estimated as a percentage. This was repeated for the five participants.
To compare the technique, data classification was also repeated for raw sEMG corresponding to the sEMG without separation, and the sEMG separated using ICA based technique reported in literature, where the unmixing matrix was generated for each experiment, similar to the technique used for separating audio and similar data.

5. Results and observations

The experiments have compared the accuracy of identifying the hand actions based on the RMS of sEMG recordings using raw sEMG, sEMG separated using standard ICA and using the unmixing matrix and weight matrix corresponding to the individual. These results have been tabulated in Table 1 in the following form:

- Method 1: Experimental results for Hand Gesture Identification using Raw sEMG (without using ICA)
- Method 2: Experimental results for Hand Gesture Identification using muscle activity separated from sEMG using traditional (matrix repeated) ICA
- Method 3: Experimental results for Hand Gesture Identification using muscle activity separated from sEMG (using fixed matrix) ICA
- Method 4: Experimental results for Hand Gesture Identification using muscle activity for amputee data from sEMG (using fixed matrix) ICA

From this table, it is observed that classification of sEMG after pre-processing using the unmixing matrix and corresponding weight matrix has 97% accuracy. This accuracy is only 65% when the unmixing matrix is generated for each set of experiments. When RMS from unseparated sEMG was used (referred to as raw sEMG), the accuracy of classification was only 60%. This indicates that using the set of unmixing matrix and weight matrix corresponding to each participant improves the system accuracy dramatically, from 60% to 97%. Similar analysis was performed for amputee data, where the classification accuracy was 90%. The discrepancy between able and amputee data could be due to more cross-talk among the amputee data, as compared to that of able bodied subjects.

Methods	Wrist flexion		Index and Middle Finger flexion		Little and ring finger flexion		Finger and wrist flexion together.	
	Day one	Day two	Day one	Day two	Day one	Day two	Day one	Day two
Method 1	60%	60%	60%	60%	60%	60%	60%	60%
Method 2	65%	65%	65%	65%	65%	65%	65%	65%
Method 3	97%	97%	97%	97%	97%	97%	97%	97%
Method 4	90%	90%	90%	90%	90%	90%	90%	90%

Table 1. Over all Experimental results (average) for Hand Gesture Identification

6. Myo electric prosthetic control using robotic hand

Most prosthetic hands, available in current market, utilize at best take sensory feedback and are dependent on muscle rather than neural control. They require an extensive training before the patients can properly use the prosthetic hand and are bounded with limited functionality. Many organizations, Defense Advanced Research Projects Agency (DARPA), for instance have set an agenda and envision of advanced prosthetic arm with various functional properties. Whereby, the data glove can be used in the training to help the end users in controlling the robotic hand. Also it would be useful for the rehabilitation program i.e. for amputees who lost the sensation of their fingers movement. The development of Robotic hand system shall provide an option for both doctors and patients (amputees and those of weak upper-limbs) to monitor their finger movements. Furthermore, the designed system 'Command for Robotic Hand' may be able to assist the elder people who are facing decrepitude; the users may be able to control the robotic hand to do the work for them which required similar training i.e. wirelessly or embedding the robotic hand on the wheel chair.

Prosthetic hand controlled by EMG has the ability to control more joints than other conventional prosthetic hands, such as functional upper extremity prostheses, which can control no more than two joints. Development of EMG prosthetic hands that can perform more humanlike and inherent movements is desired in rehabilitation engineering and welfare work, especially by amputee patients. However, the majority of the EMG prosthetic hands on the market cannot meet the demand of doctors and amputees because those products can control only one joint and perform hand-opening/closing. Current EMG prosthetic hands use two dry-type electrodes; which applied to skin surface under two types of muscles -i.e. extensor digitorum muscle and flexor digitorum superficial muscle- during EMG signals detection. The control of prosthetic hands required the amputees to generate EMG patterns which are different from the patterns before amputation. This task is quite difficult for most patients; hence the training for certain period of time is essential before the amputees can adapt themselves with the prosthetic hand.

6.1 On-line learning method for EMG prosthetic hand control

On-Line Learning Method for EMG Prosthetic Hand control enumerated their problems blocking the realization of multi-degree-of-freedom EMG prosthetic hands as follows:

- Problem of developing an effective analysis method, which can discriminate multiple motions. It is feasible because EMG is the superimposed complex signal of the electromyograms from multiple muscles.
- Problem of adapting to an operator's individual variation, which is caused by differences of physical properties, such as impedance generated by the amount of muscles and fat, and differences of the results of motor learning, that is, personal habits of motion, which result in different EMG even if in the same motion.
- Problem of developing a robotic hand which will be light enough, with high power and durability, and the same size as the human hand.
- Problem of reducing noise, which results form the fact that the EMG is a physiological signal with small amplitude (about 1mV), while the dry-type electrode has high impedances.

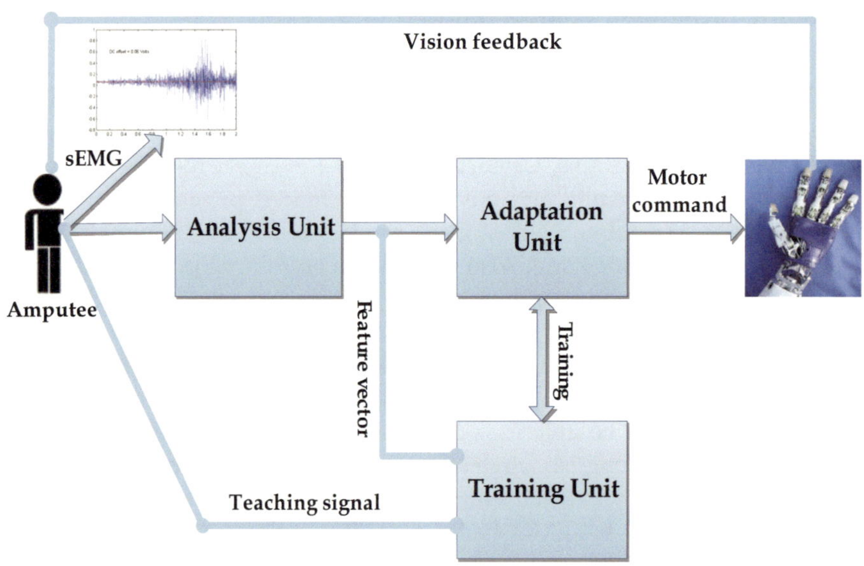

Fig. 4. A schematic diagram of the proposed prosthetic hand control system using electromyogram.

The research work of *Real-time control of a virtual hand* has been integrated a myoelectric control system for prostheses and evaluated on six healthy subjects; the system were able to control a computer-animated hand in real time with a 20Hz refresh rate. The schematic block diagram of the entire process is shown in Fig. 4. A data glove, equipped with joint angle sensors, was used to train the system and to evaluate the continuous joint predictions -prediction error. A linear envelop filter was used for EMG signal preprocessing and the recognition of muscle patterns.

EMG recognition, in nature, is indeed a complex task and the variation between different subjects is often large; one reason being for this is reported in their demographic data, but rarely been reported in related publications, is tissue thickness that heavily affect the capability to record deeply located muscles. Their purpose of the experiments was to investigate the performance of the proposed real time control system and the advantages of on-line training on a healthy homogenous study group. Different parameters that are important for practical use of hand prosthesis were calculated such as the delay between the prediction and measured hand joints and the accuracy of 4 basic hand movements.

7. Discussion and conclusions

The poor results when using raw sEMG is attributed to the closeness of the multiple active muscles resulting in high level of cross-talk. The poor accuracy in the use of ICA is attributable to the ambiguity associated with ICA and these results in inaccurate classification of the data. The proposed technique has overcome the above mentioned shortcomings of the earlier methods by having a set of unmixing matrix and weight matrix corresponding to each participant. The result is that the ambiguity associated with ICA does not affect the outcomes because while the order is not known, this remains the same during the unmixing and classification stages. The result is that the system identifies the actions accurately.

The presented work has demonstrated that use of the RMS of sEMG recorded from four channels from the forearm is not suitable to accurately identify finger and wrist flexions. The experiments have also demonstrated that if ICA is used to separate the signal to obtain muscle activity from individual muscles, there is the problem of order and scale ambiguity. To mitigate these shortcomings, the proposed system uses a set of unmixing and weight matrices that separate and classify the signal. This set is built for each individual and is referred to as ICANN. Because the ambiguities are the same during the testing and during the training phases, the test results indicate that the system accuracy is markedly better than the earlier techniques. While the results for raw sEMG was 60%, and for sEMG separated using ICA was only 65%, the method reported in this paper gives an accuracy of 97% for the same set of recordings, and classified by similar methods.

The other benefit of this technique is that it offers a system that is suitable for real time operations. During regular operations, the pre-trained system would only require two multiplication steps of 4×4 matrix to obtain the classification. While the training requires supervision, the regular operation does not require any supervision and is suitable for a lay user. The experimental results demonstrate that the system is resilient to inter-experimental variations. One of the possible shortcomings of such the proposed technique is that even after separation, background noise can make the segmentation difficult to automate and hence there is need for identifying a non-linear feature set that can replace RMS of the signal. The other shortcoming of such a technique is that while it has made a set of unmixing and classification matrices, there is the randomness associated with genetic algorithms and this set is not optimized. There is a need for optimizing such a system.

This research also proposes a visual training method for prosthetic hand control; in particular, it had succeeded in generating a command control for the Robotic hand. The success included the ability in discriminating distinct finger movements (per channel) with respect to time. As the training design aimed to assist people in the rehabilitation who have weaken limbs, encountered lost of sensation and amputees by delivering output as visual feedbacks. It is a crucial contribution to those who lost their sensation after the amputation. This experiment represents a modest foray into using supervised learning to control a prosthetic hand. It was a proof-of-concept and needs to be extended before being considered proven. To be useful, a prosthetic hand must be capable of executing considerably more than six gestures. It may not be possible to recognize directly the large number of gestures used in every day life. A better approach might be to build a small vocabulary of gesture components from which the larger set of everyday gestures can be built. Based on this paper, an intelligent EMG control scheme can be implemented to recognize the hand's gesture and grasping force simultaneously. That makes a big improvement to current multi-DOF prosthetic hands' myocontrol. Future work will concentrated on validating this method on patients.

8. References

Baker, Scheme, Englehart, K., Hutcinson, & Greger. (2010). Continuous Detection and Decoding of Dexterous Finger Flexions with Implantable MyoElectric Sensors. *IEEE Transactions on Neural Systems and Rehabilitation Engineering*.

Basmajian, & Deluca, C. (1985). *Muscles Alive: Their Functions Revealed by Electromyography*: Williams & Wilkins.

Bell, A. J., & Sejnowski, T. J. (1995). An information-maximization approach to blind separation and blind deconvolution. *Neural Comput, 7*(6), 1129-1159.

Chan, A. D. C., & Englehart, K. B. (2005). Continuous myoelectric control for powered prostheses using hidden Markov models. *Biomedical Engineering, IEEE Transactions on, 52*(1), 121-124.

Cheron, G., Draye, J. P., Bourgeios, M., & Libert, G. (1996). A dynamic neural network identification of electromyography and arm trajectory relationship during complex movements. *IEEE transactions on bio-medical engineering, 43*(5), 552-558.

Cichocki, A., & Amari, S.-I. (2002). *Adaptive Blind Signal and Image Processing: Learning Algorithms and Applications*: John Wiley \& Sons, Inc.

Comon, P. (1994). Independent component analysis, A new concept? *Signal Processing, 36*(3), 287-314.

Doerschuk, P., Gustafon, D., & Willsky, A. (1983). Upper Extremity Limb Function Discrimination Using EMG Signal Analysis. *Biomedical Engineering, IEEE Transactions on, BME-30*(1), 18-29.

Englehart, K., & Hudgins, B. (2003). A robust, real-time control scheme for multifunction myoelectric control. *50*(7), 848-854.

Farry, K. A., & Walker, I. D. (1993). *Myoelectric teleoperation of a complex robotic hand*. Paper presented at the IEEE International Conference on Robotics and Automation.

Graupe, D., & Cline, W. (1975). Functional separation of EMG signals via ARMA identification methods for prosthesis control purposes. *IEEE Transacton on Systems Man and Cybernetics, 5*, 252-259.

Hamdi, N., Dweiri, Y., Al-Abdallat, Y., & Haneya, T. (2010). A practical and feasible control system for bifunctional myoelectric hand prostheses. *Prosthet Orthot Int, 34*(2), 195-205.

Hyvärinen, A. (2000). Independent component analysis: algorithms and applications. *Neural Networks, 13*(4-5), 411-430.

Hyvärinen, A., Karhunen, J., & Oja, E. (2001). *Independent Component Analysis*: Wiley-Interscience.

Hyvärinen, A., & Oja, E. (1997a). A fast fixed-point algorithm for independent component analysis. *Neural Comput., 9*(7), 1483-1492.

Hyvärinen, A., & Oja, E. (1997b). A Fast Fixed-Point Algorithm for Independent Component Analysis. *Neural Computation, 9*(7), 1483-1492.

Jacobsen, S. C., & Jerard, R. B. (1974). *Computational requirements for control of the Utah arm.* Paper presented at the ACM Annual Conference.

Ju, P., Kaelbling, L. P., & Singer, Y. (2000). *State-based Classification of Finger Gestures from Electromyographic Signals.* Paper presented at the Seventeenth International Conference on Machine Learning.

Jung, T.-P., Humphries, C., Lee, T.-W., Makeig, S., McKeown, M., Iragui, V., et al. (1998). *Extended ICA removes artifacts from electroencephalographic recordings.* Paper presented at the NIPS '97: Proceedings of the 1997 conference on Advances in neural information processing systems 10, Denver, Colorado, United States.

Jung, T. P., Makeig, S., Humphries, C., Lee, T. W., McKeown, M. J., Iragui, V., et al. (2000). Removing electroencephalographic artifacts by blind source separation. *Psychophysiology, 37*(2), 163-178.

Kiguchi, K., Tanaka, T., & Fukuda, T. (2004). Neurofuzzy control of a robotic exoskeleton with EMG signals. *IEEE Trans. on Fuzzy Systems, 12*(4), 481-490.

Koike, Y., & Kawato, M. (1996). Human interface using surface electromyography signals. *Electronics and Communications in Japan (Part III: Fundamental Electronic Science), 79*(9), 15-22.

Lee, T.-W. (1998). *Independent component analysis: theory and applications*: Kluwer Academic Publishers.

Makeig, S., Bell, A., Jung, T., & Sejnowski, T. (1996). *Independent Component Analysis of Electroencephalographic Data.* Paper presented at the Advances in Neural Information Processing Systems.

Meyer, C. D. (2000). *'Matrix Analysis and Applied Linear Algebra.* U.K: Cambridge press,.

Nishikawa, D., Yu, W., Yokoi, H., & Kakazu, Y. (1999). *Emg prosthetic hand controller discriminating ten motions using real time learning method.* Paper presented at the IEEE/RSJ international conference on intelligent robots and systems.

Peleg, D., Braiman, E., Yom-Tov, E., & Inbar, G. (2002). Classification of finger activation for use in a robotic prosthesis arm. *IEEE transactions on neural systems and rehabilitation engineering : a publication of the IEEE Engineering in Medicine and Biology Society, 10*(4), 290-293.

Rehg, J., & Kanade, T. (1993). *DigitEyes: Vision-Based Human Hand Tracking.*

Schieber, M. H. (1995). Muscular production of individual finger movements: the roles of extrinsic finger muscles. *Journal of Neuroscience., 15*(1), 248-297.

Schlenzig, J., Hunter, E., & Jain, R. (1994). *Vision based hand gesture interpretation using recursive estimation.* Paper presented at the Signals, Systems and Computers, 1994. 1994 Conference Record of the Twenty-Eighth Asilomar Conference on.

Tenore, F., Ramos, A., Fahmy, A., Acharya, S., Etienne-Cummings, R., & Thakor, N. (2007). *Towards the control of individual fingers of a prosthetic hand using surface EMG signals.* Paper presented at the IEEE Engineering in Medicine and Biology Society. IEEE Engineering in Medicine and Biology Society. Conference.

Zardoshti-Kermani, M., Wheeler, B. C., Badie, K., & Hashemi, R. M. (1995). EMG feature evaluation for movement control of upper extremity prostheses. *Rehabilitation Engineering, IEEE Transactions on [see also IEEE Trans. on Neural Systems and Rehabilitation], 3*(4), 324-333.

7

Feedback Control and Time-Optimal Control about Overhead Crane by Visual Servo and These Combination Control

Yasuo Yoshida
Chubu University
Japan

1. Introduction

Overhead cranes are broadly used in industry. However, undesirable load swing is often induced with transportation. Therefore, various researches about control of positioning and swing suppression of crane load have been made. On the other hand, a worker operates a crane skillfully by manual-operation in transportation. Hence, support system and automation of crane operation are expected from the viewpoint of safety and production efficiency.

Feedback, time-optimal and these combination controls by visual servo for practical use are described in this section. A visual servo device has the following two types. One type is installed separately from a crane and another type is installed on a trolley. Two problems related to the crane control will be discussed from here.

The first problem is feedback control by separate type of visual servo device. Separate type device is the case that the device cannot be installed on the crane. It has to track transportation movement and measure swing of the load. Base-line stereo which has parallel optical axes of two cameras uses parallax where tracked target point is not in the center of camera image and therefore, tracking control is difficult to succeed. Gaze controlled stereo vision device has the mechanism that an intersection point of two cameras optical axes always accords with the target and can catch the target point in the center of camera image. This stereo vision device is made and used it for experiments. Using camera image data and tracking controlled camera angles, crane load can be tracked and at the same time 3 dimensional position of the load can be got. By using these data, feedback control of positioning and swing suppression was applied to crane load in experiments.

The second problem is combination control of time optimal control and feedback control by installed type of visual servo device. Installed type device is not necessary to track transportation movement, then that is why 3D-camera fixed to trolley is used. This type device can measure displacements of both swing of the load and rope hoisting. Time optimal control is feed forward control. Trolley acceleration time for swing suppression control is set using exact natural period of load swing that is beforehand estimated. However, when with lack of information of actual rope length, position of center of gravity

or moment of inertia of crane load, it is often that an incorrect natural period of load swing is used, and swing is induced. Therefore, it is practical to use combination control of both time optimal control of transportation positioning and feedback control of swing suppression of the load and it was inspected by experiments.
In addition, when rope length changes while swing suppression control, feedback control needs variable control gains of nonlinear system. Then, the method that control digital gains can be redesigned by analog gains was developed and these variable digital gains were used in experiments.

2. Crane feedback control by gaze-controlled stereo vision installed separately from the crane

2.1 Dynamic model and control of overhead crane

An overhead crane model and a stereo vision installed separately from the crane are shown in Fig. 1. This system has five-degree-of-freedom (trolley x, y, rope length l, swing angle amplitude θ, swing direction angle ϕ). Stereovision has four-degree-of-freedom from θ_1 to θ_4.

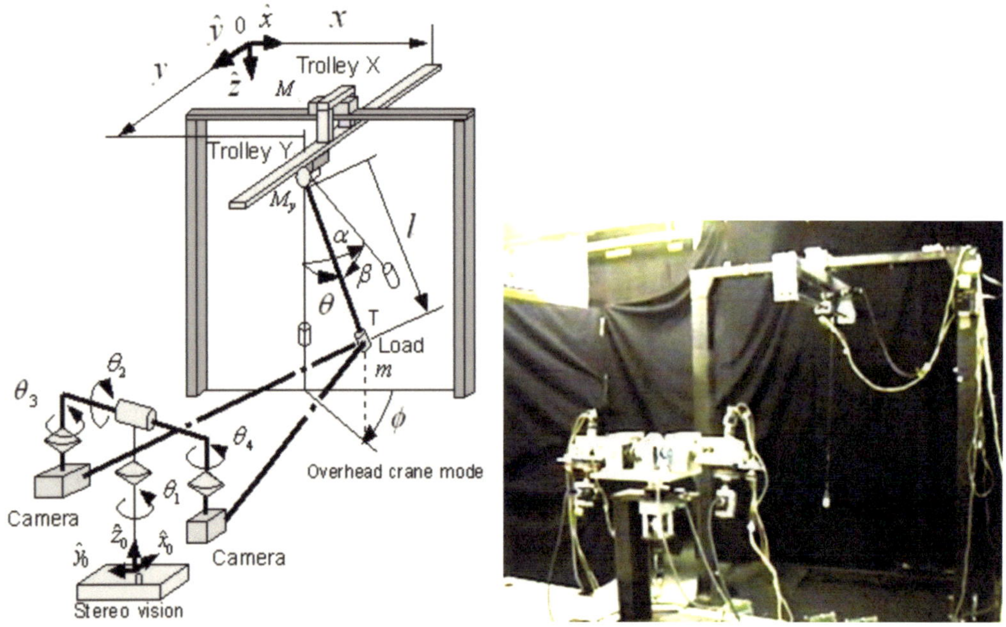

Fig. 1. Experimental model of overhead crane and stereo vision

Transportation and swing suppression are thought about the crane. The state equation associated with trolley displacement and swing angle is obtained using control input $\ddot{x} = u_x, \ddot{y} = u_y$,

$$\frac{d}{dt}\tilde{x} = A\tilde{x} + bu_x, \quad \frac{d}{dt}\tilde{y} = A\tilde{y} + bu \tag{1}$$

where

$$A=\begin{bmatrix}0&0&1&0\\0&0&0&1\\0&0&0&0\\0&-\frac{g}{l}&0&-2\frac{\dot{l}}{l}\end{bmatrix},\quad b=\begin{bmatrix}0\\0\\1\\-\frac{1}{l}\end{bmatrix},\quad \tilde{x}=\begin{bmatrix}x\\\alpha\\\dot{x}\\\dot{\alpha}\end{bmatrix},\quad \tilde{y}=\begin{bmatrix}y\\\beta\\\dot{y}\\\dot{\beta}\end{bmatrix}$$

Swing angles are supposed to be small values. Digital regulator where trolley displacements follow desired positions and swing angles decrease to zero value can be designed as,

$$u_{xk}=G_d\tilde{e}_{xk},\quad u_{yk}=G_d\tilde{e}_{yk},\tag{2}$$

where

$$\tilde{e}_{xk}=\begin{bmatrix}x_d-x_k & -\alpha_k & -\dot{x}_k & -\dot{\alpha}_k\end{bmatrix}^T,\quad \tilde{e}_{yk}=\begin{bmatrix}y_d-y_k & -\beta_k & -\dot{y}_k & -\dot{\beta}_k\end{bmatrix}^T.$$

k = 1,2,… indicates time series with sampling time. From the closed loop relation between analog state and digital state that is redesigned to digitize the analog state, digital gain $G_d=\begin{bmatrix}k_{1d} & k_{2d} & k_{3d} & k_{4d}\end{bmatrix}$ can be obtained from analog gain G_a determined by pole assignment.

$$G_d=\frac{1}{2}\left(1+\frac{1}{2}G_a q\right)^{-1}G_a\left(I+p\right),\tag{3}$$

where

$$p=\left(I-\frac{T}{2}A\right)^{-1}\left(I+\frac{T}{2}A\right),\quad q=T\left(I-\frac{T}{2}A\right)^{-1}b.$$

Velocity of the trolley can be obtained from eq.(2) by integrating with sampling time.

2.2 Gaze-controlled stereo vision

Stereo vision is shown in Fig. 2 and camera (XV1000,Keyence Corp.) coordinates in Fig. 3. Camera image positions of moving point T indicated by camera coordinate 3c, 4c are expressed as,

$${}^{3c}P_T=\begin{bmatrix}0 & y_{3c} & z_{3c}\end{bmatrix}^T,\quad {}^{4c}P_T=\begin{bmatrix}0 & y_{4c} & z_{4c}\end{bmatrix}^T.$$

Therefore, point T indicated by target coordinate 3t, 4t can be expressed with image data,

$${}^{3t}P_T=\frac{a_1}{f}\begin{bmatrix}0 & y_{3c} & z_{3c}\end{bmatrix}^T,\quad {}^{4t}P_T=\frac{a_2}{f}\begin{bmatrix}0 & y_{4c} & z_{4c}\end{bmatrix}^T,$$

a_1,a_2 are distances from the lens center to the target-point along optical axes using $c_i\equiv\cos\theta_i, s_i\equiv\sin\theta_i\,(for\ i=1\sim 4)$ and $c_{34}\equiv\cos(\theta_3-\theta_4),\ s_{34}\equiv\sin(\theta_3-\theta_4)$,

$$a_1=\frac{2b\left(\frac{y_{4c}}{f}s_4-c_4\right)-r\left\{\frac{y_{4c}}{f}(1-c_{34})+s_{34}\right\}}{\frac{y_{3c}-y_{4c}}{f}c_{34}+\left(1+\frac{y_{3c}y_{4c}}{f^2}\right)s_{34}},\quad a_2=\frac{2b\left(\frac{y_{3c}}{f}s_3-c_3\right)+r\left\{\frac{y_{3c}}{f}(1-c_{34})-s_{34}\right\}}{\frac{y_{3c}-y_{4c}}{f}c_{34}+\left(1+\frac{y_{3c}y_{4c}}{f^2}\right)s_{34}}\tag{6}$$

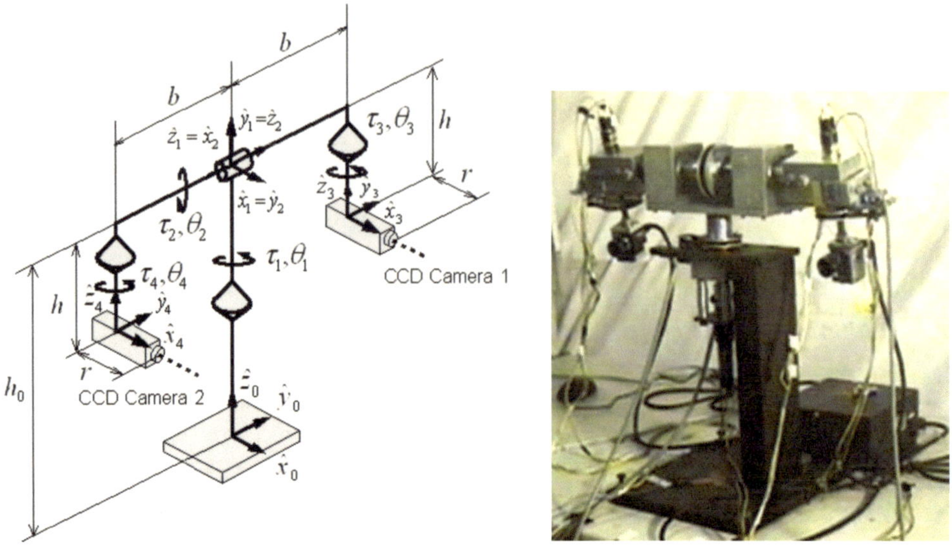

Fig. 2. Stereo vision installed separately from the crane

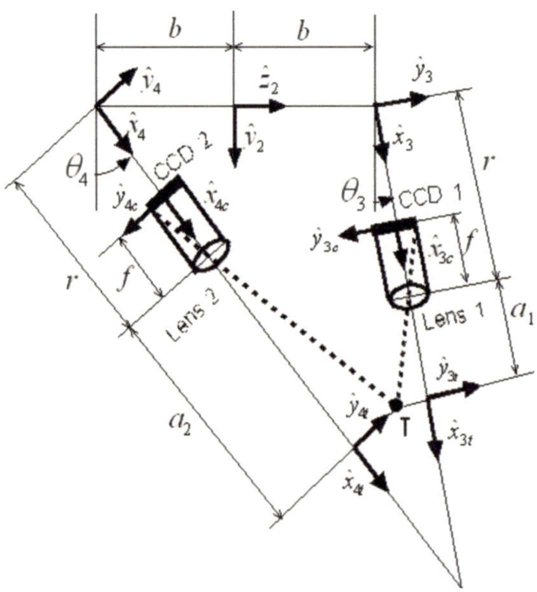

Fig. 3. Camera coordinates and optical axes

Positions by joint coordinate 3, 4 can be expressed,

$$ {}^{3}P_{T} \equiv \begin{bmatrix} {}^{3}x_{T} \\ {}^{3}y_{T} \\ {}^{3}z_{T} \end{bmatrix} = \begin{bmatrix} r + a_1 \\ (a_1/f)y_{3c} \\ (a_1/f)z_{3c} \end{bmatrix}, \quad {}^{4}P_{T} \equiv \begin{bmatrix} {}^{4}x_{T} \\ {}^{4}y_{T} \\ {}^{4}z_{T} \end{bmatrix} = \begin{bmatrix} r + a_2 \\ (a_2/f)y_{4c} \\ (a_2/f)z_{4c} \end{bmatrix} \tag{7} $$

From ${}^{3}P_{T}$, the position expressed with base coordinate 0 is given by,

$$
{}^{0}P_T \equiv \begin{bmatrix} {}^{0}x_T \\ {}^{0}y_T \\ {}^{0}z_T \end{bmatrix} = \begin{bmatrix} c_1 s_2\left({}^{3}z_T - h\right) + c_1 c_2\left(c_3\,{}^{3}x_T - s_3\,{}^{3}y_T\right) - s_1\left(s_3\,{}^{3}x_T + c_3\,{}^{3}y_T + b\right) \\ s_1 s_2\left({}^{3}z_T - h\right) + s_1 c_2\left(c_3\,{}^{3}x_T - s_3\,{}^{3}y_T\right) + c_1\left(s_3\,{}^{3}x_T + c_3\,{}^{3}y_T + b\right) \\ c_2\left({}^{3}z_T - h\right) - s_2\left(c_3\,{}^{3}x_T - s_3\,{}^{3}y_T\right) + h_0 \end{bmatrix} \tag{8}
$$

Similary, the position of the target can be obtained from ${}^{4}P_T$. Fig. 4 (a) shows the gazing state where the target is on a point of intersection of optical axes of both cameras where the position is given by putting $y_{3c} = y_{4c} = 0,\ \ z_{3c} = z_{4c} = 0$ into eq.(7). Fig. 4 (b) shows the looking state of parallel optical axes of both cameras by putting $\theta_3 = \theta_4 = 0$ into eq.(7). Therefore, expressions of (6), (7) include not only gaze but also parallel stereo.

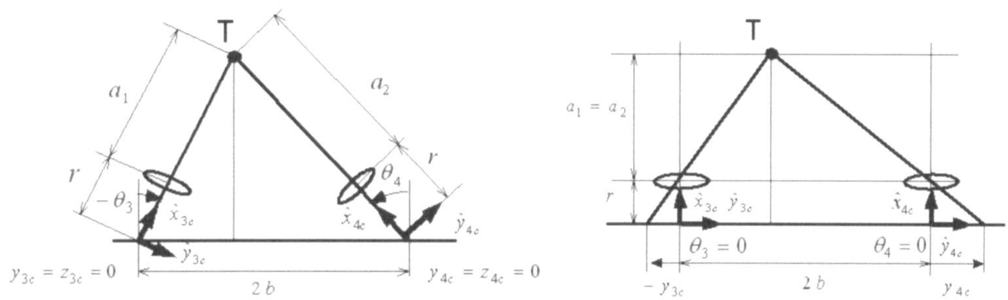

(a) Gazing the target (b) Looking using parallel optical axes

Fig. 4. Looking states using gaze and parallel optical axes of stereo cameras

We can see that the three-dimensional position of a crane load is the function with camera angles, image data, the focus distance and stereovision dimensions. It is gaze control that there is a moving point in the center of camera image, and tracking control that always catches a moving point to camera image.

When an intersection point of two cameras optical axes accords with a moving point, stereovision angles become target desired angles for the gaze/tracking control. Target desired angles as shown in Fig. 5 using measured position of eq.(8) are provided by inverse kinematics,

$$
\begin{aligned}
\theta_{1d} &= A\tan 2\left({}^{0}y_T, {}^{0}z_T\right) \\
\theta_{2d} &= -\frac{\pi}{2} - A\tan 2\left({}^{0}z_T - h_0, \sqrt{\left({}^{0}x_T\right)^2 + \left({}^{0}y_T\right)^2}\right) \\
&\quad + A\tan 2\left(\sqrt{\left({}^{0}x_T\right)^2 + \left({}^{0}y_T\right)^2 + \left({}^{0}z_T - h_0\right)^2 - h^2}, h\right), \\
\theta_{3d} &= -\theta_{4d} \\
\theta_{4d} &= A\tan 2\left(b, \sqrt{\left({}^{0}x_T\right)^2 + \left({}^{0}y_T\right)^2 + \left({}^{0}z_T - h_0\right)^2 - h^2}\right)
\end{aligned}
$$

Digital controller by which the controlled angle θ_i follows the desired angle θ_{id} is given by,

$$
u_{ik} = k_{c1d}\left(\theta_{idk} - \theta_{ik}\right) - k_{c2d}\dot{\theta}_{ik} \quad (i = 1 \sim 4) \tag{10}
$$

Manipulating velocities are obtained by integrating with sampling time.

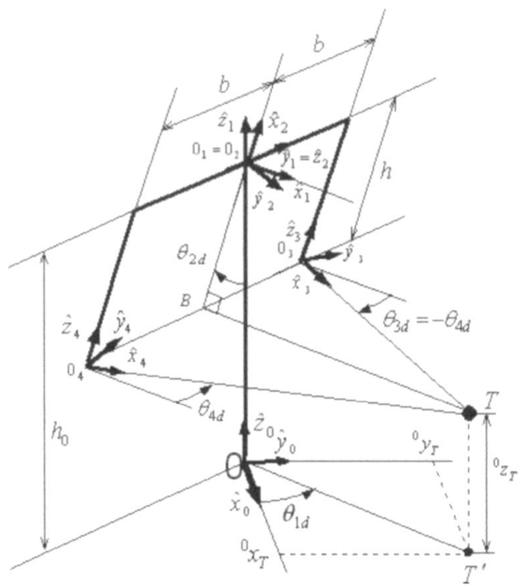

Fig. 5. Desired angles for gaze and tracking to control stereovision

2.3 Visual feedback control of overhead crane

Fig. 6 shows the block diagram of visual feedback control of overhead crane. Half of the upper part shows crane control, and half of the bottom shows stereo vision control.

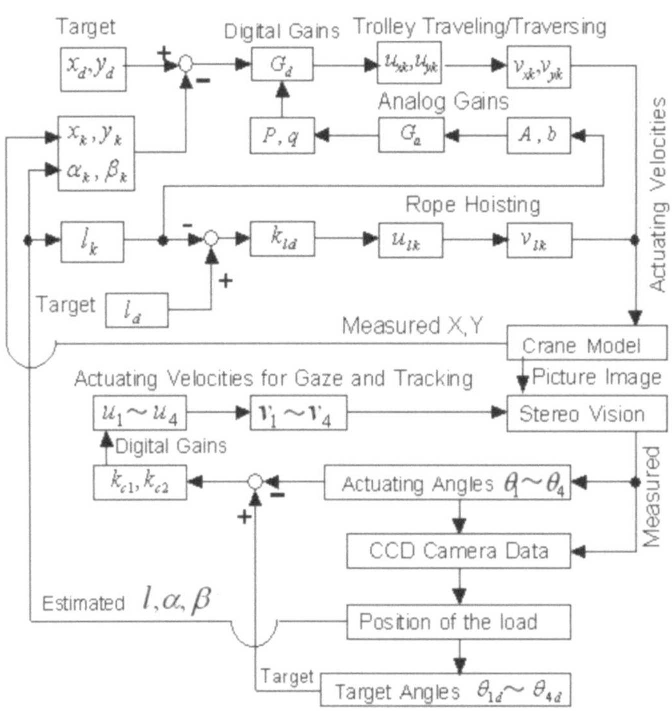

Fig. 6. Block diagram

Experimental results are shown from Fig. 7 to Fig. 11. In these figures, "with control" is defined that trolleys are controlled by visual feedback. Fig. 7 shows rope length. Hoisting the crane load is controlled where the rope length changes upward from initial length $1.25m$ to target length $0.8m$. Fig. 8 shows traveling and traversing transportation of trolley-X, Y. As for the controlled response, both of traveling of trolley-X and traversing of trolley-Y move to the target position $0.6m$ while doing swing suppression. Fig. 9 shows swing amplitude suppressed by visual feedback control though the rope length changes. Fig. 10 shows manipulating angles of stereo vision. The horizontal angle θ_1 changes largely to track the transportation of the load when trolleys are moving. And when the trolleys stay, it tracks the swing of the load, particularly large swing without control. The vertical angle θ_2 changes to track the hoisting of the load. Two camera angles θ_3, θ_4 change to gaze the load. Fig. 11 shows the changes of digital gains of trolley traveling and traversing. Gains change according to the rope length change of hosting of the load and the coefficient k_{2d} changes most, in proportion to swing angle that is sensitive for length.

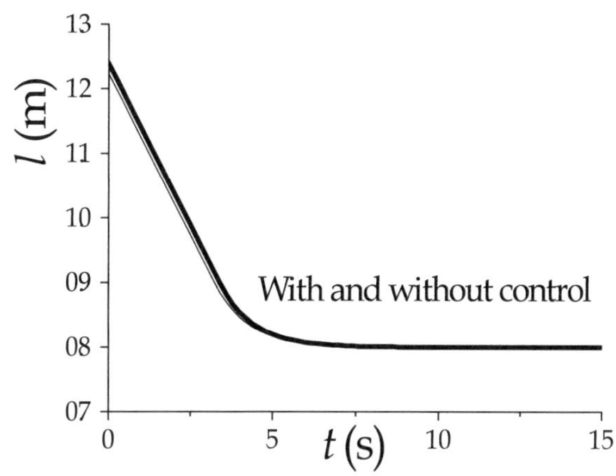

Fig. 7. Rope length with hoisting upward

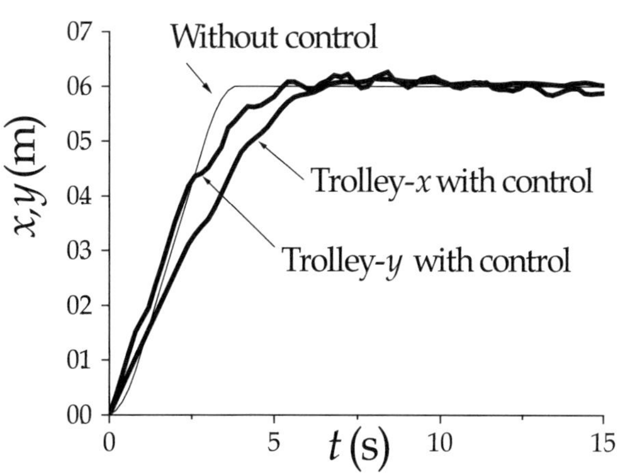

Fig. 8. Traveling and traversing transportation of trolley- X, Y

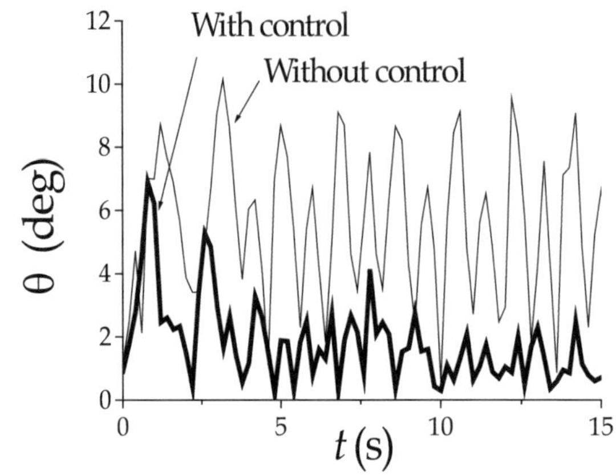

Fig. 9. Swing amplitude

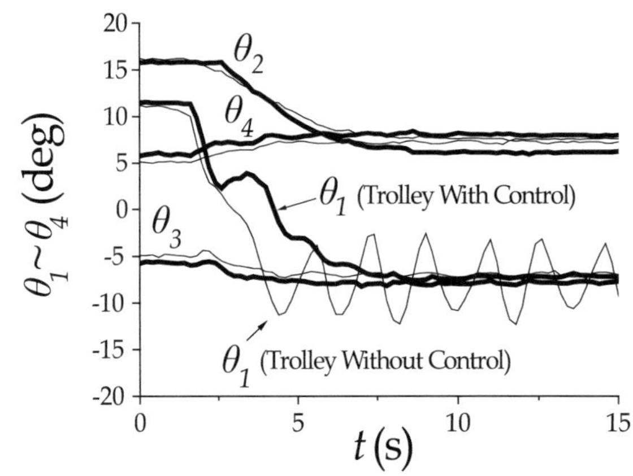

Fig. 10. Manipulating angles of stereo θvision

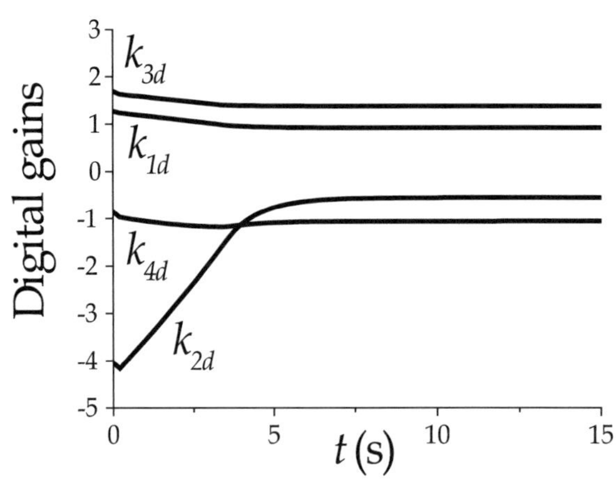

Fig. 11. Digital gains for crane control

3. Combination control of time optimal control and feedback control by 3D-camera installed on the trolley

3.1 Time optimal control as feed forward

Fig. 12 shows an overhead crane with 3D-camera (SR3000, CSEM Inc.) installed on the trolley and the marker on the crane load. Fig. 13 shows the crane load having constant rope length under the condition of trolley's constant acceleration u_0.

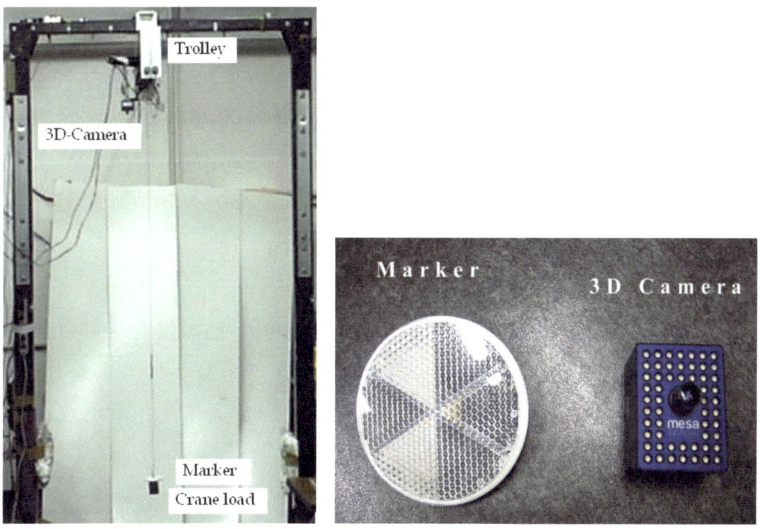

Fig. 12. Photograph of 3D-camera and marker

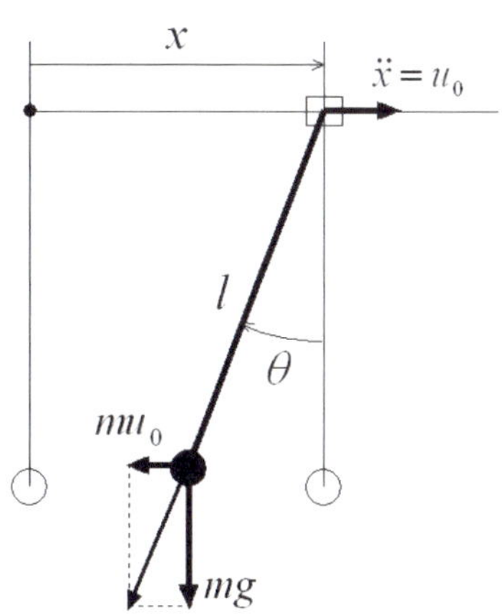

Fig. 13. Crane load under trolley's acceleration

The equations of motion is written as,

$$\ddot{\theta} + \omega^2\theta = u_0/l \tag{11}$$

where, $\omega = \sqrt{g/l}$ is natural circular frequency and $T = 2\pi/\omega$ is natural period.

Solution of eq.(11) is,

$$\theta = (1 - \cos\omega\, t)u_0/g \tag{12}$$

When initial condition is $\theta(0) = \dot{\theta}(0) = 0$, the swing angle is maximum $\theta(T/2) = 2u_0/g$ at $T/2$ and $\theta(T) = \dot{\theta}(T) = 0$ at T. Therefore, when the trolley accelerates with u_0 from $t = 0$ to T, swing angle is zero after the time T. This is well known and we call this "synchronous control" here. To make the travelling time less than natural period T, there is time-optimal control where the acceleration u_0 and $-u_0$ are switched each other using the minimum principle. Fig. 14 shows time trajectory of feedforward control. Two of synchronous control and time-optimal control are indicated in phase plane of swing angle. Synchronous control moves around P-point of circle with radius u_0/g in order O-A-C_1-B-O. Necessary time to move around the circle is natural period T. Time-optimal control moves in order O-A-C_2-B-O where A, B are switching points and center of arc A-C_2-B is Q-point. Necessary time T_{op} to move around is as follows,

$$T_{op} = 2T_\varsigma + T_\eta \ (\leq T), \tag{13}$$

where, $T_\varsigma = \varsigma/\omega, T_\eta = \eta/\omega,\ \eta = 2\tan^{-1}(\sin\varsigma/(2 - \cos\varsigma))$.
Fig. 15 shows feedforward control of trolley. (a) is acceleration and (b) is velocity of trolley

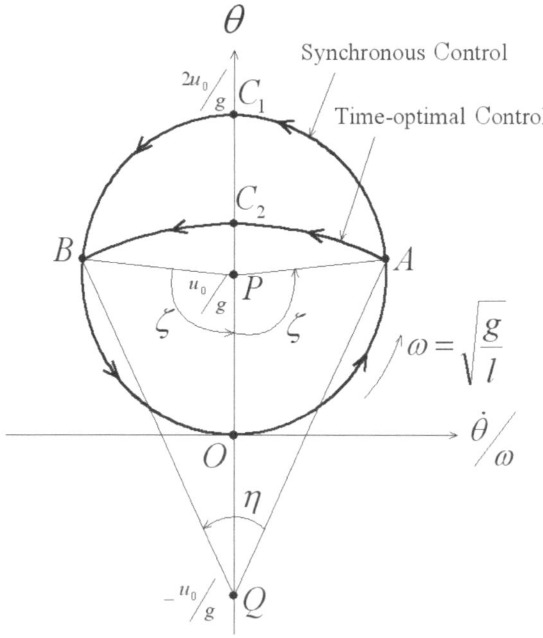

Fig. 14. Time trajectory of feed forward control

Fig. 16 shows time trajectory of feedforward control of trolley using false natural period $\omega_e = \sqrt{g/l_e}$ where l_e is error rope length. Synchronous control moves in order O-A-C_1-O-O_1 and necessary time becomes $(\omega/\omega_e)T$. Similarly, time-optimal control moves in order O-A-C_2-B-O_2 and necessary time becomes $(\omega/\omega_e)T_{op}$. As a result of these, residual swing vibration occurs as shown a circle around origin O-point with radius OO_1 or OO_2.

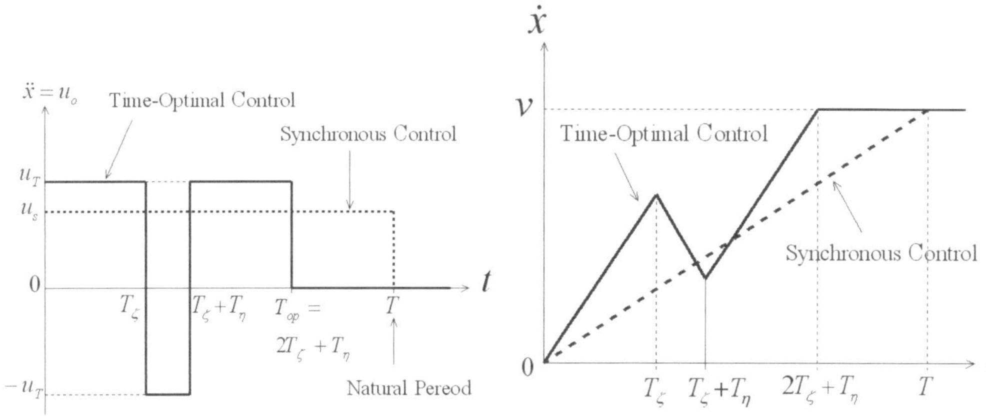

(a) Acceleration of trolley (b) Velocity of trolley

Fig. 15. Feed forward control of trolley

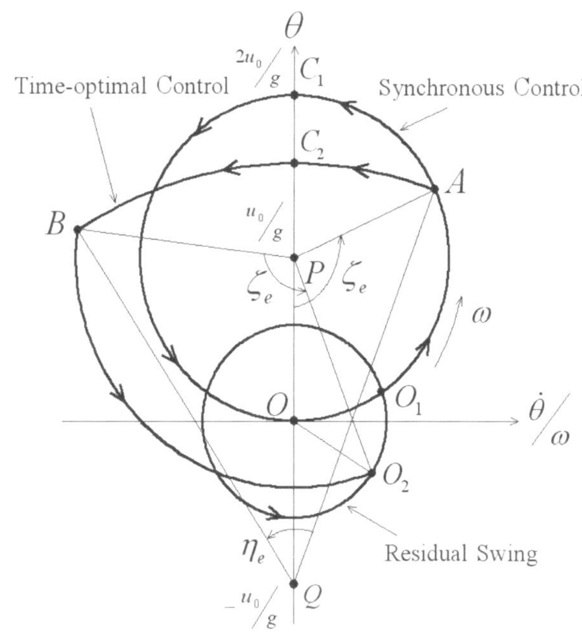

Fig. 16. Time trajectory of feed forward control using false natural period

3.2 Visual control experiments by 3D-camera

Three kinds of experiments using 3D-camera are done. These are experiment of feedback control, feed forward control and combination control.

(a) Visual feedback control experiment

Most of the same experiment as the visual feedback control by stereo vision was performed by 3D-camera. Fig. 17 shows rope length and trolley transportation. These are similar as results of Fig. 7 and Fig. 8 using stereo vision control.

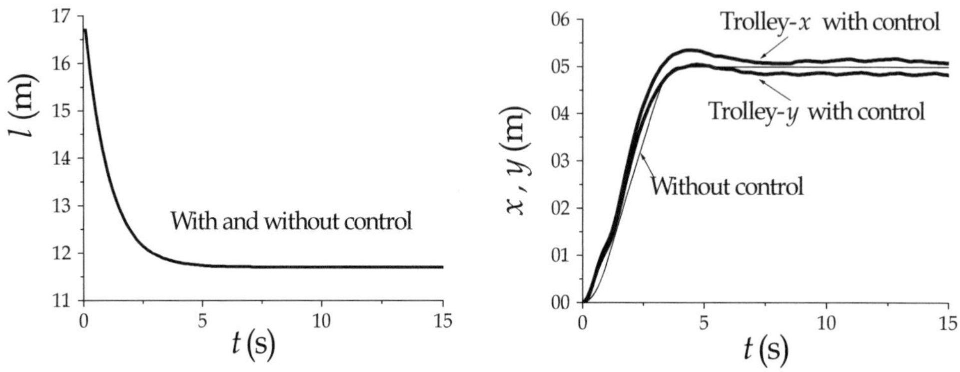

(a) Rope length with hoisting upward (b) Trolley transportation

Fig. 17. Rope length and trolley of visual feedback control using 3D-camera

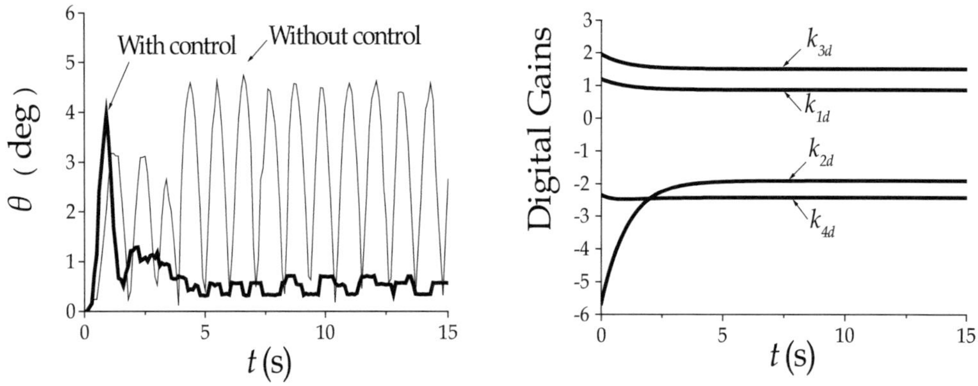

(a) Swing amplitude (b) Digital gains for crane control

Fig. 18. Swing amplitude and digital gains of visual feedback control using 3D-camera

Fig. 18 shows swing amplitude and digital gains for crane control. The swing vibration is suppressed using visual feedback by 3D-camera same as by stereo vision.

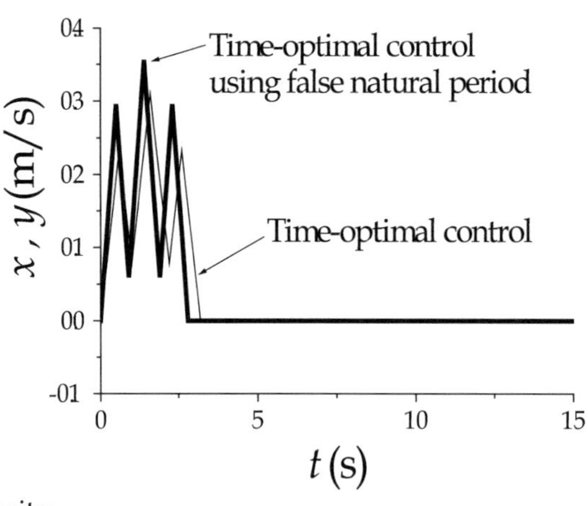

Fig. 19. Trolley's velocity

(b) Visual time-optimal control experiment

Experimental results of time-optimal control of the load suspended by rope length $l = 1.67m$ are from Fig. 19 to Fig. 21. Fig. 19 shows trolley's velocities of two patterns. One is time-optimal $T_\varsigma = 0.6s, T_\eta = 0.4s$ from correct natural period and another is error time $T_\varsigma = 0.5s, T_\eta = 0.4s$ using false natural period. Fig. 20 is the trolley's displacement controlled to the desire 0.5m. Fig. 21 is the swing amplitude. In the case of time-optimal control, the load swings with 4 deg amplitude until the time $2T_{op} = 3.2s$ and after that it stays without swing. In the case of the control using false natural period, the load swings with 5 deg until the time $2T_{op} = 2.8s$ and after that it's swing continues with 1.5 deg.

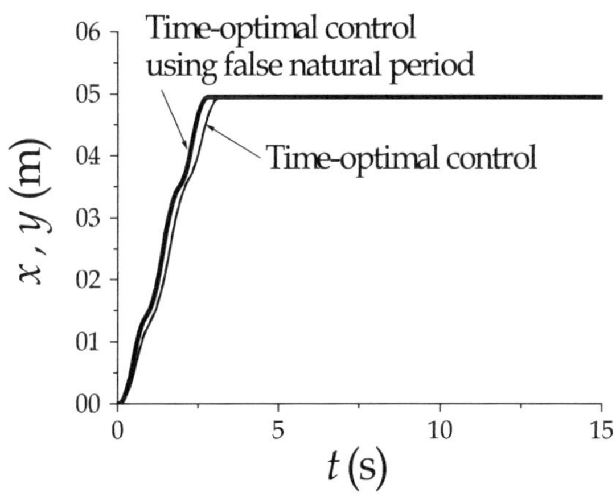

Fig. 20. Trolley transportation

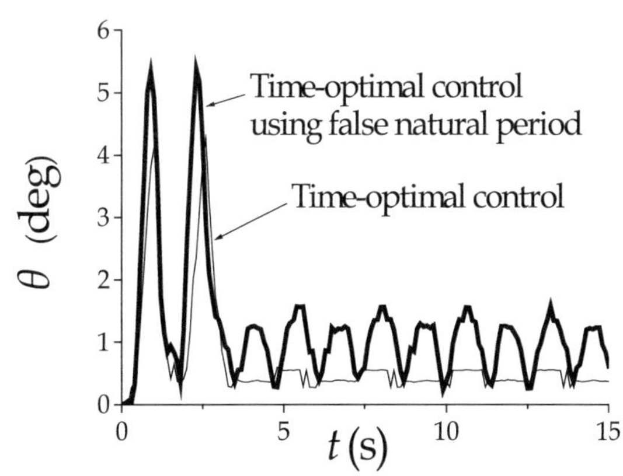

Fig. 21. Swing amplitude

(c) Combination control experiment

We tried the combination control that is the time-optimal feed forward control of transportation and the visual feedback control of swing suppression. Experimental results of combination control are from Fig. 22 to Fig. 24. Fig. 22 shows trolley's velocities. In the first part, the time-optimal feedforward control with time $T_{\varsigma} = 0.5s, T_{\eta} = 0.4s$ using false natural period is used until $2T_{op} = 2.8s$ .Afterwards visual feedback control is used.

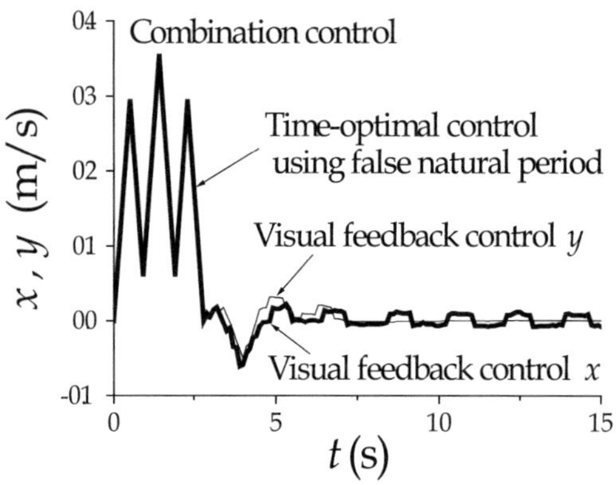

Fig. 22. Trolley's velocity

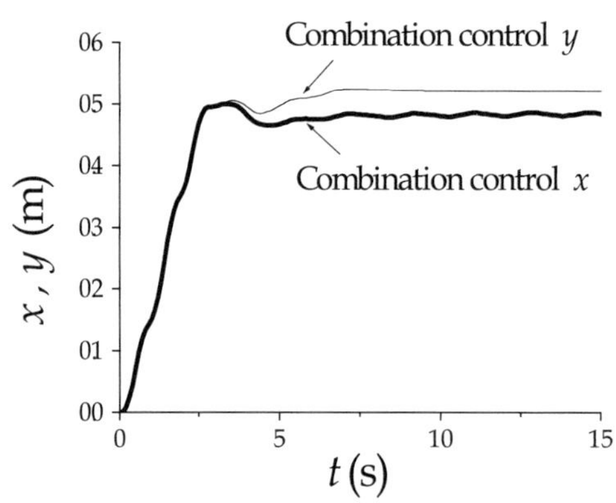

Fig. 23. Trolley transportation

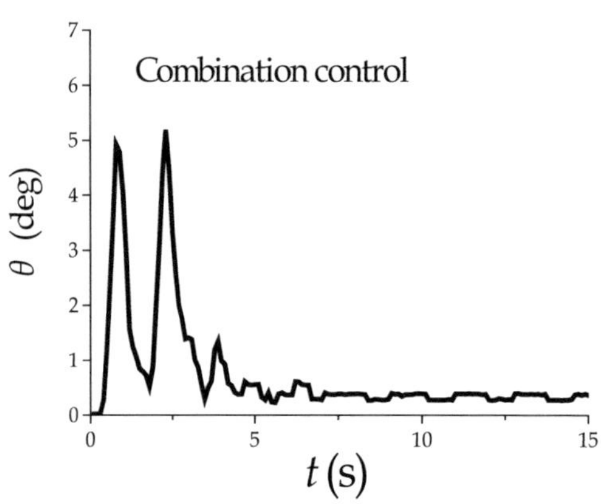

Fig. 24. Swing amplitude

Fig. 23 is the trolley's displacement controlled to the desire 0.5m. A little drift occurs in y direction. Fig. 24 is the swing amplitude. In this case, the load swings with 5 deg until the time $2T_{op} = 2.8s$, but after that it's swing decreased to zero by visual feedback control, different from Fig. 21.

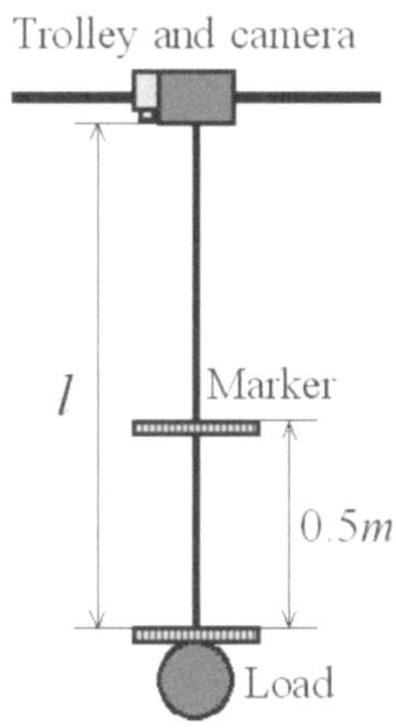

Fig. 25. Two different marker positions

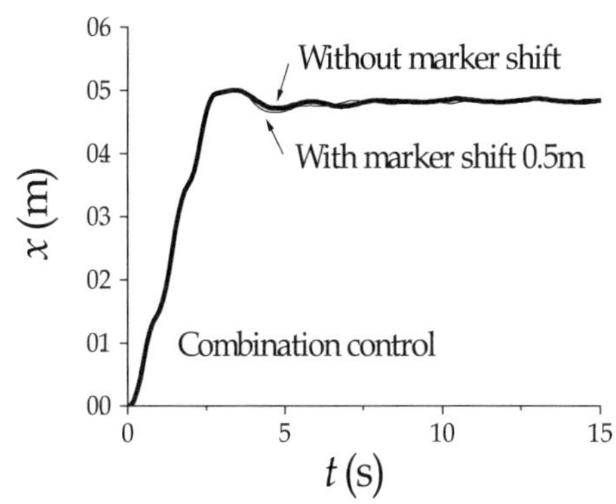

Fig. 26. Trolley transportation for arbitrary marker position

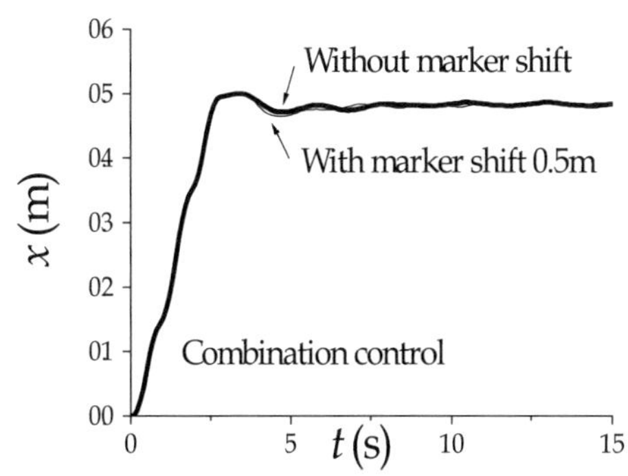

Fig. 27. Swing amplitude for arbitrary marker position

As for the load marker, it is realistic to install the marker on the crane hook or on the arbitrary position of the rope. Fig. 25 shows two different marker positions where one is installed on the load and another is located 0.5m upward. Fig. 26 and Fig. 27 are experiments using combination control for two different positions of load marker. No significant difference appeared in the response of trolley position and swing angle.

4. Conclusion

A trial stereovision device of separate type from the crane is manufactured and tried to control the overhead crane by vision feedback. Three-dimensional positions measurement and the control of tracking/gazing about the crane transportation load were carried out by the stereovision. Using measured visual three-dimensional load position, positioning and swing suppression control of the overhead crane was done. The experimental results of the overhead crane control by vision feedback showed that the used control system was effective. Next, combination control of time-optimal feedforward control for transportation and feedback control for swing suppression were carried out by a visual servo device using 3D-camera installed on the crane. The combination control is practical and effective for actual crane with various swing natural period. There is little influence which the installed location of a marker gives visual feedback control. In addition, when rope length changes while swing suppression control, feedback control needs variable control gains of nonlinear system. Then, the method that control digital gains can be redesigned by analog gains was developed and these variable digital gains were used in experiments.

5. References

Lee, H.H. (1998). Modeling and Control of a Three-Dimensional Overhead Crane, *ASME Journal of Dynamic Systems, Measurement, and Control,* Vol. 120, pp. 471-476.

Mita, T. & Kanai, T. (1979). Optimal Control of the Crane System Using the Maximum Speed of the Trolley (in Japanese with English abstracts), *Trans. of the Society of Instrumentation and Control Engineers, Japan,* Vol. 15, No.6, pp. 833-838.

Mostafa, K.A.F. & Ebeid, A.M. (1988). Nonlinear Modeling and Control of Overhead Crane Load Sway, *ASME Journal of Dynamic Systems, Measurement, and Control*, Vol. 110, pp. 266-271.

Papanikopoulos, N. P.; Khosla, P. K. & Kaneda, K. (1993). Visual Tracking of a Moving Target by a Camera Mounted on a Robot: A Combination of Control and Vision, *IEEE Trans. on Robotics and Automation*, Vol. 9, No. 1, pp. 14-35.

Yoshida, Y. (2004). Visual Feedback Control of a Three-Dimensional Overhead Crane, *Proceedings of ASME 7th Biennial Conf. on Engineering System Design and Analysis*, pp. 1-7, ISBN:0-7918-3741-6, Manchester, UK.

Yoshida, Y.; Hirano, M.; Tomida, T. & Teshima, H. (2005). Visual Feedback Control of Traveling Crane (in Japanese with English abstracts), *Trans. of the Society of Instrumentation and Control Engineers, Japan,* Vol. 41, No.6, pp. 527-532.

Yoshida, Y. & Tsuzuki, K. (2006). Visual Tracking and Control of a Moving Overhead Crane Load, *Proceedings of IEEE the 9'th Int'l Workshop on Advanced Motion Control*, pp. 630-635, ISBN:0-7803-9511-5, Istanbul, Turkey.

Yoshida, Y. & Tabata, H. (2008). Visual Feedback Control of an Overhead Crane and Its Combination with Time-Optimal Control, *Proceedings of the 2008 IEEE/ASME Int'l Conf. on Advanced Intelligent Mechatronics*, pp. 1114-1119, ISBN:978-1-4244-2495-5, Xi'an, China.

8

Intelligent Methods for Condition Diagnosis of Plant Machinery

Huaqing Wang[1] and Peng Chen[2]
[1]*Beijing University of Chemical Technology*
[2]*Graduate School of Bioresources, Mie University*
[1]*China*
[2]*Japan*

1. Introduction

In the case of condition diagnosis of the plant machinery, particularly rotating machinery, the utilization of vibration signals is effective in the detection of faults and the discrimination of fault type, because the signals carry dynamic information about the machine state. Condition diagnosis depends largely on the feature analysis of vibration signals, so it is important that the feature of the signal can be sensitively extracted at the state change of a machine (Lin et al, 2000) (Liu et al, 1999) (Matuyama, 1991) (Wang et al, 2007a). However, feature extraction for fault diagnosis is difficult because the vibration signals measured at any point of the machine often contain a strong noise.

Intelligent systems such as neural networks (NN) and support vector machine (SVM) have potential applications in pattern recognition and fault diagnosis. Many studies have been carried out to investigate the use of neural networks for automatic diagnosis of machinery, and most of these methods have been proposed to deal with discrimination of fault types collectively. However, the conventional neural network cannot reflect the possibility of ambiguous diagnosis problems, and will never converge when the first layer symptom parameters have the same values in different states. Furthermore, diagnostic knowledge is ambiguous because definite relationships between symptoms and fault types cannot be easily identified. In addition, due to the complexity of plant machinery conditions, and the number of fault states to be identified is enormous, it is very hard to find one or several symptom parameters that can identify all of those faults perfectly, simultaneously. Particularly, it is difficult to judge the relationship between fault states and the symptom parameters by a theoretical approach (Pusey, 2000) (Mitoma et al, 2008) (Wang et al, 2008a).

For the above reasons, in order to process the uncertain relationship between symptom parameters and machinery conditions, and improve the efficiency and accuracy of fault diagnosis at an early stage, the authors reviewed their recent researches on intelligent diagnosis methods for rotating machinery based on artificial intelligence methods and feature extraction of vibration signals. That is: the diagnosis method based on wavelet transform, rough sets and neural network; the diagnosis method based on sequential fuzzy inference; diagnosis approach by possibility theory and certainty factor model; the diagnosis method on the basis of adaptive filtering technique; feature extraction method based on information theory; the diagnosis method by time-frequency techniques in unsteady operating conditions; and the fault identification by support vector machine.

2. Diagnosis method based on wavelet transform, rough sets and neural network

An intelligent diagnosis method for rotating machinery is proposed using the wavelet transform (WT), the rough sets (RS) and the fuzzy neural network (NN), on the basis of the features of vibration signals (Wang et al, 2008a) (Wang et al, 2007b). The flowchart of this approach is shown in Fig. 1.

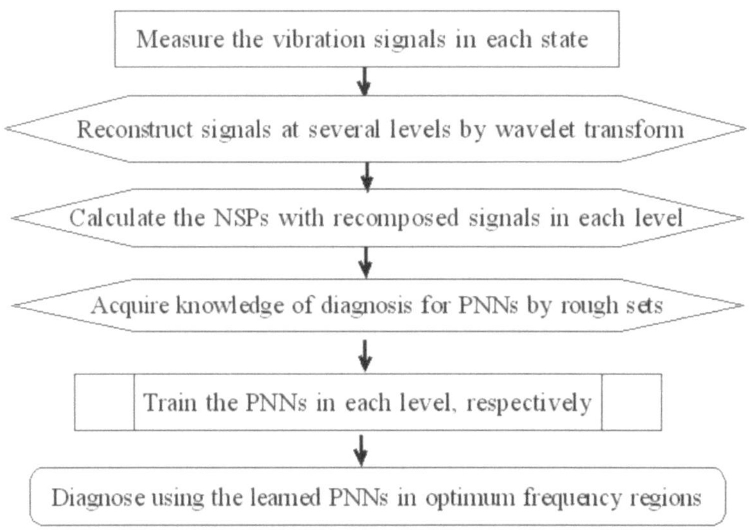

Fig. 1. Flowchart of intelligent diagnosis method

2.1 Feature extraction by WT

The WT is a time-frequency signal analysis method, which has also been used for feature extraction and noise canceling of measured signals (Lin et al, 2000) (Liu et al, 1999). Here, WT performed to extract fault features of each state from measured signals to capture the true fault information across optimum frequency regions. The signals can be decompose into some levels in approximations and details by wavelet function. After use of the wavelet reconstruction function, the signal constituents at each level of the decomposition are reconstructed.

2.2 Definition of symptom parameters

For automatic diagnosis, symptom parameters (SPs) are needed that can sensitively distinguish the fault types. A large set of SPs has been defined in the pattern recognition field (Mitoma et al,2008) (Wang et al, 2008a) (Wang et al, 2007b). In the present work, nondimensional symptom parameters (NSPs) in time domain are considered, and are calculated with the reconstructed time signals of each level in each state to be diagnosed, respectively (Wang et al, 2008a).

2.3 Fuzzy neural network

Although many studies have been carried out to investigate the use of NNs for automatic diagnosis of machinery condition, most of these methods have been proposed to deal with discrimination of fault types collectively (Saxena et al, 2007) (Samanta et al, 2003) (Li et al, 2006) (Alguindigue et al, 1993) (Samanta et al, 2006). However, the conventional NN cannot adequately reflect the possibility of ambiguous diagnosis problems, and will never converge when the first-layer parameters have the same values in different states (Bishop , 1996).
To solve this ambiguous problem, the fuzzy neural network is realized with the developed back propagation NN called as "the partially-linearized neural network (PNN)". The detailed principle of the PNN had been described in (Wang et al, 2008a).

2.4 Knowledge acquisition by rough sets

Rough sets theory (Pawlak, 1991), a mathematical tool to deal with vagueness and uncertainty, has found many interesting applications. In this case, to decrease the number of input parameters for the NN, and improve the efficiency of NN learning, the rough sets are used to acquire diagnosis knowledge.

2.5 Verification

Practical examples of condition diagnosis for a centrifugal pump system are provided to verify the method's efficiency. Faults which often occur in pump systems, such as cavitation, impeller damage, impeller unbalance and shaft misalignment between the motor and the pump are considered. Original vibration signals measured in each state are decomposed into several levels in low frequency and high frequency by the rbio2.8 wavelet function. After use of the wavelet reconstruction function, the signal constituents at each level of the decomposition are reconstructed in time domain respectively. As input data of PNNs, SPs are calculated with the reconstructed signals.
The PNNs are quickly convergent by learning the training data. We used the data that had not been learned by the PNNs to verify the diagnostic capability of the PNNs. When inputting the test data, the learned PNNs can quickly diagnose those faults with the possibility grades of relevant state.

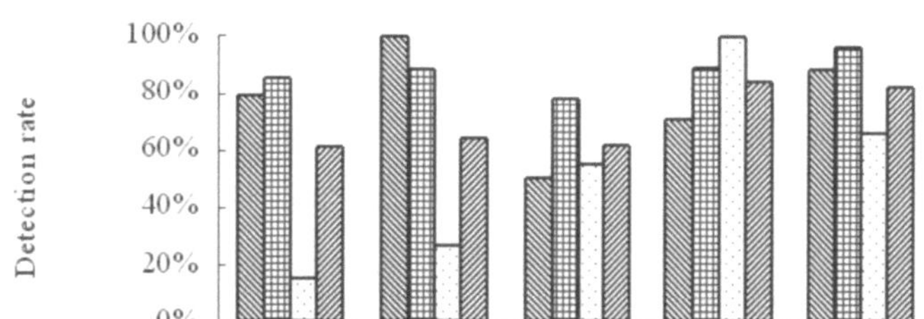

	A0	A1	D1	A2	D2
Normal	0.793	0.996	0.503	0.715	0.878
Cavitation	0.851	0.885	0.78	0.89	0.961
Misalignment	0.148	0.271	0.552	0.999	0.656
Unbalance	0.612	0.642	0.624	0.837	0.817

Level

Fig. 2. Parts of detection rates in each state

As examples, Fig. 2 shows a part of diagnostic results. It can be seen from Fig. 2, the detection rates are different for different levels. The features in the different states have appeared in different frequency levels, so we used the recomposed signals and obtained a maximum detection rate in optimum frequency regions for distinguishing relevant state from other states.
According the verification results, the efficiency of NN learning can be also improved by using knowledge acquired by the rough sets. Having the diagnosis knowledge acquired by the rough sets, the PNN can obtain a good convergence when learning. When diagnosing, the PNN is always convergent, and can quickly and automatically distinguish fault types with high accuracy in optimum frequency regions.

3. Diagnosis method based on sequential fuzzy inference

3.1 Basic conception of sequential inference

In the case of fault diagnosis, if we can find several SPs by which most of faults can be diagnosed, conditions of machinery should be easily identified. However, the number of fault states to be identified is enormous, and it is very hard to find one SP or several SPs that can identify all of those faults, simultaneously. Sometimes, an ideal SP with high detection rate does not exist. The SP to identify only two states is quite easy to find (Zhou et al, 2008). Therefore, a sequential diagnosis inference is proposed. At this time, it is just required to diagnose two states in the various diagnosis steps.
A practical example of a bearing diagnosis by sequential diagnosis is shown in Fig. 3. We must distinguish among four states, namely, the normal state, outer race flaw state, inner race flaw state, and roller element flaw state, sequentially. In those diagnostic stages, we should distinguish only two states in one diagnostic step.

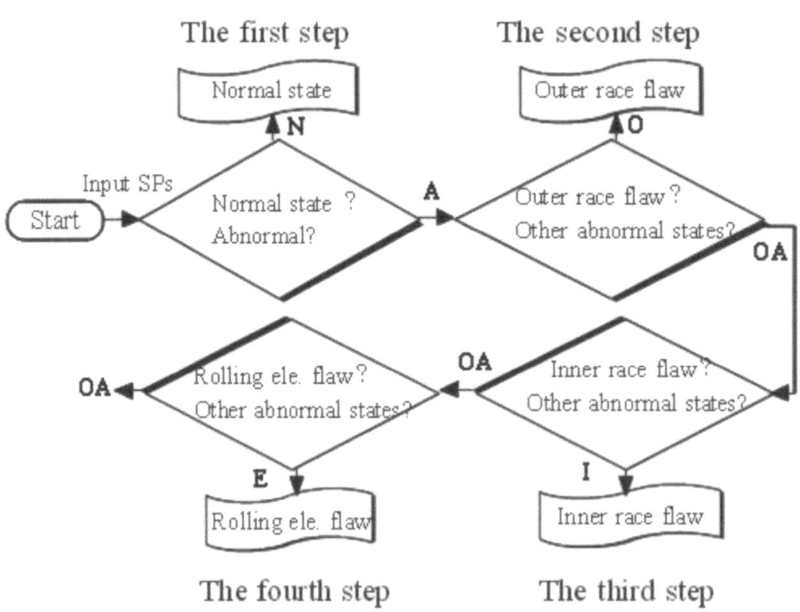

Fig. 3. Example for sequential diagnosis approach

3.2 Sensitivity evaluation for SP

Generally, the symptom parameters used for the condition diagnosis of plant machinery are selected by the following procedure (Pechon et al, 2007).

1. Measure the signals in normal state and each of the abnormal state using the sensors;
2. Define several symptom parameters;
3. Calculate the values of the symptom parameters using the obtained signals;
4. Check the sensitivity of each symptom parameter; if there are no adequate symptom parameters, return to step (2);
5. Then, adopt highly sensitive symptom parameters for fault diagnosis.

In order to evaluate the sensitivity of a SP for distinguishing two states, the distinction index (*DI*) (Wang et al, 2007) is defined as follows:

$$DI = |\mu_1 - \mu_2| \Big/ \sqrt{\sigma_1^2 + \sigma_2^2} \tag{1}$$

where μ_1, μ_2 are the mean values of the symptom parameter calculated by the data in state 1 and state 2, respectively, and σ_1 and σ_2 are their standard deviations. Here, the distinction rate (*DR*) is calculated as follows:

$$DR = 1 - \left(\int_{-\infty}^{-DI} \exp\left(-\frac{u^2}{2}\right) \Big/ \sqrt{2\pi}\right) du \tag{2}$$

It is obvious that the larger the value of the *DI*, the larger the value of *DR* will be, and the better the symptom parameter will be. Therefore, the *DI* can be used to evaluate the sensitivity of the symptom parameter for distinguishing the states of machine. The *DI* is used for selecting the better symptom parameters for each sequential diagnosis step.

3.3 Verification by PNN

We also use the PNN to verify the efficiency of this diagnosis approach. Part verification results for fault diagnosis are shown in Table 1.

(a)	p_5	p_8	g_N	g_A	Judge
	4.389	3.544	0.985	0.014	*N*
	3.767	3.789	0.996	0.005	*N*
	96.62	57.99	0.011	0.988	*A*
	109.8	75.85	0.01	0.99	*A*

(b)	p_6	p_7	g_O	g_A	Judge
	0.881	6.875	0.921	0.011	*O*
	1.224	6.885	0.755	0.059	*O*
	1.378	8.125	0.000	0.999	*A*
	1.342	9.570	0.175	0.963	*A*

Table 1. Verification results (a) in the first step, (b) in the second step

According to the test results, the possibility grades output by the PNNs show the correct judgment in each state, and fault types can be sequentially and automatically distinguished by the sequential algorithm.

4. Diagnosis approach by possibility theory and certainty factor model

Due to the complexity of plant machinery conditions, it is difficult to judge the relationship between fault types and the symptom parameters by a theoretical approach. Therefore, diagnostic knowledge is uncertain because the definite combination relationships between several symptoms and relevant machinery state cannot be easily obtained. In order to explain the uncertain relationship, a diagnosis approach based on possibility theory and the certainty factor is proposed.

4.1 Possibility theory

Possibility theory is used to solve the ambiguous problem of fault diagnosis (Chen et al, 2003). The possibility function of the symptom parameter can be obtained from its probability density function (Wang et al, 2008b) (Bendat, 1969). For example, when the probability density function of the SP conforms to the normal distribution, it can be changed to a possibility function $P(x_i)$ by

$$P(x_i) = \sum_{k=1}^{N} \min\{\lambda_i , \lambda_k\} \tag{3}$$

λ_i and λ_k can be calculated as follows:

$$\lambda_i = \int_{x_{i-1}}^{x_i} \frac{1}{\sigma\sqrt{2\pi}} \exp\left\{-\frac{(x-\mu)^2}{2\sigma^2}\right\} dx,\ \lambda_k = \int_{x_{k-1}}^{x_k} \frac{1}{\sigma\sqrt{2\pi}} \exp\left\{-\frac{(x-\mu)^2}{2\sigma^2}\right\} dx \tag{4}$$

where, σ and μ are the mean and standard deviation of the SP respectively, and $x=\mu-3\sigma\sim\mu+3\sigma$.

4.2 Certainty factor

The certainty factor (*CF*) model was introduced as a method for representation and manipulation of uncertain knowledge in the rule-based expert system MYCIN. The certainty factor can handle the problem of a combination of heterogeneous data (Binaghi et al, 1998). Therefore, in order to process the ambiguous relationship between symptom parameters and fault types, the combining possibility function of each state to be distinguished can be obtained by the MYCIN certainty factor. Practical examples should be shown in the later verification.

4.3 Verification

In the field of the condition diagnosis, it is important to first distinguish between the state of plant machinery in a normal state (*N*) or abnormal states. In this stage, the abnormal state commonly includes two levels, namely, the caution level (*C*) and the danger level (*D*). The possibility function $\mu_n(x)$ of the SP of the normal level can be easily calculated by (4) from the signal measured in the normal state. The possibility functions of the caution level and the danger level are expressed with $\mu_c(x)$ and $\mu_d(x)$, respectively. $\mu_c(x)$ and $\mu_d(x)$ are calculated by

$$\mu_n(x)+\mu_c(x)+\mu_d(x)=1,$$
$$\mu_c/_{x=\mu\pm3\sigma}=\mathrm{Max}[\mu_c(x)], \tag{5}$$
$$\mu_c/_{x\geq\mu+6\sigma \text{ and } x\leq\mu-6\sigma}=0,\ \mu_d/_{\mu-3\sigma\leq x\leq\mu+3\sigma}=0$$

Where σ and μ are the mean value and the standard deviation of the SP calculated from the signal in the normal state, respectively.
The membership function of each level is shown in Fig. 4 (a), and it is calculated beforehand and used for diagnosing the states of machinery.
If $\mu_y(x_i)$ is the possibility function calculated from the data in the state to be diagnosed, the matching degrees with relevant level are calculated as follows:

$$w_{ni}=\int_{-\infty}^{\infty}\mu_n(x_i)\cdot\mu_y(x_i)dx_i,\ w_{ci}=\int_{-\infty}^{\infty}\mu_c(x_i)\cdot\mu_y(x_i)dx_i,\ \ w_{di}=\int_{-\infty}^{\infty}\mu_d(x_i)\cdot\mu_y(x_i)dx_i \tag{6}$$

where w_{ni}, w_{ci}, and w_{di} express the possibility degree of a state in normal level, the caution level, and the danger level, respectively. These degrees are normalized by

$$w_{ni}+w_{ci}+w_{di}=1 \tag{7}$$

The condition of machinery should be judged as the state where the value of the possibility degree is maximum.

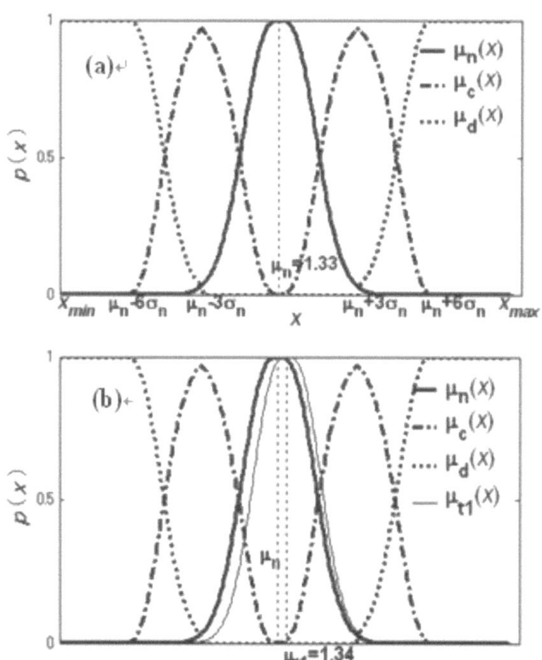

Fig. 4. Membership functions for the first diagnostic step (a) The pre-calculated membership function of each level, (b) Test example

In order to verify the diagnosis capability, all of the verification signals that are measured in the relevant known states had not been used for the pre-calculated membership function. The practical diagnosis example is shown in Fig. 4 (b). Here, t1 is the possibility distributions of the SP that are calculated from the verification signals under the normal state.

The matching degrees (possibility degrees) of relevant state are obtained by (Pusey, 2000) (Wang et al, 2008a), and the verification results are shown as follows:

Possibility of the normal state (*N*):w_{ni}=77.48%;

Possibility of the caution state (*C*):w_{ci}=22.49%;

Possibility of the danger state (*D*):w_{di}=0.03%.

Conclusion: the state can be judged as "Normal state".

According to the verification results, the condition of a bearing can be judged correctly.

5. Condition diagnosis method based on adaptive filtering technique

This section introduces a diagnosis method using adaptive signal processing and fuzzy neural network for raising the accuracy of the fault diagnosis in reciprocating machine (Wang et al, 2008c). The flowchart of this diagnostic method is shown in Fig. 5.

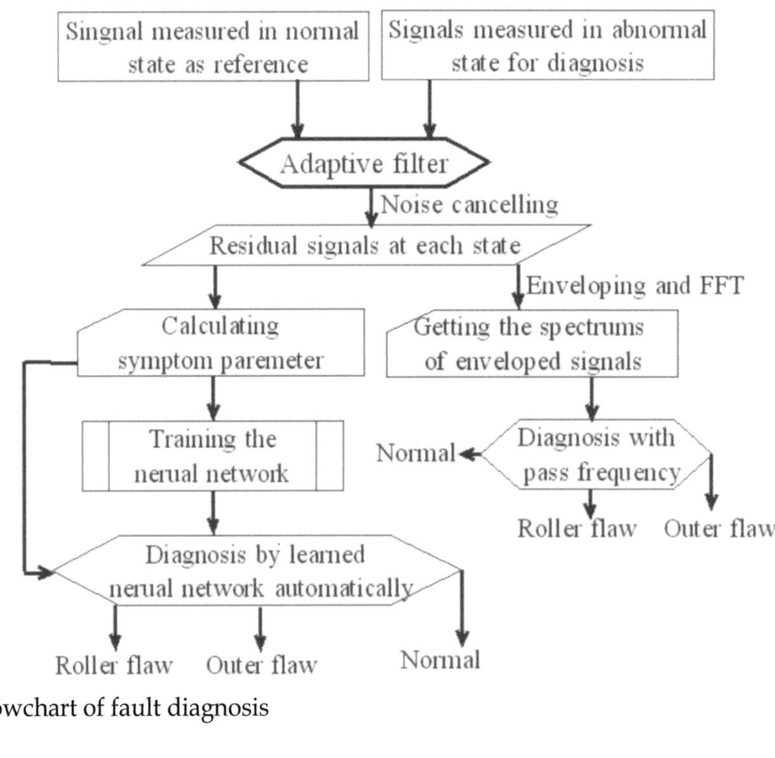

Fig. 5. Flowchart of fault diagnosis

5.1 Adaptive noise cancelling

Adaptive filters have been applied in signal processing and control, as well as in many practical problems (Alexander et al, 1986). As the signal continues into the filter, the adaptive filter coefficients adjust themselves to achieve the desired results, such as identifying an unknown filter or cancelling noise in the input signal.

An illustration for adaptive noise canceling is shown in Fig. 6. The signal S is the fault signal from the bearing and N_0 is the noise of the vibration signal measured from normal bearing. A reference noise N_1, which is related to noise N_0 in some unknown way but not correlated with signal S, is received by the reference sensor in normal state. The output filter y is then adaptively filtered to match N_0 as close as possible. Then the filter output is then subtracted from the primary input $S+N_0$ to produce the system output ε called residual signal.

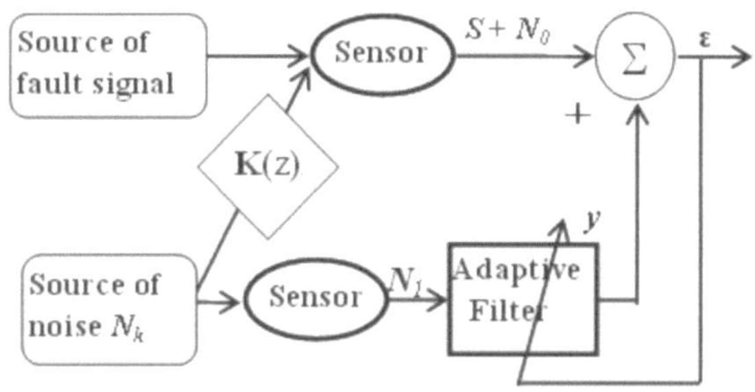

Fig. 6. Adaptive noise cancelling

5.2 Verification and discussion

In this case, roller bearings are utilized in a rice husking machine, and the simulated flaw is located at the roller and outer race of the rolling bearing for the tests of the condition diagnosis.

5.2.1 Enveloped spectrum analysis

As an example, Fig. 7 shows the enveloped signal spectrums under outer-race flaw with high-pass filtering (cut-off frequency is 3 kHz) and adaptive filtering, respectively. The spectrum of enveloped signal with adaptive filtering clearly shows the peak at the pass frequency and exhibits many harmonics that correspond to the pass frequency. It is clearly evident that the bearing flaw can be detected more easily after adaptive filtering.

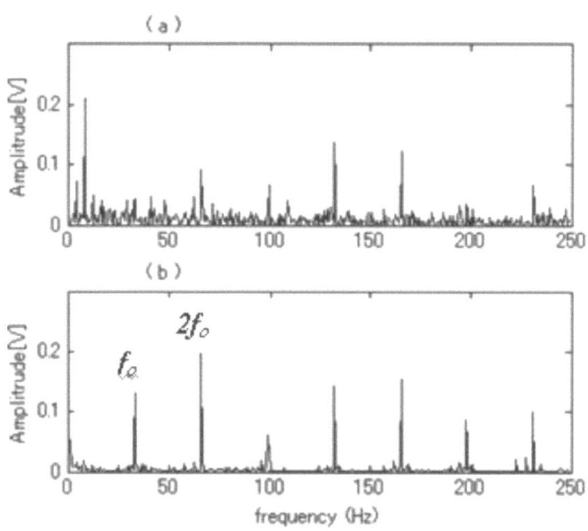

Fig. 7. Enveloped spectrums in outer-race flaw, (a) high-pass filtering, (b) adaptive filtering

5.2.2 Diagnosis by PNN

As input data of PNN, three symptom parameters in the time domain are considered (Wang et al, 2008c). Values of those symptom parameters are calculated using the signals with high-pass filtering and adaptive filtering for training the fuzzy neural network, respectively. Using the symptom parameters calculated from the high-pass filtered signals, the fuzzy neural network cannot converge. It can be explained that the symptom parameters calculated form the high-pass filtered signals are poor, and cannot discriminate the fault types. Contrarily, using the symptom parameters calculated with adaptive filtered signals, the neural network can get a good convergence.
Table 2 shows the diagnosis results for each state. According to the test results, the probability grades output by the PNN show the correct judgment in each state.
By these diagnosis results, this method is efficient for condition diagnosis of rolling bearing used in the reciprocating machine.

P1	P2	P3	N	O	R	Judge
0.83	1.74	0.75	0.99	0.01	0.00	*N*
0.80	1.34	0.72	0.99	0.00	0.00	*N*
1.41	12.16	1.47	0.00	0.77	0.01	*O*
1.06	8.18	1.07	0.01	0.99	0.00	*O*
2.13	16.7	2.26	0.01	0.02	0.97	*R*
1.775	16.38	1.72	0.00	0.45	0.72	*R*

Table 2. Diagnosis results

6. Feature extraction method based on information theory

A method of feature extraction is proposed based on information theory (Wang et al, 2007c) (Wang et al, 2009), and a flowchart of this diagnosis approach is shown in Fig. 8.

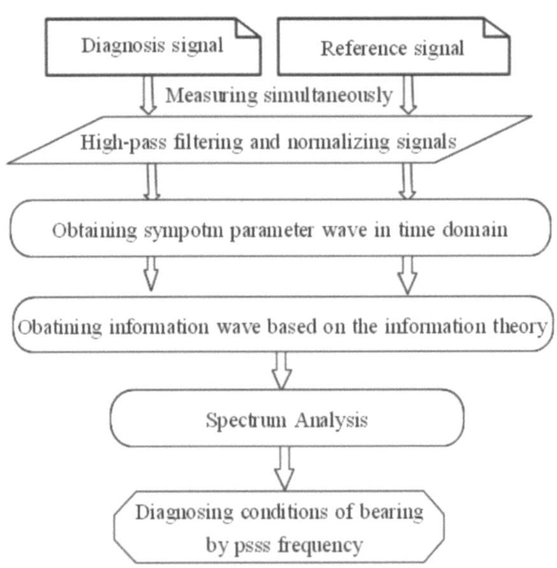

Fig. 8. Flowchart of diagnosis approach

6.1 Obtaining method for a symptom parameter wave

To acquire the signal feature in the time domain, an obtaining method of the symptom parameter wave is proposed, as shown in Fig. 9. The whole data of a discrete signal is divided into some smaller regions. The value of the symptom parameter can be calculated using the data in those regions. Connecting those points of the symptom parameter, the symptom parameter wave (*SP(j), j=1~L*) can be derived.

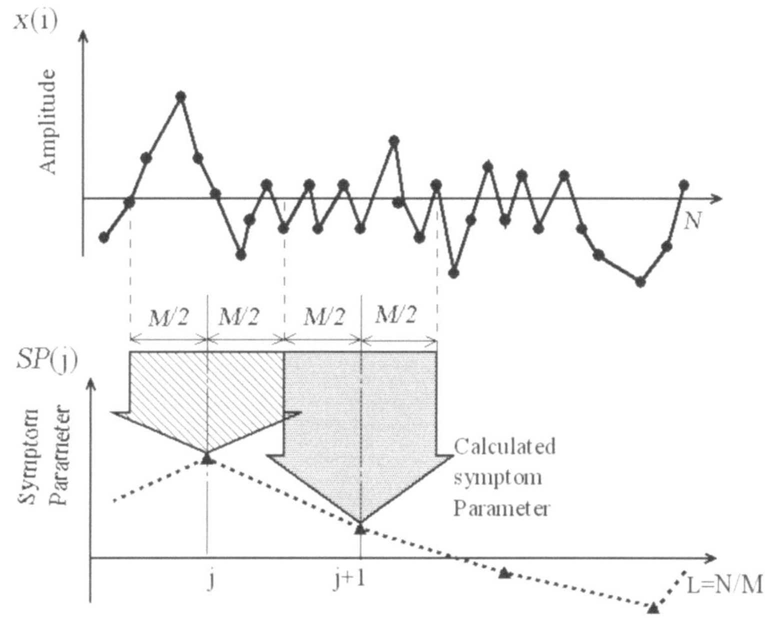

Fig. 9. A method to obtain a symptom parameter wave

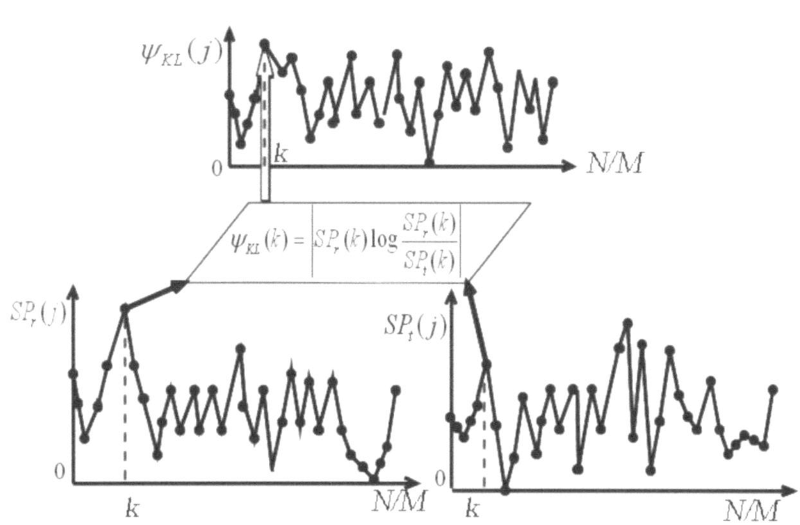

Fig. 10. Derivation of information wave

6.2 Feature extraction by information theory

Kullback-Leibler divergence (*KL*) plays a central role in the information theory of statistical inference. In case of fault diagnosis, the information theory had been applied for comparing the unknown distribution to be diagnosed with the known reference distribution. Those diagnosis method based on the information divergence has been used for the simple diagnosis of machinery (Toyata et al, 1996) (Liu et al, 1996). However, its diagnostic capability for the precise diagnosis is lower. Information of a symptom parameter wave shortened as "information wave" are proposed in the work based on the information theory. An illustration for the derivation of the information waves (*KL* information wave: Ψ_{KL}) is shown in Fig. 10. Here, a symptom parameter wave $SP_t(j)$ is calculated with the signal measured in the location to be diagnosed, and a symptom parameter wave $SP_r(j)$ is calculated with the signal measured in the reference point which is faraway from the diagnosis position.

6.3 Spectral analysis for information wave

"Spectral difference of information wave" is used for identifying the fault types and is defined as follows.

$$Q_A^{KL}(f_i) = P_A^{KL}(f_i) - P_N^{KL}(f_i) \tag{8}$$

where, Spectral difference Q is a difference between spectral of information wave in the abnormal state and the normal state, and fi is frequency of spectral. P_N and P_A are spectra of information wave in the normal state and the abnormal states, and can be obtained by FFT technique respectively.

6.4 Practical application

Practical diagnosis example for a bearing used in the diesel engine is given to verify the efficiency of this method.

Fig. 11 shows the spectrums of enveloped diagnosis signal in each state after high-pass filtering with a 5 kHz cut-off frequency. The feature frequency (f_{ki}, f_{ko}, f_{kn} and f_{kr}) in each state is the shock frequency of a piston. Fault frequency of bearing under abnormal states cannot appear, and therefore, the fault types cannot be identified.

As examples, Fig. 12 shows spectral differences of information waves $Q_O(f_i)$ under the outer race flaw state. Fig. 12 (a), (b) and (c) show $Q_O{}^{KL}(f_i)$ of absolute mean value, root mean square, and shape of wave, respectively. Where, f_O is a pass-frequency of the outer race flaw. It is clearly evident that the outer flaw of a bearing can be detected easily by those spectral differences. According to the verification results, the fault types of a bearing can be identified effectively.

7. Diagnosis method by time-frequency analysis in unsteady operating conditions

Time–frequency analysis (Cohen, 1995) has proven to be an effective tool for analysing the behaviour of nonstationary signals. There are several time-frequency analysis methods, such as the short-time Fourier transform (STFT), wavelet analysis (WA), and the Wigner-Ville distribution (WVD), which may be used for condition diagnosis of rotating machinery in unsteady operating conditions. Many studies have been carried out with the goal of diagnosis of machinery condition. Those studies identify machinery faults by using the distribution of the spectrum in the time-frequency domain collectively, the symptom parameters are not considered, and the fault types cannot be identified automatically.

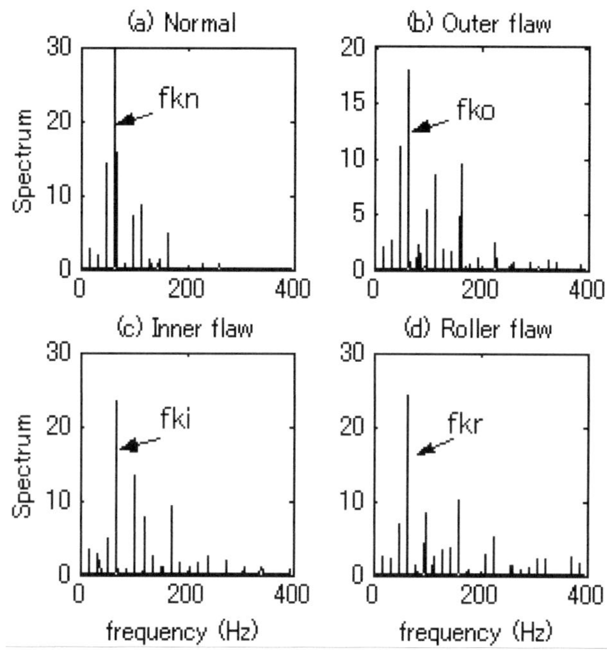

Fig. 11. Spectrums of enveloped diagnosis signal

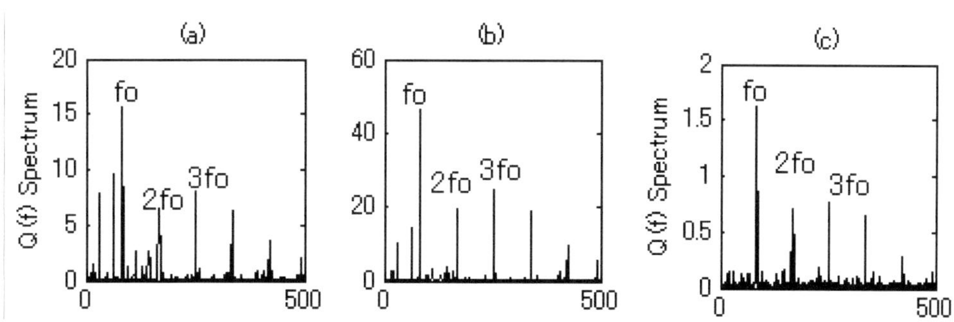

Fig. 12. Spectral differences of information waves with outer race flaw

This section introduces a method that is to integrate time-frequency analysis techniques with the automatic feature extraction method, sequential inference and possibility theory, and apply them to the condition diagnosis of a machine at the variable rotating speed (Wang et al, 2010a).

7.1 Feature spectra extracted by relative crossing information (RCI)

In this section, we propose a method for extracting the feature spectra from the time-frequency analysis techniques using the relative crossing information (RCI). The RCI is used to automatically extract the feature spectra from time-frequency analysis by computer in order to distinguish each state. The $q_i(t)$ is the number of crossings over some level i of the vertical coordinate of the spectrum $P(t,\omega)$ with a positive slope in unit time, and it can be calculated from the spectrum P(t,ω), shown as follows(Chen et al, 2001):

$$q_i(t) = \frac{\sigma_v(t)}{2\pi\sigma_x(t)}\exp\left(-\{i/2\sigma_x(t)\}^2\right) \tag{9}$$

$$\sigma_x(t)^2 = \int_0^\infty P(t,\omega)d\omega\,,\quad \sigma_v(t)^2 = \int_0^\infty (\omega)^2 P(t,\omega)d\omega \tag{10}$$

where $P(t,\omega)$ is a spectrum calculated from the time signal by the time-frequency methods. The vertical coordinate axis of a spectrum in the normal state is divided into M equal sections from maximum amplitude to minimum amplitude of the spectrum. $q_{ni}(t)$ and $q_{ai}(t)$ can be calculated by (9) in section i $(i=1{\sim}M)$ in the normal state and the state to be detected, respectively.

The RCI is expressed by the $I_q(t)$ and is defined as

$$I_q(t) = \sum_{i=1}^{M}\left|\log\left(q_{ai}(t)/\overline{q_{ni}}\right)\cdot\left(q_{ai}(t)-\overline{q_{ni}}\right)\right| \tag{11}$$

where $\overline{q_{ni}} = \frac{1}{T}\int_0^T q_{ni}(t)dt$ and T is the sampling time.

I_q can be used to express the difference in spectra between the normal state and abnormal states, by which the feature spectra can be extracted in the time-frequency domain. Some real examples of $I_q(t)$ (RCI) are shown in Fig. 13.

We can extract the feature spectra of the outer flaw state in the position A-A where the value of I_q is maximum.

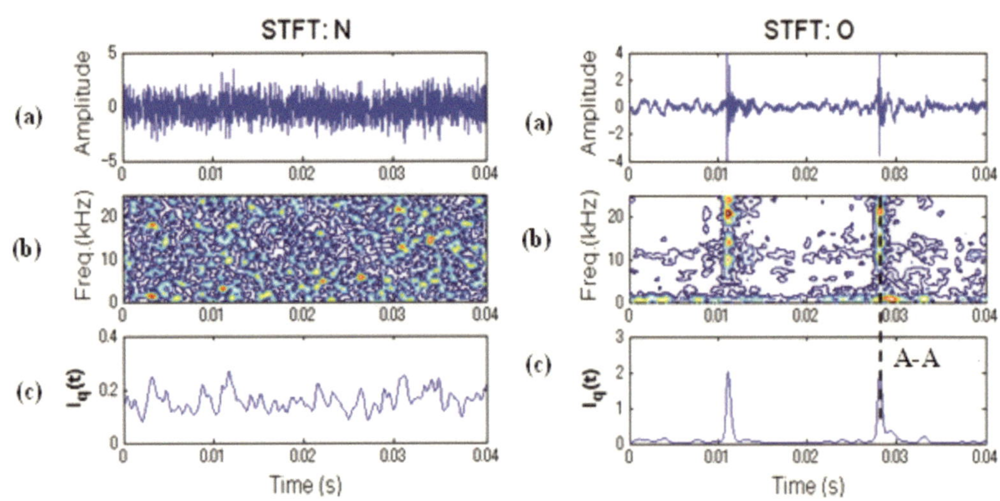

Fig. 13. Examples in normal state (N) and outer race flaw(O), (a) normalized signal, (b) the contour graphs of spectra processed by the STFT, (c) the RCI of spectra expressed by the Iq.

7.2 Synthesizing symptom parameter by least squares mapping (LSM)

After extracting the feature spectra by the RCI, the symptom parameters can be calculated using the feature spectra. In this section, in order to improve the diagnosis sensitivity, we propose a method by projecting the symptom parameters into discrimination space using least squares mapping (Chen et al, 1995). The number of symptom parameters ($\boldsymbol{Y}_k$) is K, and the category number of states is M. In the coordinate space of K dimension, the endpoint of the vector $\boldsymbol{Y}_{ij}$ expresses state i. $\boldsymbol{Y}_{ij}$ is

$$\mathbf{Y}_{\mathbf{ij}}=\left\{y_{ij1},y_{ij2},\ldots,y_{ijK}\left|i=1\sim M,j=1\sim N_i\right.\right\}^T \tag{12}$$

where N_i is the number of symptom parameters in the state i.
$\boldsymbol{Y}_{ij}$ can be projected into a new space **L**, and the new vector $\mathbf{L}_{ij}$ in the space **L** can be calculated as follows:

$$\mathbf{L}_{\mathbf{ij}}=\mathbf{A}\mathbf{Y}_{\mathbf{ij}} \tag{13}$$

where $\mathbf{L}_{\mathbf{ij}}=\left\{I_{ij1},I_{ij2},\ldots,I_{ijK}\left|i=1\sim M,j=1\sim N_i\right.\right\}^T$.
According to LMS algorithm, transformation matrix **A** can be obtained as

$$\mathbf{A}=\left[\sum_{i=1}^{M}\sum_{j=1}^{N_j}\left\{\mathbf{V}_{\mathbf{i}}\mathbf{Y}_{\mathbf{ij}}^{'}\right\}\right]\left[\sum_{i=1}^{M}\sum_{j=1}^{N_i}\left\{\mathbf{Y}_{\mathbf{ij}}\mathbf{Y}_{\mathbf{ij}}^{'}\right\}\right] \tag{14}$$

Fig. 14 shows an illustration of the projection by the LSM, where K=2 and M=2. Namely, the two states (state 1 and state 2) should be classified using two symptom parameter series.
According to the projected results shown in Fig. 14 (b), the points in state 1 and state 2 are congregated to vector $\mathbf{V}_1$ and $\mathbf{V}_2$, respectively. The two states in the space **L** can be distinguished more easily than in the space **Y**.

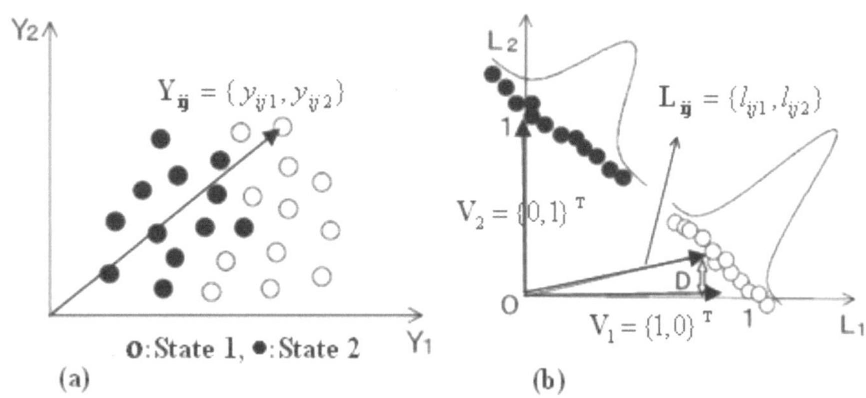

Fig. 14. Projected examples by the LSM, (a) Before projection, (b) After projection

7.3 Performance comparison of time frequency techniques

The performance of this approach is evaluated using three time-frequency transformation techniques. Fig. 15 shows the flowchart for performance comparison.

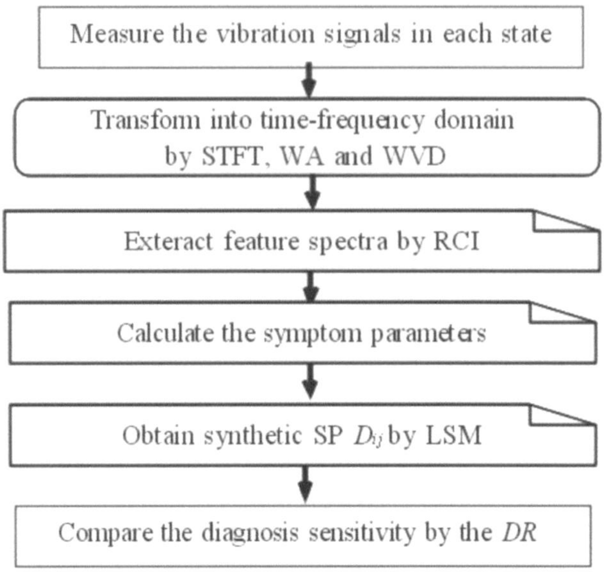

Fig. 15. Flowchart of performance comparison

In order to raise the diagnosis sensitivity of the symptom parameters, the new synthetic symptom parameter should be obtained by the LSM algorithm. In this case, the synthetic symptom parameter is defined as follows:

$$D_{ij} = \sqrt{(l_{i1} - 1)^2 + l_{i2}^2 + \cdots + l_{i7}^2} \tag{15}$$

Fig. 16 shows the values of the D_R and D_I of the synthetic symptom parameter in each diagnostic stepby the time-frequency techniques. It is obvious that the diagnosis accuracy of the WVD is the best, because all of the D_{Rs} (or D_{Is}) obtained by the WVD method are largest. The calculation results of D_I also show that the sensitivity of the new synthetic symptom parameter is higher than the original symptom parameters.

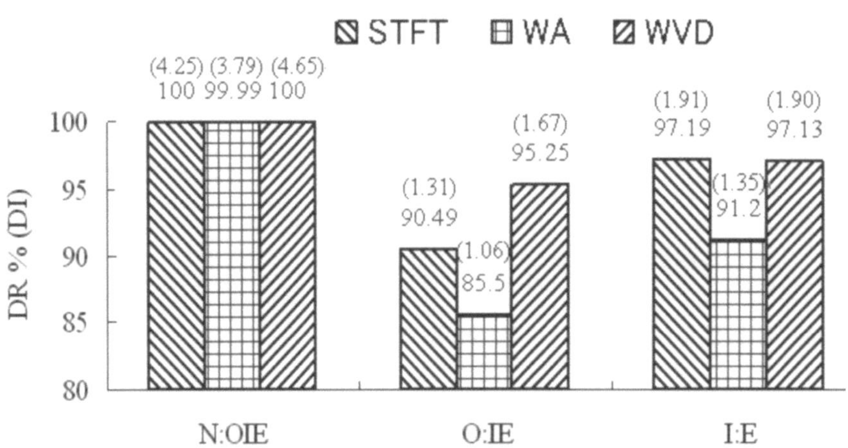

Fig. 16. Comparison result for the DR (DI) of each method.

8. Application of support vector machine in condition diagnosis

Most pattern recognition methods used in condition diagnosis of rotating machinery are studied that the sufficient samples are available. However, it is hard to obtain sufficient fault samples in practice. Support vector machine (SVM) can solve the learning problem with a small number of samples. This section reports a condition diagnosis method for a centrifugal blower using a multi-class classification technique, such as SVM to identify fault types. The statistic feature parameters are also acquired in the frequency domain for classification purposes, and those parameters can reflect the characteristics of vibration signals. The effectiveness of the method is verified by the application to the condition diagnosis for a centrifugal blower. The result shows that multi-class SVM produces promising results and has the potential for use in fault diagnosis of rotating machinery (Wang et al, 2010b).

8.1 Brief of support vector machine (SVM)

Support vector machine (SVM) is a relatively new computational learning method based on the statistical learning theory. SVM is based on Vapnik-Chervonenkis theory (VC-theory) that recently emerged as a general mathematical framework for estimating dependencies from finite samples. The basic idea of applying SVM to pattern classification can be stated as follows: first, map the inputs vectors into one features space, possible in higher space, either linearly or nonlinearly, which is relevant with the kernel function. Then, within the feature space from the first step, seek an optimized linear division, that is, construct a hyperplane which separates two classes. This technique can also be extended to multi-class classification. SVM training seeks a global optimized solution and avoid over-fitting, so it has the ability to deal with a large number of features (Wang et al, 2010b).
In the linear separable case, there exists a separating hyperplane whose function is

$$\mathbf{w} \cdot \mathbf{x} + b = 0 \tag{16}$$

which implies,

$$y_i(\mathbf{w} \cdot \mathbf{x} + b = 0) \geq 1, i = 1,\ldots,N \tag{17}$$

By minimizing $\|\mathbf{w}\|$ subject to this constrain, the SVM approach tries to find a unique separating hyperplane. Here $\|\mathbf{w}\|$ is the Euclidean norm of w, and the distance between the hyperplane and the nearest data points of each class is $2/\|\mathbf{w}\|$. By introducing Lagrange multipliers αi, the SVM training procedure amounts to solving a convex quadratic problem. The solution is a unique globally optimized result, which has the following properties,

$$w = \sum_i^N \alpha_i y_i \mathbf{x}_i \tag{18}$$

Only if corresponding $a_i > 0$, these x_i are called support vectors.
When SVM are trained, the decision function can be written as

$$f(\mathbf{x}) = sign(\sum_i^N \alpha_i y_i (\mathbf{x} \cdot \mathbf{x}_i) + b) \tag{19}$$

For a linear non-separable case, SVM performs a nonlinear mapping of the input vector x from the input space RN into a higher dimensional Hilbert space, where the mapping is determined by kernel function. According to the different classification problems, the different kernel function can be selected to obtain the optimal classification results (Widodo et al, 2009).

8.2 Practical application

Practical diagnosis example for a bearing used in a centrifugal blower is given to verify the efficiency of this method. The diagnosis approaches are shown as follows. A high-pass filter with a 5 kHz cut-off frequency was used to cancel noise from the measured signals. The filtered signals were processed by the Hilbert-Huang Transform, and the spectra of the enveloped waveforms were obtained by the FFT. Then, the feature parameters were calculated with the envelope spectra, and used for SVM classification. Lastly, fault diagnosis was carried out by the SVM. In this study, seven of these parameters (Wang et al, 2010b) in frequency domain, commonly used for the fault diagnosis of plant machinery, are acquired for classification purposes.

The number of train samples and the number of test samples in each state are 10. The optimized kernel parameters, C and γ are 1.533 and 3.526, respectively. We used which data measured in each state had not been trained by the SVM in order to verify the diagnostic capability of the SVM. The training results and test results are shown in Figs. 17 and 18. In Fig. 17 and 18, class 1, 2, 3 and 4 mean normal, inner-race flaw, and outer-race flaw and roller-race flaw states, respectively. In this case, the test accuracy is 100%, and the test error is zero.

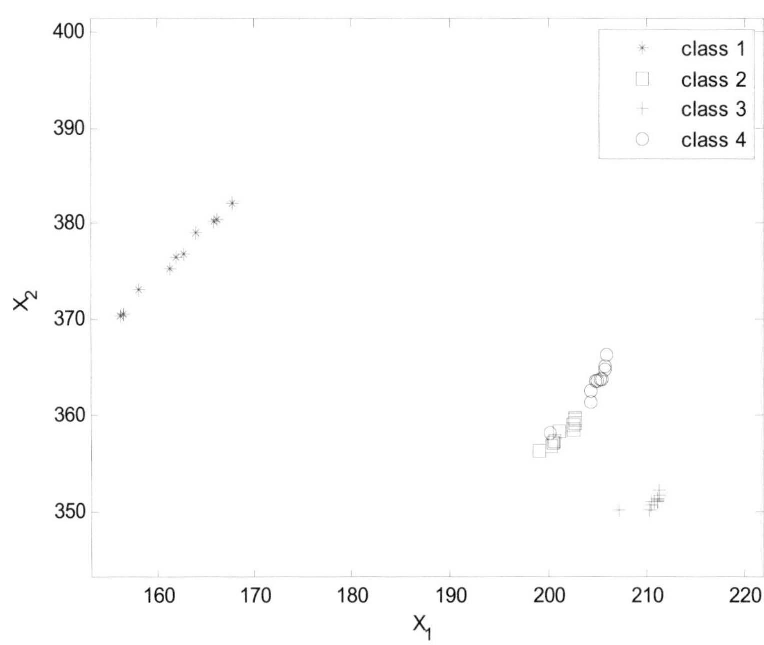

Fig. 17. Training result

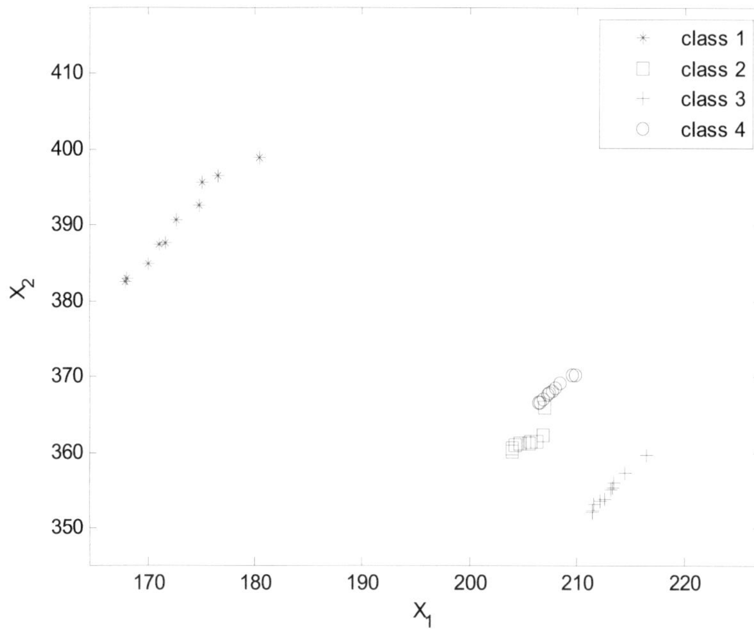

Fig. 18. Test result

The classification results show that the multi-class LS-SVM can identify the fault types with high accuracy, and SVM has the potential for use in fault diagnosis of rotating machinery.

9. Conclusion

This chapter reported several intelligent diagnostic approaches for rotating machinery based on artificial intelligence methods and feature extraction of vibration signals. Main conclusions are described as follows:

1. An intelligent method for condition diagnosis of rotating machinery was proposed, which constructed on the basis of the wavelet transform, the rough sets and the fuzzy neural network realized by the PNN. The wavelet transform performed to extract fault features of each state from measured signals to capture the true fault information across optimum frequency regions. Having the diagnosis knowledge acquired by the rough sets, the efficiency of NN learning was also improved, and the PNN accurately and quickly identified conditions of machine.
2. A fuzzy diagnostic method was also advanced, which was achieved through a sequential diagnostic approach and constructed on the basis of possibility theory, certainty factor and a fuzzy neural network. This diagnostic approach manipulated the ambiguous relationship between the symptom parameters and machinery conditions, and identified the states sequentially.
3. A diagnosis method was also adopted based on the adaptive signal processing technique for reciprocating machine. Diagnosis results showed that the diagnostic approach was very effective even for cancelling highly correlated noise, and for automatically discriminating the fault types.

4. A feature extraction method was present based on the information theory. An obtaining method of a symptom parameter wave was defined, and a concept of spectral difference was derived. "Information wave" was also proposed by the information theory. By spectral difference of information wave, the feature spectra can be extracted clearly, and conditions of a machine can be discriminated effectively.
5. A new extraction method of feature spectra for rotating machinery in unsteady operation conditions was also proposed using the relative crossing information (RCI), by which the feature spectra from time-frequency analysis can be automatically extracted. The diagnosis sensitivities of three time-frequency analysis methods (STFT, WA, and WVD) were investigated for the condition diagnosis of machinery. The obtaining method of the synthetic symptom parameter was also projected by the Least-Squares Mapping (LSM) approach in order to improve the diagnosis sensitivity of the symptom parameter.
6. A condition identification method for a centrifugal blower was proposed using a multi-class classification technique, such as SVM. The effectiveness of the method is verified by the application to the condition diagnosis for a centrifugal blower. The result shows that the multi-class SVM produces promising results and has the potential for use in fault diagnosis of rotating machinery.

These proposed diagnostic methods had been successfully applied to condition diagnosis of practical rotating machinery, such as, a centrifugal pump, a centrifugal blower, reciprocating machine such as a diesel engine, etc.

10. Acknowledgment

Publication of this chapter is supported by the project of National Nature Science Foundation of China (51075023)

11. References

Alexander, S,T. (1986). *Adaptive signal processing-theory and applications*, Springer-Verlag, ISBN:0-387-96380-4, New York.

Alguindigue, I,E. ; A. Loskiewicz-Buczak & R. E. Uhrig. (1993). Monitoring and diagnosis of rolling element bearings using artificial neural networks, *IEEE Transactions on Industrial Electronics*, Vol. 40,No.2, pp.209-216. ISSN :0278-0046.

Binaghi, E. ; Luzi.,L. ; Madella, P. ; Pergalani, F. & Rampint, A. (1998). Slope instability zonation: a comparison between certainty factor and fuzzy dempster–shafer approaches, *Natural Hazards*, Vol. 17, pp. 77–97.

Bishop, Christopher,M.(1996). *Neural networks for pattern recognition*, Oxford University Press, ISBN :0 19 853864 2,New York.

Bendat, J,S. (1969). Probability Function for Random Processes: Prediction of Peak. Fatigue Damage, and Catastrophic Failure, *NASA Report CR-33.*

Chen, P. & Toyota,T. (1995). Method for raising diagnosis accuracy by Least-squares mapping, *Journal of the Society of Plant Engineers Japan*, Vol. 7, No. 3, pp. 162-166.

Chen, P. & Toyota,T. (2003). Sequential fuzzy diagnosis for plant machinery, *JSME International Journal*, Series C, Vol. 46, No. 3, pp. 1121-1129.

Chen, P. ; Toyota, T. & He, Z,J. (2001). Automated function generation of symptom parameters and application to fault diagnosis of machinery under variable operating conditions, *IEEE Transactions on Systems Man and Cybernetics Part A-Systems and Humans*, Vol. 31, No. 6, pp. 775-781.

Cohen, L. (1995). *Time-Frequency Analysis*, Prentice-Hall, New Jersey.

Liu, B. & Ling, S,F (1999). On the selection of informative wavelets for machinery diagnosis, *Mechanical Systems and Signal Processing*, Vol. 13, No. 1, pp. 145-162.

Li, R,Q. ; Chen, J. & Wu,X. (2006). Fault diagnosis of rotating machinery using knowledge-based fuzzy neural network, *Appl. Math. Mech-Engl.*, Vol. 27, No. 1, pp. 99-108.

Lin, J. & Qu, L. (2000). Feature extraction based on Morlet wavelet and its application for mechanical fault diagnosis, *Journal of Sound and Vibration*, Vol. 234, No. 1, pp. 135-148.

Liu, X. ; Toyota, T. & Chen, P. et al.(1999). Study on deterioration trend control for rotating machinery by information theory, *Journal of Reliability Engineering Association of Japan*. Vol. 21, No. 8, pp. 507-514.

Matuyama, H. (1991). Diagnosis algorithm, *Journal of JSPE*, Vol. 75, No. 3, pp. 35-37.

Mitoma, T. ; Wang, H. & Chen, P. (2008). Fault diagnosis and condition surveillance for plant rotating machinery using partially-linearized neural network, *Computers & Industrial Engineering*, Vol. 55, No. 4, pp. 783-794.

Pawlak, Z.(1991). *Rough sets*. Kluwer Academic Pubishers, ISBN :0-7923-1472-2,Thw Netherlahds.

Pechon, R. ; Jinyama, H. & Wang, H. et al.(2007). Fault diagnosis of rolling bearing using partially-linearized neural network, *Journal of the Society of Plant Engineers Japan*, Vol. 19 No. 2, pp. 133-141.

Pusey, H,C. (2000). Machinery condition monitoring, *Journal of Sound and Vibration*, Vol. 34, No. 5, pp. 6-7.

Saxena, A. & Saad, A. (2007). Evolving an artificial neural network classifier for condition monitoring of rotating mechanical systems, *Applied Soft Computing*, Vol. 7, No. 1, pp. 441-454.

Samanta, B. & Al-Balushi, K,R. (2003). Artificial neural network based fault diagnostics of rolling element bearings using time-domain features, *Mechanical Systems and Signal Processing*, Vol. 17, No. 2, pp. 317-328.

Samanta, B. ; Al-Balushi, K,R. & Al-Araimi, S,A. (2006). Artificial neural networks and genetic algorithm for bearing fault detection, *Soft Computing*, Vol. 10, No. 3, pp. 264-271.

Toyota, T. ; Fukuda, K. & Chen, P. (1996). Failure diagnosis of machinery by Kullback information theory, *Technical Report of Institute of Electronics, Information and communication engineers*, R96-13, pp. 13-16.

Wang, H. & Chen, P. (2007a). Sequential condition diagnosis for centrifugal pump system using fuzzy neural network, *Neural Information Processing-Letters and Reviews*, Vol. 11, No. 3, pp. 41-50.

Wang, H. & Chen, P. (2007b). Fault diagnosis of centrifugal pump using symptom parameters in frequency domain, *Agricultural Engineering International: The CIGR E-journal*, Manuscript IT 07 005. Vol. 11, pp.1-14.ISSN :1618-1130.

Wang, H. ; Chen, P. & Toyota, T. (2007c). Condition diagnosis for a bearing used the diesel engine based in information theory,*Proceedings of the Society of Plant Engineers Japan*, Japan, Nov.

Wang, H. & Chen, P. (2008a). Intelligent diagnosis method for a centrifugal pump using features of vibration signals, *Neural Computing & Applications*, Vol. 8, No. 4, pp. 397-405.

Wang, H. & Chen, P. (2008b). Sequential Fuzzy Diagnosis for Condition Monitoring of Rolling Bearing Based on Neural Network, *Proceedings of Fifth International Symposium on Neural Networks, LNCS*, Vol. 5624, pp. 284-293, Springer.

Wang, H. & Chen, P. (2008c). Fault diagnosis for a rolling bearing used in a reciprocating Machine by adaptive filtering technique and fuzzy neural network, *WSEAS Transactions on Systems*, Vol. 7, Issue 1, pp. 1-6.

Wang. H. & Chen. P. (2009). A Feature Extraction Method Based on Information Theory for Fault Diagnosis of Reciprocating Machinery, *Sensors*, Vol. 9, pp. 2415-2436.

Wang, H. & Chen, P. (2010a). Fuzzy diagnosis method for rotating machinery in variable rotating speed, *IEEE Sensors Journal*, Doi :10.1109/JSEN.2010.2049103.

Wang, H. ; Guo, Y. & Yuan, H. et al.(2010b). Application of Support VectorMachine in Condition Diagnosis of Centrifugal Blower, *Proceedings of 23rd Condition Monitoring and Diagnostic Engineering Management- COMADEM2010*, pp. 199-204.

Widodo, Achmad ; Eric Y. Kim ; Jong-Duk Son ; Bo-Suk Yang; Andy C.C. Tan ; Dong-Sik Gu; Byeong-Keun Choi ; Joseph Mathew. (2009). Fault diagnosis of low speed bearing based on relevance vector machine and support vector machine. *Expert Systems with Applications*, Vol. 36, pp. 7252-7261.

Zhou, X ; Wang, H ; Chen, P. & Song, J. (2008). Diagnosis method for gear equipment by sequential fuzzy neural network, *Proceedings of Fifth International Symposium on Neural Networks, LNCS*, Vol. 5624, pp. 473-482, Springer.

Part 3

Physical System Modelling and Real Time Applications

9

Methodology for Reusing Real-time HiL Simulation Models in the Commissioning and Operation Phase of Industrial Production Plants

Sebastian Kain[1], Frank Schiller[1], and Sven Dominka[2]
[1]*Institute of Information Technology in Mechanical Engineering, Automation Group, Technische Universität München,*
[2]*School of Electrical and Computer Engineering, Royal Melbourne Institute of Technology,*
[1]*Germany*
[2]*Australia*

1. Introduction

Automation Industry is dealing with productivity, costs, quality and flexibility in the manufacturing process. In addition to the classic requirements of increasing productivity, reduction of costs and increasing quality, flexibility plays a significant role in the modern manufacturing process. These requirements are fulfilled by development, adaptation, and application of advanced engineering processes and methodologies improving mechanics, electronics and software development. One methodology that is widely applied in the development of complex mechatronic systems is HiL Simulation. By definition, a Hil (HiL) Simulation is a system that combines parts of a real system with parts of its simulated counterpart (Stoeppler et al., 2005; Grega, 1999). It is one of the best-known methods for testing embedded systems (Yan et al., 2005; Boot et. al., 1999) by providing the possibility of validating software execution on the original hardware without the dangers that are evoked by a malfunction of the controller in combination with the real mechanic system (Pfister et al., 1999).

In the early stages of the production plant lifecycle, ranging from Basic Design to Plant Engineering, discrete event simulation models are created reproducing the behavior of the mechanic system. Supporting the engineering of the control system, these models are executed in a real-time environment and connected to the control logic for desk checks. At the end of the Engineering Phase, the application of HiL Simulation shifts to the virtual commissioning of the overall process control system (Deutel & Suhm, 2002; Rohrlack, 2008).

The commissioning of modern mechatronic production plants is a critical cost intensive phase. At the beginning of the commissioning phase, the controller and its embedded control software and its correct interaction with the production plant are validated by HiL Simulation. These techniques strongly base on accurate simulation models.

Because of model inaccuracies between model and specification as well as between model and real plant, it is unlikely to find all errors in the controller software solely by controller testing with HiL Simulation. But even one failure in the controller may cause critical

machine states and damaging parts of the production plant. Thus the authors present a new approach for improving the commissioning of production plants. This approach, called Hybrid Commissioning, is an incremental procedure for commissioning that begins with a HiL Simulation, goes on with a mixture between reality and simulation, and ends up with the commissioning of the real production plant.
For achieving an optimal operation of the production plant, it is unavoidable to consider varying production conditions, e.g. in the flow of material or required retooling of components in the plant. Monitoring and diagnostics can support the operation of complex mechatronic systems and have to be provided efficiently. The authors present a methodology for reusing HiL Simulation models during the operation phase of industrial production plants. HiL Simulation models are used for determining the nominal state of the plant. Therefore, the HiL Simulation has to run in parallel to the real plant and has to be stimulated simultaneously by the control system. Initialized with the current state of the plant and the control system, a HiL Simulation can be applied predicting the behavior of the controlled plant. Therefore, the discussed application scenarios comprise the reuse of HiL simulation models for the purpose of monitoring and diagnostics as well as prognostics and optimization.
This chapter is structured as follows: In Section 2, fundamentals of industrial production plants are presented: production plants are structured according to their functionality and a summary over the lifecycle of production plants is given. The methodology of HiL is presented in the Engineering Phase of production plants and applications to mechatronic systems are discussed in general. In Section 3, the application of HiL Simulation methods in the commissioning phase by Hybrid Commissioning is introduced. In Section 4, two approaches for supporting the operation of manufacturing systems are presented: running a simulation model in parallel to the manufacturing system in Synchronous Simulation for purposes of diagnostics and using a simulation model in Forward Simulation for prognosis. Afterwards, application scenarios are discussed in Section 5. In Section 6, conclusions of the reuse of HiL Simulation in the Commissioning and Operation Phase of industrial production plants are given.

2. Fundamentals

This chapter deals with some methods and techniques that are recommended to apply during different stages of the lifecycle of production plants or production machines. These methods and techniques are based on Hil Simulation or rather Hil Simulation models. To gain a common understanding of these innovative approaches under the readership, this chapter is dedicated to general description of 'production plants', a generalized 'lifecycle of production plants' and the well known method called 'Hil' Simulation or 'Hil' Testing.

2.1 Industrial production plants

Industrial production plants represent complex mechatronic systems in factory automation machinery. There is a wide range of applications where production plants are used. Examples are wood-processing, packaging and varies kinds of manufacturing of all sorts of products. Although automation industry is dominated by a few suppliers, industrial production plants are mainly built by small and medium-sized enterprises. Considering unique specifications, they are mainly built in individual production. Regarding this constraint, the costs for development and commissioning are much more considerable than in markets with mass production and, hence, a cost efficient usage of engineering methods and techniques is essential.

Production plants are typically composed of modular sub-systems, so called production machines. In these machines, goods are processed in varies ways. The machines are connected via feed systems, which move the products or goods from one unit to the next, c.f. Fig. 1.

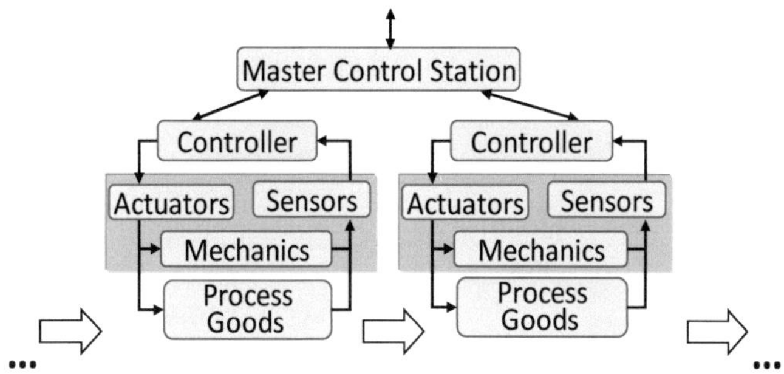

Fig. 1. Structure of a production plant

Production plants, comprising machines and feed systems, are built up of sensors and actuators, which typically represent mechatronic systems themselves, mechanical parts and different types of controllers. The type of controller depends on the functionality to be achieved. The functionality ranges from planning purposes at the top, e.g. realized in an operations control computer on process control level, to operational purposes at the bottom, e.g. a programmable logic controller on plant-floor level.

For structuring the functionality and the tasks of components in manufacturing systems, the Automation Pyramid is often applied (e.g. (Brown & Lean, 1987; IEC 2008), c.f. Fig. 2. In higher levels, real-time requirements are less important, whereas planning aspects gain in importance at the same time. In the Process Control Level, the process realized in the manufacturing system is planned in an optimal way regarding production constraints like availability of resources or priority of orders. In the Cell Level, the main task consists in evaluating and controlling the signals in the subordinated levels, e.g. several logic controllers or intelligent field devices. In the Actuator/Sensor Level, information about the process is gained via sensors and transmitted to the control system in the Plant-floor Level. There the information is processed according to the specified behavior and actuators are activated, while the data processing has to fulfill real-time requirements. In the Actuator/Sensor Level and Plant-floor level, real-time requirements play an important role. In regard of the designated process, dynamic or discrete properties dominate. Consequently, the task of the control system on plant-floor level range from executing step sequences up to advanced closed-loop control algorithms.

Depending on these different main tasks, different types of controllers are used. The programmable logic controller (PLC) is typically used event-driven or rather logic-driven step sequences. Numerical controllers (NC) are used for processes with very dynamical and highly precise motions, typically in machine tools (CNC machines). A controller type that combines the qualities of a PLC and a NC is a motion controller (MC). Such controllers are used for processes with fast and precise motions and logic-driven step sequences. A very flexible controller that has the potential to substitute the PLC, the NC as well as the MC is the industrial PC (IPC): a PC system that is tailored to industrial needs and is based on a real-time operation system for that reason.

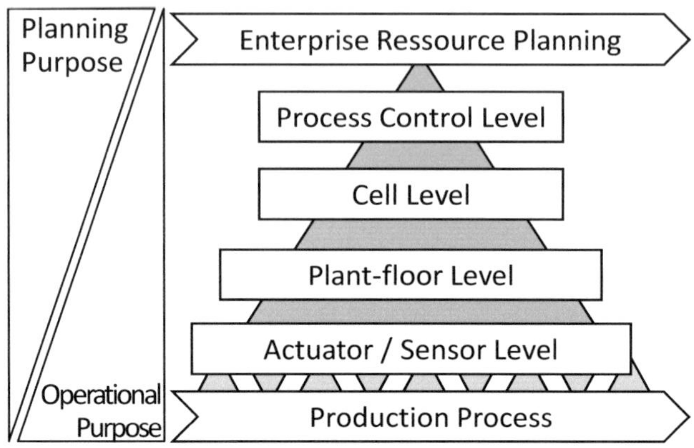

Fig. 2. Functional range of production plants in the Automation Pyramid according to (Brown & Lean, 1987)

Nowadays, industrial production plants have to feature high availability. Simultaneously, they have to produce as many goods as possible with the required quality. For cost-efficient production, the plant needs to be capable to manufacture different goods without any huge efforts in refitting (Wiendahl et al., 2007). Regarding optimal operation, it is necessary to consider changing conditions. To fulfill such requirements, production plants become more and more complex. The installed sensors and actuators are mechatronic systems themselves that embed controller with significant processing power. Examples are modern electric drives. Drives must fulfill requirements that lead to the above mentioned highly precised and speedy motions. Here, speeds of 10.000 rpm (rounds per minute) and accelerations of 12.000 rpm/s are common. To fulfill such requirements, drives are equipped with modern CPUs that are capable to run advanced control strategies.

One main enabler for the above mentioned functionality of production plants (and its varies sub-components) is software. The amount of software in a production plant is enormous and still increasing. And with the increasing amount of software, the effort for software development increases as well.

Production plants are of different qualities, each machine has its own efficiency. For evaluating the production plants, the Overall Equipment Effectiveness (OEE) was introduced merging capacity, availability, and product quality (Hansen, 2005; Nakajima, 1988). Parts of the OEE can be determined and fulfilled by the assembly during the development process, others, like availability are determined in the operation phase by several effects, e.g. improper handling, delays in the flow of material, abrasion or even the breakdown of components in the plant.

2.2 Life cycle of industrial production plants

Goal of this chapter is to introduce advanced methods for reusing Hil Simulation models to a broad audience. The introduced methods are applied in different phases of production plant life cycle. Hence, a brief summary over the (earlier) stages of the production plant lifecycle is given, c.f. Fig. 3:

- In basic design, the essential functionality of the plant is defined. An optimal design of the plant requires the analysis of the production process as well as the required capability of the plant. Attributes like capacity, reliability, maintainability, and follow-up costs as well as the resulting quality of the produced goods have to be defined (Lenord & Brohm, 2007). By using rough geometric and approximated functional models, proofs of concepts are realized.
- During the phase of plant engineering, the basic design is extended and detailed. The plant engineering starts typically with the specification of the mechanics, is continued with the application of sensors and actuators, and leads to the development of logic controllers and control systems. The controller design can be highly supported by use of application-specific functional models (Dominka & Kausler, 2006; Deutel & Suhm, 2002; Haufe, 2006).
- During the procurement and manufacturing phase, the designed assembly of the plant is finalized.
- The commissioning phase comprises the start-up of the machine and the validation of its full functionality.
- During the operation phase, the plant produces the goods it was designed for.

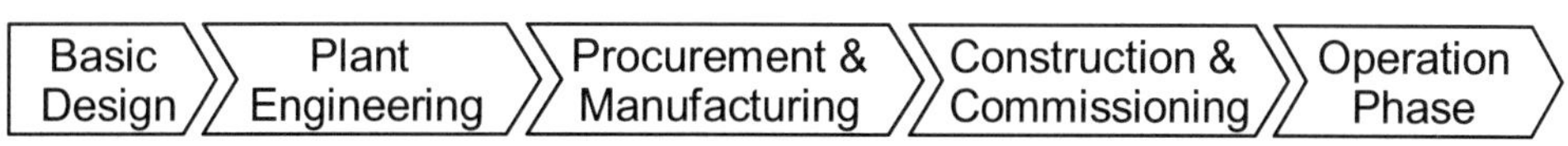

Fig. 3. Lifecycle of production plants ranging from Basic Design to Operation Phase (Dominka & Kausler, 2006).

2.3 Methodology of HiL testing in the engineering phase of industrial production plants

Hil Simulation is a well known method that is applied during the engineering phase of mechatronic systems. It is generally used as a test method. By definition, a Hil Simulation is a system that consists of real hardware components and simulated components that are connected to each other and show a (closed-) loop character (Grega, 1999; Stoepler et al., 2005).

In industrial automation, there are different use cases for HiL Simulations. One use case is the testing of industrial components (system under test), like motors and drives. Here the real component is integrated in a simulation test-bed (test-rig), which integrates HiL Simulation models and additional real parts like counter-motors, emulating the torque load and the load against a real motor (Bouscayrol, 2008; Maclay, 1997; Oh, 2005; Lin & Zhang, 2008; Murakami, 2005), c.f. Fig. 4.

Another application of HiL Simulation is testing of the control software for modern industrial production plants. In this case, the system under test, typically a programmable logic controller or a motion controller, is connected to a simulation model of the technical process, consisting of the mechanical and electrical parts (sensors and actuators) of a production plant and the process good, which is also an essential part of the technical process. In other words, the physical parts like the mechanical components, the actuators, the sensors, and the process goods are modeled in a real-time simulation system that is connected to the controller (Haufe, 2006), c.f. Fig. 5.

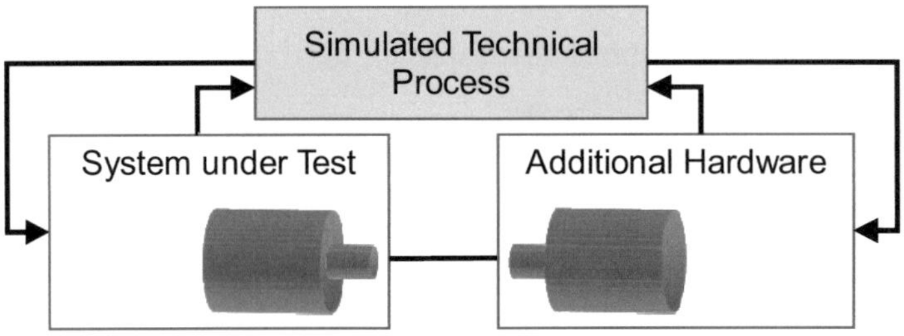

Fig. 4. Hil setup.

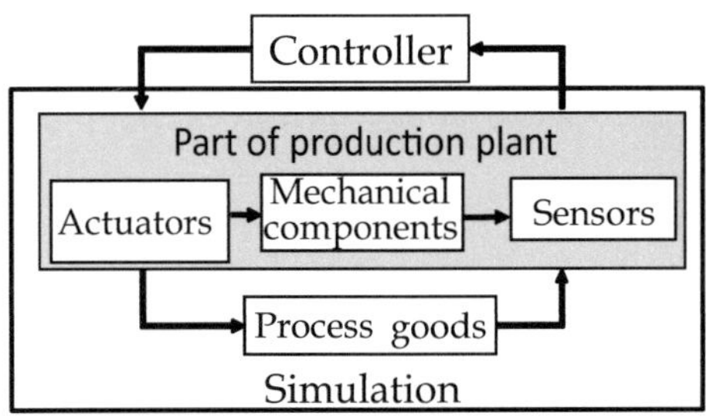

Fig. 5. HiL Setup in the realm of industrial production plants for controller testing

To be able to use Hil Simulation for controller and software testing, some very important requirements need to be fulfilled by the HiL Simulation models:

- Realistic functional behavior: The idea behind HiL Testing is that the simulation behaves in such a realistic way so that the controller does recognize the simulation as real machine. This means that the functional behavior of the simulation needs to be accurate enough, so that the controller works in its regular way.
- Realistic timing behavior: Technical processes, run by production plants or machines, are high dynamical, and so the controllers need to fulfill hard real-time requirements. To be able to test these controller characteristic properly, the HiL Simulation needs to run in real-time as well.
- Communication interface: The controller communicates with the machine via various communication interfaces (e.g. field busses). For testing, the controller must not be manipulated, i.e. the controller with its original hardware and software needs to be used. Hence, the HiL Simulation must provide the identical communication interface as its real counterpart.

HiL Simulation is one of the best-known methods for testing PLCs and motion controllers. It can emulate the plant behavior before the real plant is available. It provides the possibility of validating software without the dangers that are evoked by a malfunction of the controller in combination with the real mechanic system. Because this procedure can be considered as a commissioning of the controller by means of a simulated plant, it is known under the expression 'virtual commissioning' or 'soft commissioning' (Auinger et al., 1999; Wischnewski & Freund, 2004).

2.4 Application of HiL models in mechatronic systems

Hil Simulation is applied in various industries, like in the automotive and aerospace sector and in industrial automation. Especially for the development or rather testing of electronic control units (ECU), HiL Simulation is a very powerful and important technique that is widely used by automotive manufacturers and automotive component suppliers. Because of the high number of manufactured pieces in this industry, higher engineering costs can be better apportioned. This is also true for the modeling efforts, which consist in time and costs for building up the models that are necessary to prepare and run a HiL Simulation.

Unfortunately, this is not true for the realm of industrial automation. As mentioned earlier, production plants are themselves built in job production, which leads to the fact that engineering budgets are much more limited than e.g. in the automotive industry. The development of Hil Simulation models is therefore too costly for many companies in the industrial automation, and although its benefits are proven, HiL Simulation is often not applied.

Fortunately, the use of HiL Simulation models is not only restricted to software testing. By (re-)using such models, benefits can be achieved in the commissioning and in the operation phase of industrial production plants, and, furthermore, a justification for the modeling efforts can be achieved. In the following sections, some of such methods are presented

3. Application of HiL simulation methods in the commissioning phase

The commissioning of production plants is a very complex as well as a critical process in the lifecycle of production plants. Commissioning is defined as the transfer from the non-operating state of the production plant after assembling into the operating state of production. "The goal of any commissioning effort is to energize the equipment safely, in the time frame dictated by construction completion and process startup plans." (Bowen, 2000). Depending on the production plant and its dimension, assembling for the purpose of testing at the engineers place is not possible or very uneconomical. Thus the production plant is often assembled for the first time at the customers place.

Due to the increasing interdisciplinary between mechanical, electrical, and software components, it is hard to test one component without the availability of the remaining components. Consequentially during commissioning phase, the whole production plant is activated almost at once. Issues with parts of the plant or interfaces between components must be resolved one after the other. As a result the commissioning phase often takes an unjustified long time and costs a lot (McGregor, 2000).

Due to the increasing ratio of software in modern production plants, also the commissioning of these software components and the control unit/embedded system get more time-consuming. The costs for commissioning the controller and the embedded software exceed the costs for commissioning the mechanical components and are continuously increasing. The duration of the complete commissioning phase to the engineering phase takes about 15 to 25%. Thereof the commissioning of the electrical and software components takes about 90%. The reason for this high rate is in 70% due to software failures. The controller software therefore could take 60% of the commissioning time and 15% of the complete engineering time (VDW, 1997). Hence, there is a high industrial demand to increase the efficiency of commissioning the controller software.

3.1 Techniques for reducing the efforts of commissioning

There are several techniques how the efforts of commissioning could be reduced. These techniques can be classified as direct and indirect techniques and also constructive and analytical techniques. Direct techniques are used directly in the phase of commissioning contrary to indirect techniques, which are used in former phases and affect the commissioning indirectly by increasing the quality of the production plant or its components.

Analytic techniques are applied for the detection and localization of failures. All testing and checking methods are therefore analytical ones. Constructive techniques by contrast aim on high quality during the engineering. That means that constructive techniques avoid the appearance of failures.

In (Rieger, 1995) some of these techniques are described exemplary:

- A systematic requirements analysis for avoiding expensive aberrations (indirect, constructive)
- A systematic software-engineering with the phases 'concept', 'implementation' and 'test' should reduce errors during the development (indirect, constructive as well as analytic)
- Reuse of software-components for combining the advantages of high quality software and low development efforts (indirect, constructive)
- Pre-checking of machine and plant components before assembling the machine for recovering failures previously to the commissioning. (indirect, analytic)

Approaches during the commissioning phase comprise structured, checklist-based procedures and various tools, like monitoring, parameterization, and diagnostic tools.

3.2 Hybrid commissioning – incremental procedure from virtuality to reality

The main goal of HiL Simulation is to test the controller and its embedded software, and to find remaining software errors. In spite of the big advantages of HiL Simulation, it is not guaranteed that all software bugs can be detected. Because of discrepancies between the real and the simulated plant, it is unlikely to find all mistakes in the controller software with HiL Simulation. But even one failure in the controller may cause critical machine states and damage parts of the production plant. Because of this fact, the procedure of commissioning has to be executed very carefully. Nevertheless, it is possible that a failure in the controller occurs, so the machine could be damaged and the commissioning phase could be hold up further more. In addition, it might burry the risk to endanger human life (Schludermann et al., 2000).

This risk can be reduced by applying a technique called Hybrid Commissioning. Hybrid Commissioning is the stepwise transfer from the non-operating state of the production plant after assembling into the operating state of production by use of HiL Simulation (Dominka et al., 2007a; Dominka et al., 2007b; Dominka et al., 2007c). The rate of the real to the simulated components increases in the course of the commissioning procedure. In the beginning, a completely simulated plant is connected to the controller, whereas at the end the ratio of reality has been increased up to 100%. By applying this procedure, damages to the production plant and the process goods during commissioning can be avoided, and the commissioning can be performed less risky, more safely, and less time-consuming.

The procedure of Hybrid Commissioning is classified and presented in detail: 1) Hil Simulation, 2) partial commissioning of actuators and sensors, 3) Simulation-of-Goods, 4) commissioning of the real plant.

Phase 1: Hil Simulation

Phase 1 of the Hybrid Commissioning is the complete Hil Simulation, c.f. Fig. 6. The purpose of this phase is to detect most software bugs in the control unit and mainly to check the controller for correctness. The big advantage of this phase is the avoidance of real danger to field service personal, machine or process good in the case of a wrongdoing of the controller. However, a disadvantage arises. It is only possible to check the behavior of the controller while communicating with the simulation. Due to discrepancies between reality and simulation, it is not possible to obtain exactly the same behavior while connected to the real plant. This phase is finished, when the controller connected to the simulated production plant works as demanded.

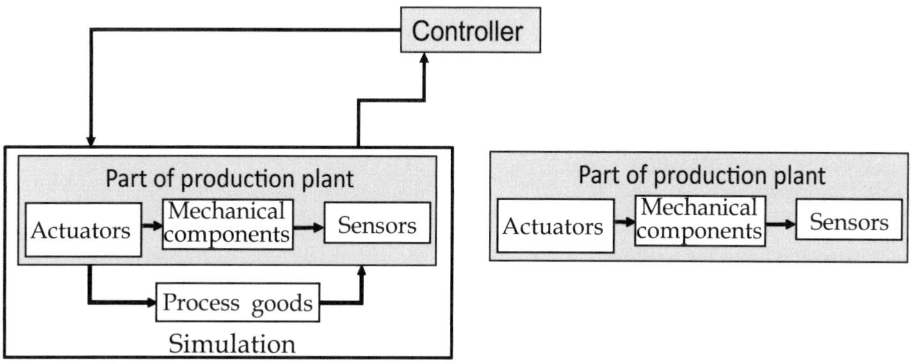

Fig. 6. Phase 1: Hil Simulation

Phase 2: Partial commissioning of actuators and sensors

In phase 2 of the Hybrid Commissioning, the actuators and sensors of the production plant are started up stepwise. The goal of this phase is to check, whether the single components themselves work correctly. In the beginning, the Hil simulation is connected to the controller like in the 1st phase. After that, control commands are additionally send to particular real actuators step by step. That means, that the control unit communicates with the Hil simulation, but the control commands are also sent to selected actuators. In this manner, it is possible to check the communication and operability between the control unit and real actuators. Malfunction could point to an error in the actuator, e.g. a wrong configuration or an error in the control software, c.f. Fig. 7.

Another goal of this phase is the check of sensors. Although the control unit still communicates with the Hil Simulation, it is possible to check the state of certain sensors and for example compare them with the simulated ones. As there is still no good to be processed in the real production plant only these sensors could be checked that are triggered mechanically by the production plant itself and not by the good. Light barriers are for example often used to be triggered by the good whereas encoders are triggered by a mechanical component. After verifying these sensors, it is possible to send the state of the real instead of the simulated sensors to the controller.

The advantage of this phase of Hybrid Commissioning is obvious: Every actuator and a big part of the sensors can be checked individually. So the operability of the controller and the plant components can be proved. A concurrent switch-on of the all actuators, however, is dangerous, because an undetected malfunction of a component could cause damage to the plant, and thus increase the costs and the time of commissioning.
Phase 2 is finished, when all actuators of the real plant are receiving control commands and all sensors triggered mechanically send their status back to the controller.

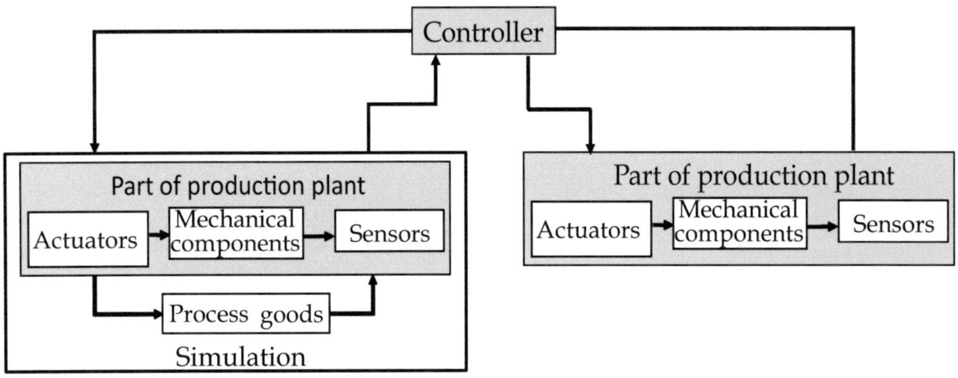

Fig. 7. Phase 2 and 3: Partial Commissioning and Simulation-of-Goods

Phase 3: Simulation-of-Goods

The Focus of this phase is the Simulation-of-Goods, which is defined as follows. Simulation-of-Goods is the simulation of the good and its behavior, which is processed in a production machine or plant. For simulating the good and its behavior, it is also necessary to replicate the actuators, which influence the good, and the sensors, which are influenced by the good.
The outstanding difference of Simulation-of-Goods to the Hil simulation is, that only these sensors and actuators are reproduced in the simulation system that have direct contact to the good. Operating controls, safety mechanisms, like safety light barriers, automatically closing safety doors, and final position switches, for example, need not to be simulated. The Simulation-of-Goods could be regarded as a real commissioning with simulated instead of real process good. For the Simulation-of-Goods, the models of the Hil simulation could be completely reused.
In this third phase of Hybrid Commissioning, Augmented Reality visualization is very promising. In this case, the process good, as bottles, cars, food, etc., is presented virtually. This virtual process good is then overlaid to the real production plant. The advantage of this kind of visualization is that the interaction between real plant and virtual simulated process good can be monitored ideally. It could be recognized very easy, whenever a process good is positioned directly in the scope of a sensor or an actuator.
The advantage of Simulation-of-Goods is that a big part of the real plant could be commissioned without the danger of damage of the real plant or of the real product by crashes. By this way the waste of production could be reduced as well.
This phase is finished, when the by then integrated part of the real production plant works as demanded.

Phase 4: Commissioning of the real plant

Phase 4 of the Hybrid Commissioning is finally the commissioning of the complete real production plant. After checking the operability and correctness of the control unit in interaction with the real plant with the simulated good, the simulation could be disconnected. In this phase, the control unit gets all its sensor signals from the real production plant processing real good, c.f. Fig. 8.

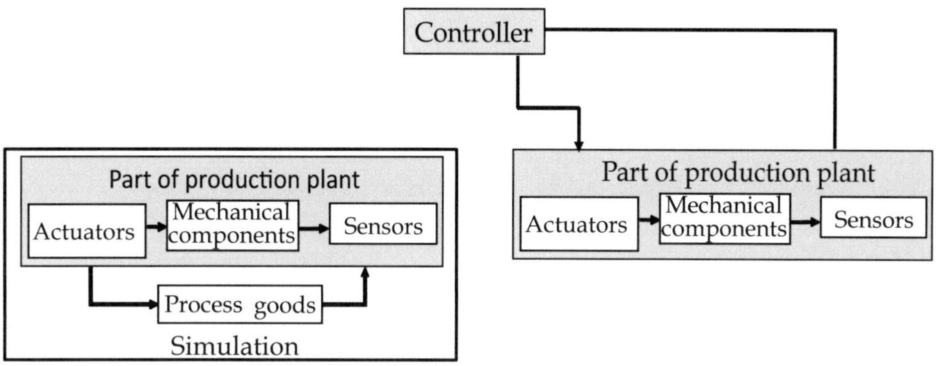

Fig. 8. Phase 4 Commissioning of the real plant

By this procedure, damaging of the production plant and the process good during commissioning can be avoided. This procedure allows a systematic, stepwise commissioning of the mechatronic system without the risk of injuring the operating personnel and of damaging the production plant and the process good. The waste of production and the efforts can be reduced.

4. Application of HiL Simulation in the operation phase

By (re-)using HiL Models, modeling efforts can be apportioned. In this section, the advantages of applying HiL Simulation models during the operation phase of industrial production plants are discussed. These comprise the reuse of HiL Simulation models for the purpose of monitoring and diagnostics as well as prognostics and optimization.

4.1 Supporting the operation of production plants by simulation

In the operation phase of production plants, the OEE (section 2.1.) can be influenced by several effects, e.g. changing production conditions, improper handling, and delays in the flow of material, abrasion, or even the breakdown of components in the plant. Therefore, condition-based monitoring and machine diagnostics represent an important topic of application and research (Williams et al. 1994). On component-level, several diagnostic functionalities and adaptive control strategies can be applied for detecting and reacting to these influences (e.g. self-tuning, system identification, parameter estimation or state observer approaches) (Isermann, 2005).

Production plants realizing challenging processes, consisting of numerous sensors and actuators, represent complex mechatronic systems. They can be characterized with mainly event-driven behavior. Thus, the application of observer methods from control engineering requires considerable efforts, causes noticeable efforts for modeling, and leads to complex and specialized solutions. The reuse of already existing simulation models is highly preferable for reducing efforts. In the following sections, methods for realizing monitoring and optimization are discussed.

4.2 Monitoring and diagnostics with HiL models

In the operation of production plants, failures in components (like abrasion or defects) can be indicated in marginal deviations of the plant behavior. Identifying these deviations allows to repair or change the component before the failure effects the operation totally, e.g. by break down of components. HiL Simulation models can be used for determining the nominal state of the plant and its components. This enables monitoring and in case of deviations also diagnostics of the plant. Therefore, the HiL Simulation has to run in parallel to the real plant and has to be stimulated simultaneously by the control system. This simulation setup is named Synchronous Simulation.

4.3.1 Methodology

The HiL Model of the plant emulates the behavior of the plant on component level or the field floor respectively. It can be used for determining the nominal state of the plant. If the simulation is operated in parallel to the plant, the measurable states of the real plant and of the simulation can be compared and deviations can be detected (Kain et al., 2008; Kain et al. 2009; Kain et al. 2010; Kain & Schiller, 2009). Thus, the signals in the simulation can be used for realizing diagnostics. Furthermore, information calculated in the simulation can be provided to the control system, c.f. as virtual sensor.
For realizing this comparison, different setups can be utilized: In open-loop control systems, variations of output variables do not affect the control system variables directly. As there is no closed loop between the control system and the real system, the control outputs can be applied to the simulation as well. By this approach, the simulation of the plant represents the nominal operation. Deviations detected by comparison of the signals of the real plant with the signals of the simulation indicate disturbances in the real plant.
In closed-loop control systems, variations of control variables directly lead to changes in the controller output. Connecting this controller output to the simulation model does not lead to a nominal operation, if the disturbance is not applied to the simulation as well. Thus, for obtaining the nominal operation in the simulation, the HiL Simulation has to be controlled by a separately simulated or additional hardware control system. Thereafter, the nominal behavior of the simulation with its own control system can be used for comparison with the real plant. These basic concepts are presented in Fig. 9.

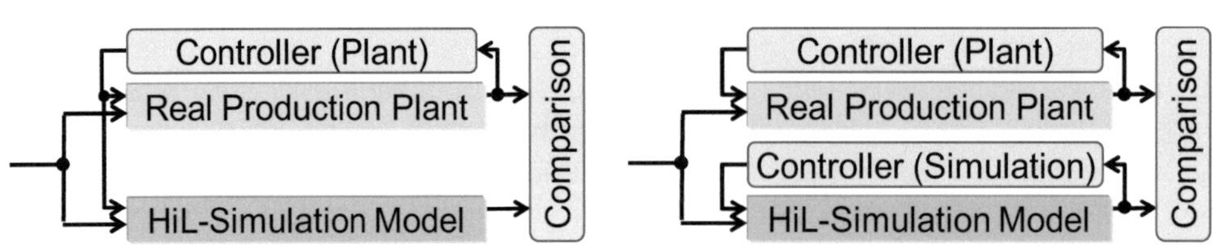

Fig. 9. Basic concepts for Synchronous Simulation

Depending on the level of detail of the HiL Simulation model, different statements can be made. If it comprises the behavior of the plant with its components in detail, the behavior of the plant according to input signals can be calculated. Assuming the same behavior of the plant and its simulation and an identical initial state, the simulation can calculate the states of the real plant when stimuli are applied to both systems. Stimuli that are not applied to the simulation, but to the real plant and vice versa, lead to deviations that can be detected by comparison of the signals.

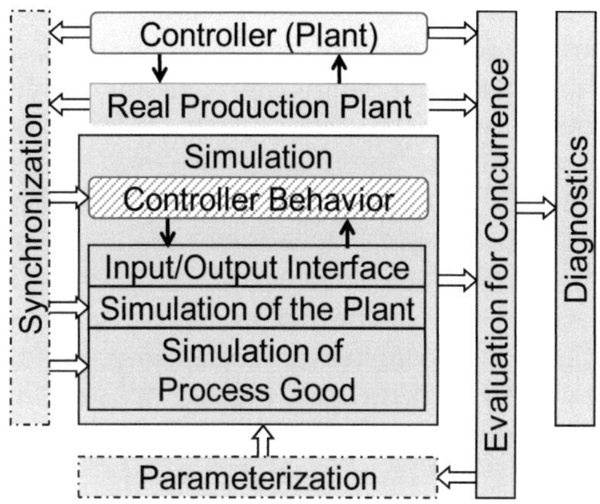

Fig. 10. Simulation architecture realizing Synchronous Simulation

In case of deviations between simulation and real plant, non-intended behavior of the plant can be detected and an appropriate reaction can be initialized. Of course, deviations can also be effected by imprecision of the simulation model.

As the real production plant is available in operation phase, the parameters of the simulation model can be verified easily. Nevertheless, parameterization of the HiL-Models has to be supported. Reducing deviations by stochastic characteristics of the realized process or by average determination of model parameters, deviation between the real plant and the simulation model can be reduced, by adapting the state of the simulation model to measured states of the plant. Thus, an architecture for Synchronous Simulation has to comprehend following parts: mechanisms enabling parameterization of the simulation model, synchronization of the simulation with the real plant and evaluation of the real and calculated behavior, c.f. Fig. 10.

4.3.2 Use cases in Synchronous Simulation

Synchronous Simulation provides knowledge about the behavior and state of the components in the production plant. It enables the online monitoring of the production plant and diagnostics of faults. Condition monitoring is defined as continuous or periodic measurement and interpretation of data for indicating the condition of an item to determine the need for maintenance (BS, 1993).

Dealing with production plants, condition monitoring can be detailed in Component Monitoring and Process Monitoring (Weck, 2001). In Component Monitoring, the components of the production plant are analyzed. The objective is the identification of faults in the components, before failures in the production plant are caused. For receiving

information about the components, this method in general requires sensors mounted directly to or stimulated by the monitored components, e.g. (Christer & Wang, 1995).
In contrast, Process Monitoring focuses on the analysis of the production process realized in the plant. The objective is to analyze, whether the process fulfills the requirements. Usually, information about the process is received by sensors in the plant stimulated by the process.

Component Monitoring

Realizing component monitoring requires information about the current state of the component in regard of the monitoring goal (availability, life period, current state, current dynamics ...). Depending on the complexity of the component, this can be realized by measuring the required data in the real component, if it is equipped with sensors.
If there is no data available, e.g. if the real component is not equipped with adequate sensors, it is possible to gain the information required by Synchronous Simulation. Therefore, the HiL Model of the component is connected to the plant and stimulated by the real signals in Synchronous Simulation. Representing the behavior of the real component (functional, timing, communication) the state of the HIL-component represents the state of the real component. Thus, the required information can be measured in the HiL Model. For realizing sophisticated Component Monitoring functionality, the functional behavior of the HiL model can be extended for providing the required information. Exemplarily, gaining information about energy consumption or friction in a real component might be very costly due to additional sensors or accessibility of the signal values.

Process Monitoring

Hybrid Commissioning requires the Hil Model to represent the component behavior and the flow of material. This characteristic also enables Process Monitoring.
Therefore, the HiL Model, representing the functional and timing behavior and the communication interface, connected to the real plant is stimulated by the real control system. Representing the chain of effects (actuators, flow of material, sensors) the calculated sensor signals can be compared with the real measured signal values.
This enables the online monitoring of the production process realized in the production plant and diagnostics of process faults. Therefore, the behavior of the realized process has to be analyzed, whether it is within the boundaries calculated by the simulation.
Reusing HiL Models in Synchronous Simulation enables monitoring and diagnostics of the production plant. Providing information about the state of the components, the HiL models enable Component Monitoring. Extending the models, e.g. with dynamic attributes, allows comprehensive conclusions about the components in the plant, e.g. energy consumption. Representing the flow of material, the HiL models enable Process Monitoring. Furthermore, information about the components and the process can be processed on the control level of the production plant for sophisticated control strategies.

4.4 Forward Simulation

Many influences to the production plant during the operation phase were presented above. If influences are determined to be systematic, their effect on the OEE of the plant can be predicted by simulation, e.g. changing production conditions, or delays in the flow of material. Additionally, if the control system was optimized in the engineering phase according to assumed process parameters divergent from the current situation, changes of the parameters in the control system can improve the plant behavior. Therefore, the models of HiL Simulation can be applied predicting the behavior of the controlled plant, called

Forward Simulation (Kain et al., 2008; Kain et al. 2009; Kain et al. 2010; Kain & Schiller, 2009).

4.4.1 Methodology

The behavior of the controlled plant consists of the functional behavior of the plant, represented by the HiL Models, and the behavior of the controller, represented by a model of the controller. Validation of the simulation results can be realized by comparison with the real plant behavior. This enables a detailed predictive simulation, called Forward Simulation, cf. Fig. 11.

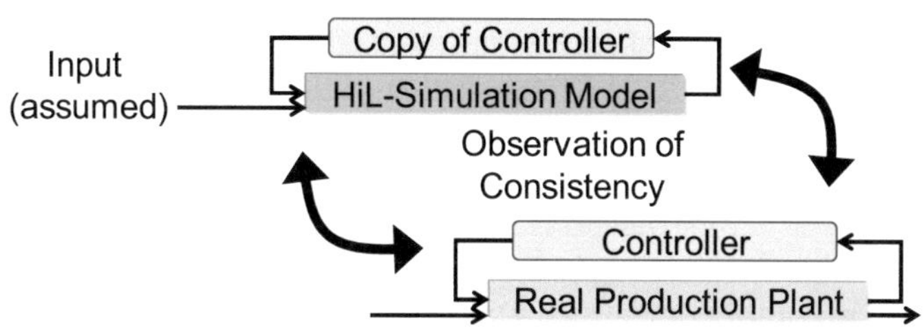

Fig. 11. Concept Forward Simulation

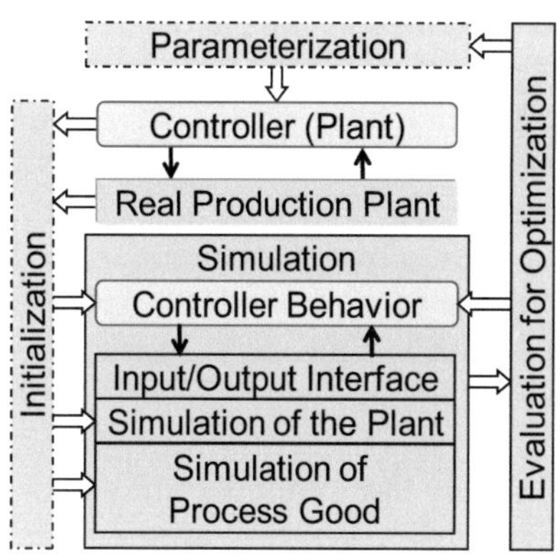

Fig. 11. Forward Simulation: concept and simulation architecture

Optimizing the control system offline based on verified HiL Simulation models representing the behavior of the plant in detail offers several advantages: In a first step, the optimal parameters can be determined based on simulation experiments with diverse constraints without influencing the real plant. This process can be repeated time or event triggered.
In a second step, the control system can be adjusted to the parameters and the optimized control system can be applied to the real plant consequently. Thereby changes in the control system might range from simple parameterization to adaptation of the controller software.
The basis for this optimization is a HiL Simulation model of the plant representing the behavior of the real plant and a control system. In addition to Synchronous Simulation, this simulation approach requires mechanisms for parameterization of model parameters and mechanisms for evaluation of the simulation results. The Forward Simulation is initialized with the current state of the plant and the controller. For predicting the behavior, inputs

applied in the future to the production plant and optimization parameters could be considered as well, if available. The resulting architecture of Forward Simulation is presented in Fig. 11.
Due to restrictions of real hardware controllers, this can be only realized by means of a simulation of the controller behavior. The HiL Simulation has to be detailed sufficiently to enable the evaluation of application specific optimization criteria.

4.4.2 Use cases Forward Simulation

Many applications exist for applying simulation methods in the operation phase of production plants. Besides production planning issues and supporting the production planning and control, also detailed functional models are applied in the ramp up and the prediction of production capability. In the following, these use cases are presented and compared.

Ramp up

The efficient ramp up of industrial production plants poses a big challenge due to short product life cycle and numerous product variants, e.g. in automotive industry. For economic reasons, the production process should be operated at the desired production rate and output as soon as possible. After commissioning the plant, e.g. by Hybrid Commissioning, there are still adjustments necessary. Therefore, in (Lanza & Fleischer, 2005) an approach is discussed, supporting this adjusting process with an extended functional simulation of the production plant. Therein, the capability of producing products with components of plant is modeled. This enables the prediction of the product quality in the plant by different action alternatives. Thus, the best action alternative can be chosen and applied to the plant.
Depending on the size of the simulated production plant and the expected simulation results, the level of abstraction of the simulation model has to be chosen. In case of detailed conclusions about significant parts of a production plant, reusing the functional part of HiL-models and functional control logic can represent the behavior of the plant sufficiently.

Forecast simulation

Predicting the plant behavior is very important for production planning and control. Thus, there exist many simulation systems, providing information about the plant behavior in different levels of detail. (Hotz et al., 2006) discuss a system in automotive industry, predicting the behavior of large production plants/production factories. Therefore, the basis is a simulation model, representing the in-firm logistic and the production times and capacities of parts in the production factory in discrete event models. Initialized with the current state of the production plant (allocation of material, future orders), the quantity produced in a time horizon in the future can be calculated. Also, conclusions about the influence of cumulated lags and dynamic behavior of storages can be analyzed. Another application of forecast simulation is the support of the detailed production planning and control. In case of identified need of action, the best action can be chosen by calculating the effects on the plant behavior.
These applications require predicting the behavior of the controlled plant. Thus, models representing the controlled behavior of the components in the plant are required. Due to the rough level of detail required in this use case, detailed HiL Simulation models are not appropriate, but discrete event representations might be suitable.

Forward Simulation

Detailing the simulation model representing the functional and timing behavior of the components in the plant and the control logic, enables Forward Simulation. Therein, the models of the components represent the uncontrolled behavior. For representing the model of the controlled plant, also the behavior of the control system has to be considered.

This detailed model of the production plant enables the calculation of the plant behavior on process level. Including the signal flow between the components and the controller enables the detailed analysis of the plant behavior on signal level. Consequently, conclusions about the behavior of the components on process level are possible (Kain et al., 2010; Kain & Schiller, 2009).

This simulation model requires the representation of the functional and timing behavior of the components in the plant. Depending on the model of the control system, the communication interface has to be defined.

In Fig. 12, the discussed use cases are arranged according to the criteria level of component detail and size of the focused plant. The use case Forward Simulation requires the most detailed models of the components in the plant and focuses the most restricted part of the production plant. In contrast to the presented use cases of Forecast Simulation, Forward Simulation is suitable for reusing HiL-Simlation Models in the operation phase of industrial production plants for prediction and optimization of the plant behavior. This approach provides detailed information about the components and the process on field floor level.

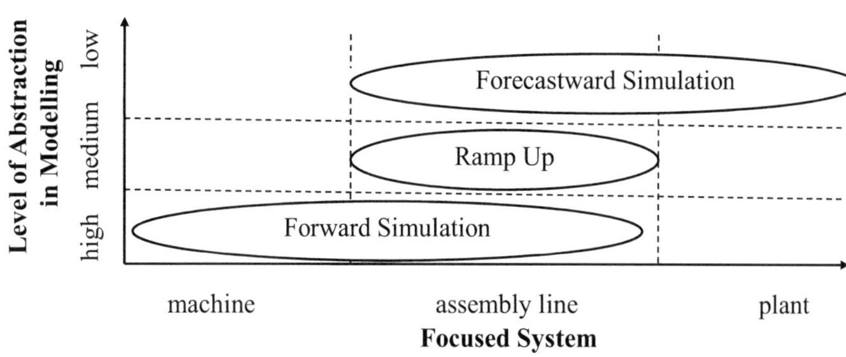

Fig. 12. Overview simulation methods for prognostics

5. Showcase realization

In this section, the application of the presented concepts is shown in practical use. Thereby, a demonstrator for production plants represents all characteristics of a real production plant.

5.1 Hybrid Commissioning

The application of Hybrid Commissioning poses some challenges. As described before, it is required that the simulation of a plant and the plant itself run in parallel. Moreover, it is necessary, that control commands could be sent to the real and the simulated actuators at the same time, and that single sensor signals could be taken either by the real or the simulated plant without big efforts of reconfiguration. Hence, Hybrid Commissioning is based on forwarding and redirecting of signals; either sensor signals to the controller or

actuator signals from the controller. It is also a requirement that the original controller interface is used. If the controller communicates with the plant, i.e. the sensors and actuators via a field bus, then this interface must be used for the Hybrid Commissioning as well.
The type of interface has a significant impact on the actual realization of Hybrid Commissioning, and so its realization cannot be generalized. As an example, however, a realization of PROFIBUS DP driven production plants is given. Anyway, the realization could also be achieved with other communication technologies like Profinet, EtherCAT, CAN (Thomesse, 2005) or Flexray (Froschauer & Auinger, 2009).
Fig. 13 shows a simplified system architecture of Hybrid Commissioning for a plant, which communicates with the field bus PROFIBUS DP. In this example, a Programmable Logic Controller (e.g. with bus address 1), is connected to a real decentralized input-output periphery (e.g. with bus address 13), and its simulated counterpart. To be able to fulfill the mentioned requirements, a so called Merge-Tool is necessary. This Merge-Tool is interconnected between the controller and the real and the simulated plant components, which act as slaves according to the Master-Slave principle in case of PROFIBUS DP.

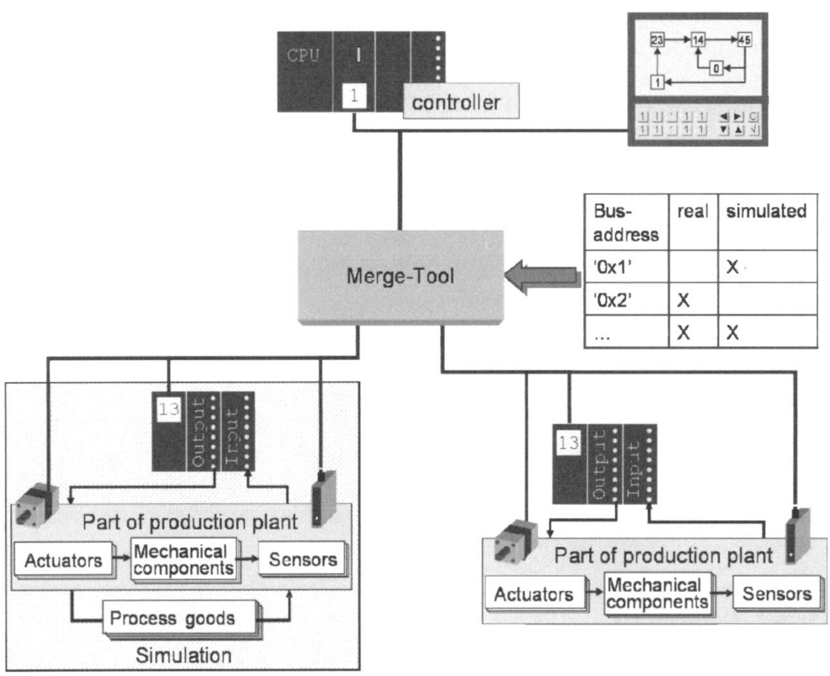

Fig. 13. System architecture with Merge-Tool

The main purpose of the Merge-Tool is to control the execution of Hybrid Commissioning, i.e. the Merge-Tool determines the forwarding and redirecting of signals and hence determines the phase of Hybrid Commissioning. Via the Merge-Tool, the real actuators are activated and step-by-step the real sensor signals are taken into the control-loop.
The Merge-Tool has, however, some more tasks that are rooted in the characteristics of the field bus PROFIBUS DP. PROFIBUS DP is a reliable field bus that detects several errors, like transmission errors or unwanted multiple devices with the same bus address. Adding a

simulation system to the real system (consisting of controller and plant) and changing signals that are transferred via the field bus would generally lead to an interruption in the PROFIBUS communication. Thus, additional task of the Merge-Tool is to keep up the field bus communication, despite additional bus components and changing of bus signals.
The tool has the task to merge the real slave and its simulated counterpart for pretending the master (controller) the existence of only one slave.
In case of PROFIBUS DP, the communication between master and slave can be divided into two parts. One part consists of the master-requests, i.e. the polling of the slaves by the master. The other part consists of the slave-responses, i.e. the responses from the slaves to the master. Master-requests are telegrams which are transmitted from the controller to the slaves. These can be telegrams for parameterization and configuration, commands to actuators or telegrams for polling the sensors. The Merge-Tool has to duplicate or copy these telegrams, respectively, in order to send them to the real and to the simulated slave. Therefore, a main aspect of the Merge-Tool is the copy functionality.
Another aspect of the Merge-Tool is to process the slave-responses. For Hybrid Commissioning, both simulated and real slaves, send telegrams to the master. However, the master can only handle one telegram for each bus address. Thus, the Merge-Tool has the task to merge real and simulated telegrams and forward new composite telegrams to the master. This is the primary purpose of the Merge-Tool, as for Hybrid Commissioning, sensor data from the real plant as well as sensor data from the simulated plant should be provided for the controller. For example, from a peripheral with 8 bits input, 4 bits should be taken from the real plant and 4 bits should be taken from the simulation. This pivotal merge-functionality led to the name Merge-Tool.
Hybrid Commissioning has so far been implemented in several demonstrators. An example of such a demonstrator is illustrated in Fig. 14. This demonstrator consists of a bottling plant, which fills up little bottles, a HiL Simulation and a control unit. Additionally, the Merge-Tool can be seen, which is interposed between the controller, the real production plant and its simulated counterpart. The bottling plant consists of four separate stations which are controlled by programmable logic device Siemens SIMATIC S7 317-2 DP and connected via PROFIBUS DP communication decentralized peripheries Siemens ET200M, to which several actuators, like driving motors, and sensors, like light-barriers, are connected.
The HiL Simulation (functional and timing behavior) of the production plant is modeled with the simulation tools Siemens SINUMERIK Machine Simulator. As the real components are connected to the controller via Profibus DP, the communication interface of the simulation is realized by SIMBApro, an interface card for Profibus communication. The Merge-Tool copies the control commands and sends them, depending on the actual configuration, to the real production plant, to the simulation, or to both of them. According to the configuration, either simulated or real sensor signals are sent back to the control unit.
The Merge-Tool has so far only been developed for PROFIBUS DP driven production plants. However, a similar device that can handle other field bus systems or point-to-point connections could be (easily) developed as well. The Merge-Tool can be reused for every other PROFIBUS DP driven plant and just needs to be reconfigured for each use case. So, Hybrid Commissioning can be considered as a very powerful and low cost approach by reusing existing Hil models.

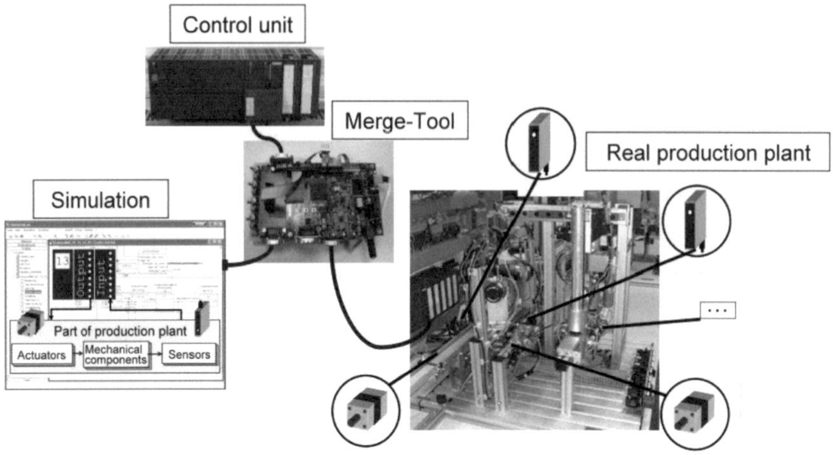

Fig. 14. Demonstrator for Hybrid Commissioning

5.2 Synchronous Simulation

First, background for the application of Synchronous Simulation is discussed, and then the application to a demonstrator for automation systems is presented. Monitoring with Synchronous Simulation require accurate simulation models. Assuming a constant behavior of the components, the parameters of the HIL models can be determined by established verification and validation techniques, e.g. (Sargent, 2007).

In the case of shifting component behavior, as induced by abrasion, the parameters of the HIL models have to be adapted for representing the behavior of the real component furthermore. Compared to observer methods in control engineering, deviations between the real and the simulated component have to be considered in parameterization of the HiL-model. Depending on the characteristic of the modeled component, this can be realized by many algorithms, ranging from simple PID feedback to sophisticated optimization algorithms, c.f. e.g. Kalman filter in control engineering (Nise, 2006). As result, the parameters of the HiL-models are adapted over time to represent the behavior of the real component in a best manner.

In a demonstrator scenario, the Synchronous Simulation model of the system and the process is built up from HiL-models. The simulation model contains the sensors and actuators of the demonstrator as well as the realized technical process of gripping and moving bottles.

The signals between the plant and the controller are available via a PROFIBUS DP slave emulation with Siemens Simba Pro. The control signal of the real controller is input for the Synchronous Simulation of the system, cf. Fig. 15.

The simulation tool Sinumerik MS enables the programming of user specific functions in a simplified version of the programming language C. The monitoring and analyzing units are also implemented in the simulation tool Siemens Machine Simulator. Therein, the output signals are calculated and compared to the real output signals. Therefore, detected deviations are analyzed and compensated on signal level in the parallel HiL Simulation. The monitoring components include the in- and outputs of both simulations and the real system. The diagnostics component enables the analysis of deviations.

In the scenario presented, component monitoring is possible by tracing the state of the actuators in the simulation. In case of exceeding defined limits, the controller of the plant is informed by the field bus interface card.
Process monitoring is realized by comparing the signals of the real sensors and the calculated sensors in time and value. For reducing the propagation of deviations, application specific intervals are defined for synchronizing the simulated flow of material with the real one permitting discrete synchronization.

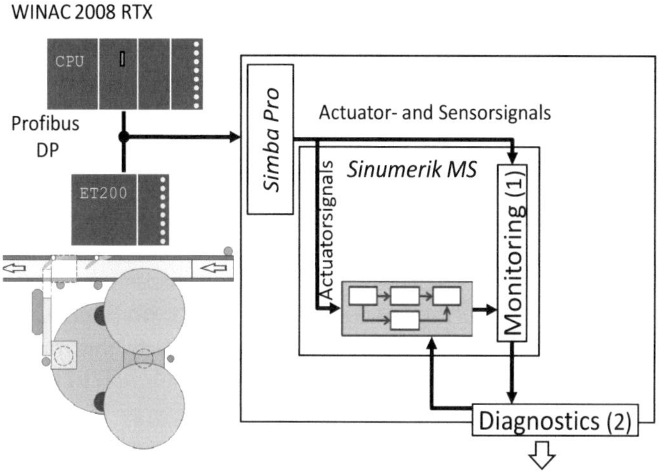

Fig. 15. Framework for simulation based monitoring and diagnostics for hybrid automation systems.

5.3 Forward Simulation

For representing the nominal behavior of a closed-loop system in the Forward Simulation, the simulation system has to fulfill certain requirements:

- Consideration of the behavior of the controller, e.g. by proposing an additional controller simulator. Depending on the chosen controller model, interfaces to a platform, executing the controller code, are necessary.
- For receiving simulation and optimization results in-time before appearing in the real plant, the simulation models must be executed faster than the reality.
- Discrete-Event-Systems can be calculated quickly, when only few events occur in the system. Therefore simulation velocity depends on the modeled behavior. This poses the challenge to consider the behavior of the real operating controller.
- The Forward Simulation is started periodically for evaluating the influence of changing production constraints. Thus an initialization method transferring the respective state of the controller and the plant to the simulated controller and the HiL Simulation is required. Therefore, a real-time simulation-plant interface has to be available.
- After evaluating the simulation result according to optimization criteria, the behavior of the real controller has to be adjusted, e.g. by parameterization.

The reuse of HiL Simulation models for Forward Simulation was applied for optimization issues. First, a software realization is discussed, then the application to a demonstrator for automation systems is discussed exemplarily.

Forward simulation requires calculating the behavior of the plant before it happens in the real plant. Afterwards the behavior is evaluated and interventions to the plant are derived. Therefore, the Forward simulation reuses the functional and the timing behavior of HiL Simulation models, the communication interface might be neglected.

In the following, a software realization is presented, which executes the HiL Models in an event discrete simulator. Therein, the communication interface of the models is reduced to the in- and outputs if the components. This enables a rapid calculation of the plant behavior and enables the application of different optimization strategies.

Forward Simulation is realized in a software framework, c.f. Fig. 16. Industrial software tools were integrated to the framework for reducing integration efforts in the industrial engineering flow. The behavior of the plant is calculated in Siemens Plant Simulation. The models represent the functional and timing behavior of the HiL Simulation models. For considering the behavior of the controller system, Siemens PLCSim is integrated. This enables the reuse of the original controller code. Obviously, also an abstraction of the controller code could be used. Depending on the representation of the controller code, a execution environment has to be chosen and integrated. For efficient calculation of the system behavior, the time-discrete behavior of PLCSim is encapsulated for event-discrete calculation. This enables the efficient event-based interaction with Plant Simulation.

Initialized with the state of the real plant (consisting of the plant, the controller system and expected production orders), this setup enables the rapid calculation of the system behavior. The evaluation of the system behavior is realized in Mathworks Matlab. This enables the application of several optimization strategies adapted to the plant characteristics.

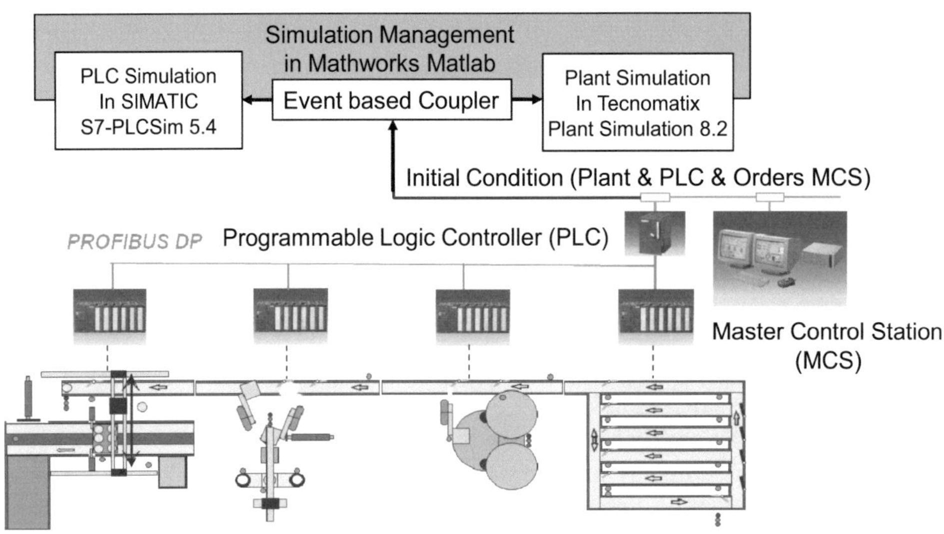

Fig. 16. Framework for Forward Simulation

Based on a first shot calculation, the system behavior is evaluated according to the chosen criteria. These might include production time, energy consumption or the abrasion of components.

The research in the application scenarios of HiL Models in the operation of production plants was supported by the project AVILUS ("Angewandte Virtuelle Technologien im

Produkt- und Produktionsmittellebenszyklus" - Applied Virtuality in the Lifecylce of the Product and the Production Facility), funded by the German Federal Ministry of Education and Research (BMBF 01 IM 08 001 X). Special thanks goes to our partner Siemens AG.

6. Conclusion

Hil Simulation is a very well known method that is applied during the engineering phase of mechatronic systems. By definition, a Hil Simulation is a system that consists of real hardware components and simulated components that are connected to each other and show a (closed-) loop character. In Automation Industry, HiL Simulation is typically applied in the development phase of the lifecycle of industrial production plants. In this chapter, the authors proposed the reuse of HiL Simulation Methods in the commissioning and the operation phase by the applications Hybrid Commissioning, Synchronous Simulation and Forward Simulation.

Hybrid Commissioning is a stepwise procedure starting from a simulated production plant to a real production plant. By this approach, it is possible to reduce the danger of damages at the process of commissioning. At the same time, the commissioning phase could be passed more systematically, and so the partially high efforts of commissioning could be reduced.

For increasing the benefits of HiL-Models, a methodology for supporting the operation phase of industrial production plants has been proposed afterwards. Synchronous and Forward Simulation enable monitoring and diagnostics as well as prognostics and optimization based on reused HiL Simulation models.

Obviously, the quality of the Synchronous Simulation is suitable to allow conclusions about the states of the plant in component and process monitoring and diagnostics and can strongly support the control system of the plant by providing virtual sensors. The Forward Simulation enables the optimization of the control system due to varying operation conditions. Mechanisms for synchronization with the real plant enable efficient modeling of the Synchronous Simulation, whereas the Forward Simulation demands for detailed modeling and for robust criteria for evaluation respectively.

It was shown that each of these approaches can easily be implemented and that a (re-)usage of HiL Simulation models in different phases of plant lifecycle can increase their benefits. The insertion of the proposed approach showed that it could be successfully used in industrial applications and hereby decreases the efforts in the commissioning and the operation phase of industrial production plants.

7. References

Auinger, F.; Vorderwinkler, M. & Buchtela, G. (1999). Interface Driven Domain-Independent Modeling Architecture for „Soft-Commissioning" and „Reality in the Loop", *Proceedings of IEEE Winter Simulation Conference*, pp. 798-805, ISBN 0-7803-5780-9, Phoenix Arizona, December 1999

Boot, R.; Richert, J.; Schutte, H. & Rukgauer, A. (1999). Automated test of ECUs in a Hil simulation environment. *Proceedings of the 1999 IEEE International Symposium on Computer Aided Control System Design*, pp. 587-594, ISBN 0-7803-5500-8, Kohala Coast, HI, August 1999

Bouscayrol, A. (2008). Different types of Hil simulation for electric drives, *Proceedings of IEEE International Symposium on Industrial Electronics*, pp. 2146-2151, ISBN 978-1-4244-1665-3, Cambridge, June-July 2008

Bowen, J. (2000). Substation Commissioning and Turnover Planning, *IEEE Industry Applications Magazine*,Vol. 6, No. 2, March/April 2000, pp. 8-15, ISSN 1077-2618

Brown; P. F. & Mac Lean, C. R. (1987). The architecture of the NBS factory automation, IFAC Congress, Munich, Germany, 1987

British Standard BS 3811 (1993). British Standard Glossary of Maintenance Management Terms in Terotechnology, British Standards Institution,

Christer, A. H. & Wang, W. (1995). A simple condition monitoring model for a direct monitoring process. *European Journal of Operational Research*, Vol. 82, pp. 258-269, ISSN 0377-2217

Deutel, F. & Suhm, C. (2002). Simulation based engineering. *A&D Newsletter*, No. 10 October 2002, pp. 14-16, ISSN 1618-2898

Dominka, S. & Kausler, B. (2006). Reuse of HiL Simulation Models in Later Phases of Production Plant Life cycle, *Proceedings of* 10th International Research/Expert Conference *Trends in the Development of Machinery and associated Technology (TMT)*, pp. 705-708, Barcelona-Lloret de Mar, Spain, September 2006, ISBN 9958-617-30-7

Dominka, S.; Schiller, F.; Kain, S. & Bröcker, E. (2007a). Modification of Field Bus Telegrams for Hybrid Commissioning, *Proceedings of IEEE Second International Symposium on Industrial Embedded Systems*, pp. 170-177, ISBN 1-4244-0840-7, Lisbon, Portugal, July 2007, IEEE

Dominka, S.; Schiller, F.& Kain, S. (2007b). Hybrid Commissioning - From Hil Simulation to Real Production Plants, *Proceedings of 18th IASTED International Conference on Modeling and Simulation*, pp. 544-549, ISBN 978-0-88986-664-5, Montreal Quebec, Canada, May June 2007, ACTA Press Anaheim, CA, USA

Dominka, S.; Schiller, F. & Kain, S. (2007c). Hybrid Commissioning - Speeding-up Commissioning of Field Bus Driven Production Plants, *Proceedings of IEEE International Conference on Mechatronics, ICM 2007*,pp. 1-6, ISBN 1-4244-1183-1, Kumamoto, Japan, May 2007

Froschauer, R. & Auinger, F. (2009). A Survey on the Integration of the FlexRay Bus in Distributed Automation and Control Systems, *Proceedings of 2nd International Symposium on Logistics and Industrial Informatics*, pp. 1-6, ISBN 978-1-4244-3958-4, Linz, September 2009, IEEE

Grega, W. (1999). Hil simulation and its application in control education, *Proceedings of Frontiers in Education Conference, 1999. FIE '99. 29th Annual*, pp. 12B6/7 - 12B612, Vol. 2, ISBN 0-7803-5643-8, San Juan, November 1999

Hansen, R. C. (2005). *Overall Equipment Effectiveness*. Industrial Press, ISBN 0831131381, New York

Haufe, J. (2006). Simulation based Prototyping of PLC-Programs, In: *Annual Report 2006 Fraunhofer Institute for Integrated Circuits IIS*, Fraunhofer IIS, pp. 141-142, Erlangen

Hotz, I., Hanisch, A. & Schulze, T. (2006). Simulation-based early Warning Systems as a practical Approach for the Automotive Industry, Proceedings of the IEEE 2006 Winter Simulation Conference, pp. 1962-1970, ISBN 1-4244-0500-9, Monterey CA, December 2006

IEC (2008). Enterprise-Control System Integration, IEC 62264, 2008

Isermann, R. (2005). Model-based Fault Detection and Diagnosis - Status and Applications, *Annual Review in Control*, Vol.29, No. 1, pp 71-85, ISSN 1367-5788

Kain, S.; Schiller, F.& Dominka, S. (2008). Reuse of Models in the Lifecycle of Production Plants, Using HiL Simulation Models for Diagnosis, *Proceedings of IEEE International Symposium on Industrial Electronics*, pp. 1802 - 1807, ISBN 978-1-4244-1665-3, Cambridge, June-July 2008

Kain, S.; Merz, M.; Schiller, F. & Dominka, S. (2009). Reuse of HiL Simulation Models in the Operation Phase of Production Plants, *Proceedings of International Conference on Industrial Technology*, pp. 1-6, ISBN 978-1-4244-3506-7, Gippsland, February 2009

Kain, S. & Schiller, F. (2009). Supporting the Operation Phase of Manufacturing Systems by Synchronous and Forward Simulation, *Proceedings of 3rd International Conference on Changeable, Agile, Reconfigurable and Virtual Production (CARV 2009)*, ISBN 978-3-8316-0933-8, Munich, Germany, October 2009, Herbert Utz, München

Kain, S.; Schiller, F. & Frank, T. (2010). Monitoring and Diagnostics of Hybrid Automation Systems based on Synchronous Simulation, *Proceedings of 8th IEEE International Conference on Industrial Informatics - IEEE INDIN 2010*, pp. 260 - 265, ISBN 978-1-4244-7298-7, Osaka, Japan, July 2010

Lanza, G. & Fleischer, J. (2005). OEE-simulation in production ramp-up, OEE-Prognose im Produktionsanlauf, *wt Werkstattstechnik online*, Vol 95, No. 7/8, pp. 554-558, ISSN 1436-4980

Lenord, O. & Brohm, M. (2007) Maschinensimulation - Eckstein des Simultaneous Engineering im Bereich Steuerungstechnik, SPS/IPC/DRIVES 2007, pp. 537-546, ISBN 3800730693, VDE Verlag, Berlin

Lin, C. & Zhang, L. (2008). Hil simulation and its application in electric vehicle development. *Proceedings of IEEE Vehicle Power and Propulsion Conference*, pp. 1-6, ISBN 978-1-4244-1848-0, Harbin, September 2008

Maclay, D. (1997). Simulation gets into the loop. *IEE Review*, Vol. 43, No. 3, May 1997, pp. 109-112, ISSN 0953-5683

McGregor, I. (2000). Reduce Your Commissioning Costs Through Emulation, Auto Flash, Vol. 13, No. 9, pp. 1-8, AutoSimulations, a Division of Brooks Automation, Bountiful, USA

Murakami, K. (2005). Hil Testing Of Distributed Electronic Systems. *Proceedings of JSAE Annual Congress*. Vol.17 , No. 5; pp. 9-13, ISSN 0919-1364

Nakajima, S. (1988). *Introduction to TPM: Total Productive Maintenance*. Productivity Press. ISBN 0915299232, Cambridge

Nise, N. (2006). Control Systems Engineering, ISBN 0471794759, John Wiley & Sons, New York

Oh, S. C. (2005). Evaluation of motor characteristics for hybrid electric vehicles using the Hil concept, *IEEE Transactions on Vehicular Technology*, Vol. 54, No. 3, May 2005, pp. 817 - 824, ISSN 0018-9545

Pfister, F., Reitze, C. & Schmidt, A. (1999). Hardware in the Loop - The Technologie for Development and Test of Vehicle Control Systems, Proceedings of the "Congrès International - La Dynamique du Véhicule Automobile et Ferroviaire", Lyon, France, June 1999, SIA R-99-04

Rieger, B. (1995). *Systematische Funktionsprüfung bei der Inbetriebnahme feldbusvernetzter Automatisierungssysteme / Systematic Functional Test in the Commissioning of fieldbus driven Automation Systems*, VDI, ISBN 3-18-316320-9, München

Rohrlack, T. (2008). The Digital Factory From Concept to Reality. Factories Solution Executive, Technical Report, September 2008, Bentley Solutions

Sargent, R. (2007). Verification and validation of simulation models. *Proceedings of the 37th Conference on Winter Simulation*, pp. 124-137, ISBN 978-1-4244-1306-5, Washington DC, December 2007

Schludermann, H.; Kirchmair, T. & Voderwinkler M. (2000). Soft-Commissioning: Hil-Based Verification of Controller Software, Proceedings of *IEEE Winter Simulation Conference*, pp. 893-899, ISBN 0-7803-6579-8, Orlando Florida, December 2000

Stoeppler, G.; Menzel, T. & Douglas, S. (2005). Hil simulation of machine tools and manufacturing systems. IEEE *Computing and Control Engineering*, Vol. 16, No. 1, Feb.- March 2005, pp. 10-15, ISSN 0956-3385

Thomesse, J. (2005). Fieldbus technology in industrial automation, *Proceedings of the IEEE*, , Vol. 93, No. 6, pp. 1073–1101, ISSN 0018-9219

VDW (1997). *Abteilungsübergreifende Projektierung komplexer Maschinen und Anlagen / Interdivisional Engineering of Complex Machines and Production Plants*, Technical Report, Aachen, German Machine Tool Builders´ Association, 1997

Weck, M. (2001). Werkzeugmaschinen 3 / Machine Tools 3. ISBN 3540225064, Springer, New York

Wiendahl, H.-P.; Elmaraghy, H.; Nyhuis, P.; Zäh, M.; Wiendahl, H.-H.; Duffie, N. & Brieke, M. (2007). Changeable Manufacturing - Classification, Design and Operation. *CIRP Annals - Manufacturing Technology*, Vol. 56, No. 2, pp. 783-809, ISSN 0007-8506

Williams, J. H., Davies, A., & Drake, P. R. (1994). *Condition-based maintenance and machine diagnostics*. ISBN 0412465000, Springer, Netherlands

Wischnewski, R. & Freund, E. (2004). COSIMIR Transport: modeling, simulation and emulation of modular carrier based transport systems, *Proceedings of IEEE International Conference on Robotics and Automation. ICRA '04*, pp. 5171-5176, ISBN 0-7803-8232-3, New Orleans, April 2004

Yan, Q.; Oueslati, F.; Bielenda, J. & Hirshey, J. (2005). Hardware in the loop for a dynamic driving system controller testing and validation. In: *Controller System Software Testing and Validation*, Warrendale, Pa.: Society of Automotive Engineers, ISBN 978-0-7680-1573-7, SAE International

10

Hybrid Planning for Self-Optimization in Railbound Mechatronic Systems

Philipp Adelt, Natalia Esau, Christian Hölscher,
Bernd Kleinjohann, Lisa Kleinjohann, Martin Krüger, Detmar Zimmer
University of Paderborn
Germany

1. Introduction

Self-optimizing mechatronic systems with inherent partial intelligence are the research objective of the Collaborative Research Centre - Self-optimizing concepts and structures in mechanical engineering (CRC 614, 2010). A mechatronic system is called a self-optimizing system in this context, if it is not only able to adapt the system behavior to reach a set of given objectives or goals but also can adapt the objectives themselfes (or their weighting) on the basis of an analysis of the actual situation.

Hence, the self-optimization approach promises to leave degrees of freedom in choosing objectives for the system open until runtime. This means that the system can decide upon internal objectives based on external user input and current environmental conditions while the system is running. This is in contrast to a system where all internal objectives are set before the system is started. Having such fixed parameters leads to complicated and overly pessimistic approximations of the parameters that are needed to be set over the course of action that the system will take.

Using self-optimization, leaving that decision open is the key idea of our approach. The external objectives like quality of control, comfort or total energy consumption along with constraints (e.g. maximal peak powers or average power consumption) are still embedded or entered into the system. But settings like distribution of energy usage among subsystems can be determined during runtime based on the actual system conditions.

One application of this approach to mechatronic system design is a novel transportation system that is developed in close collaboration with CRC 614. The core of the system consists of railbound vehicles called RailCabs that enable groups of up to 12 passengers to travel directly without intermediate stops. A test track in a scale of 1:2.5 and two RailCabs in the same scale have been built at the University of Paderborn under the name Neue Bahntechnik Paderborn (NBP; see Figure 1). The RailCab is equipped with a doubly-fed linear motor as well as several innovative subsystems that feature inherent intelligence. Among these subsystems are an Air Gap Adjustment System (AGAS) and an active suspension system described more closely below (see Section 3). Operating parameters of these (sub)systems need to be adapted depending on environmental conditions and properties of the track sections that are being travelled.

Fig. 1. Autonomous shuttles traveling on the NBP test track at the University of Paderborn

This paper will present a hybrid planning approach for realizing self-optimization concepts in complex mechatronic systems using the RailCab as application example. The following scenario will explain the tasks fulfilled by the hybrid planner and the focus of this paper. As the presented mechatronic system is used in a railbound environment, a RailCab has the task to travel from one point on a given track network along several track sections to a destination point. Finding a path along this track network is a discrete planning problem of the hybrid planner that we are not going into more detail here. Our starting point for this paper is a consecutive set of track sections to travel. Along with this comes a power budget that assigns a maximum peak and mean power to each section that must not be exceeded. The task of the hybrid planner is to choose a parametrization or configuration of the control systems that does not violate these constraints and follows a global goal like maximizing passenger comfort. Candidate controller configurations for different mechatronic subsystems like e.g. the active suspension system are determined before system operation using hierarchical optimization (see Section 5). The planner then constructs an initial discrete plan by selecting controller parametrizations for different mechatronic subsystems for each track section to be travelled. The discrete planning part is described elsewhere (Adelt & Klöpper (2009)). Since the controller configurations have been calculated offline, their optimality cannot be assured in any cases due to environmental changes or changes to subsystems happening. Hence, it is necessary to change the configuration and controller respectively, see (Geisler & Trächtler (2009)) for an exemplary model-based approach. However, these runtime changes, that have to be continuously monitored, often result in deviations from the initial plan that will lead to a violation of the overall system objectives and constraints (e.g. minimum comfort or available energy budget in our case). In such situations a replanning is necessary that may also adapt the objectives of the system or its parts. For this purpose the hybrid planner will use simulations of the system and environement model to predict the continuous effects of different controller parametrizations (see Section 6.2). In contrast to traditional approaches the hybrid planning approach presented in this paper, also takes into account the actual system and environment status and is able to adapt system objectives or goals accordingly during runtime.

The paper is structured as follows. After introducing the mechatronic subsystems, air gap adjustment system and active suspension system, used as application example in Section 3, the architecture of the hybrid planner and its use within the application example is explained in Section 4. Subsequently in Sections 5 and 6 the system models needed for optimization, monitoring, control or prediction of system behavior are described. Section 5 introduces the optimization approach by example of the active suspension system while Section 6 concentrates on the measurement, control and prediction of continuous effects using the

AGAS as example. An evaluation of our approach and some concluding remarks follow in Sections 7 and 8.

2. Related work

The term *hybrid planning* has been used in different contexts that describe a combination of general purpose planning with domain specific reasoning (Kambhampati et al. (1993)), combining hierarchical domain properties with other planning concepts (Kambhampati et al. (1998)) or matching planning techniques to business processes (Nutt (1982)). We use the term hybrid planning for the integration of discrete and continuous domains in planning. Techniques used in classical planning like hierarchical task networks and the refinement of plans are usable as part of a hybrid planning system too (Ghallab & Traverso (2004); Russell & Norvig (2003)). Planning of continuous actions, e.g. the planning of trajectories, is not in the focus of this approach to hybrid planning as we deliberately use discrete actions and choose to work on the granularity of the actions. Continuous planning as a constant interleaving of planning and execution is similar to our approach of simulating planned actions with current system state as the basis. Dearden (2005) uses a probabilistic hybrid automaton for diagnosis after initial planning. Thus a discrete plan is checked and evaluated on basis of a hybrid model and afterwards changed or even discarded. This relates to the simulation part in our hybrid planning approach.
The active suspension system of the RailCab is used for hybrid planning in this work. Such active suspension systems for automobiles and also for trains have been extensively studied in recent years. A survey of different strategies could be found in (Li & Goodall (1999)), for example. Active suspension systems are often used to enhance the contradicting objectives comfort and safety simultaneously, which is not possible by means of passive systems. Actual research mostly deals with novel control strategies and methods. Methods related to artificial intelligence as Fuzzy-based controller (Chiou & Huang (2007), Kou & Fang (2007)) or neural networks (Jin et al. (2007)) are used as well as model-based methods as H^2 and H^∞ controller, that can be found in (Yousefi et al. (2006)) for example. Approaches, that are also dealing with energy aspects as we do in our application example, can be found in (Zhang et al. (2007)) and (Stribrsky et al. (2007)).
A description of an air gap adjustment system developed at the CRC 614 is described in (Adelt et al. (2008); Schmidt, Adelt, Esau, Kleinjohann, Kleinjohann & Rose (2008)) and (Esau et al. (2008)).The adjustment of the air gap will be achieved by the usage of additional actuators, working against return springs. An alternative concept of an air gap adjustment system and the related measurement of the air gap are presented in (Gabel (2009)). This concept realizes an air gap control by the usage of the system-inherent normal force.

3. Mechatronic system

At the University of Paderborn a new, innovative traffic and transport concept has been developed (Henke et al. (2008), RailCab (2010)). The basic idea of this concept is to use the existing railway infrastructure with small autonomously driven vehicles, called RailCabs, which are adjusted to the needs of the passengers. These RailCabs accelerate by means of a doubly fed asynchronous linear motor laying in the middle of the existing railway and are equipped with an active guidance as well as an active suspension system.
Implementation of this kind of propulsion requires some kind of information processing at the track and fast communication hardware between the RailCab and the track hardware.

Hence, the track is separated into small sections each one equipped with its own information processing, the so called *Track Sectioning Controls (TSC)*. This TSC is necessary to realize the propulsion but it is also beneficial for self-optimization of the RailCab. It offers the possibility to store track section specific data that can be accessed by an arriving RailCab, see Figure 2. In this paper we concentrate on two kinds of information that are directly related to our hybrid planning approach. On the one hand the TSC provides an actual excitation profile that has been learned previously from former RailCabs. So we are in the situation, that a RailCab knows the future disturbances due to the excitation. On the other hand the measured air gap as well as the calculated air gap, that are both required for air gap modeling and prediction, are delivered by the TSC, too. More details concerning usage and relation to hybrid planning are described in the following sections. All information is updated by the measurements of the leaving RailCab.
In order to continuously enhance the RailCab system, a number of test benches for different components are used to evaluate new methods separately. Therefore we illustrate the hybrid planning approach on the basis of two test benches: the air gap adjustment system and the active suspension system. Their physical functionality is described in detail in the following two subsections at first. Afterwards we present the theoretical methods that we used for the hybrid planning approach.

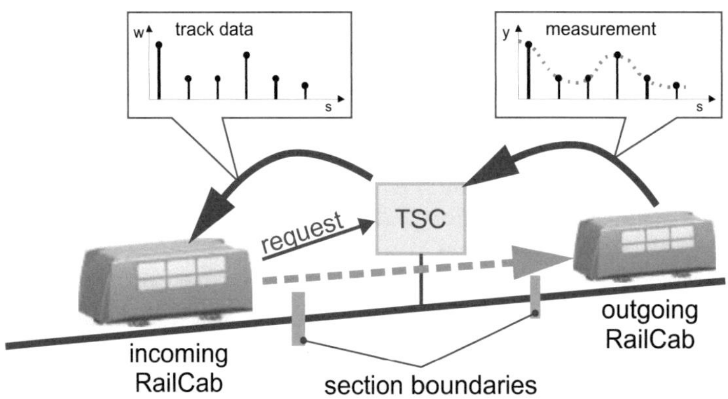

Fig. 2. Data transmission between Track Sectioning Control (TSC) and passing RailCabs

3.1 Active suspension system

The active suspension system performs the task of compensating bumps and other excitations of the railway in order to increase comfort for the passengers. The controller doing this task consists of two parts: a so-called Sky-Hook controller (Li & Goodall (1999)) tries to minimize the absolute coach body acceleration and a relative controller introduces additional virtual spring-damper forces to the coach body. More Details of the controller realization can be found in (Vöcking & Trächtler (2008)). To set up the controller there exists a Hardware-in-the-Loop (HiL) test rig, which emulates the active suspension system of a RailCab, see Figure 3(a).
The structure of the test bench can be partitioned into nine mechatronic modules arranged at three different levels as shown in Figure 3(b). On the top level there is the coach body, which can move in vertical, horizontal and rotational (body roll) degrees of freedom. Beneath the coach body there are two symmetrically allocated actuator groups, each one consisting of a nonlinear guide kinematic, which is connected to a glasfiber reinforced polymers (GRP) spring and three separately controlled hydraulic cylinders that form the third level of the

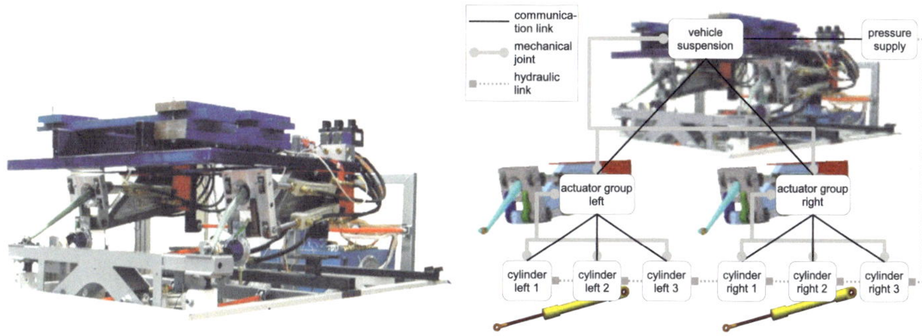

(a) Test rig (b) Partitioning into mechatronic subsystems

Fig. 3. HiL Test rig for the active suspension system

structure. The main function of the actuator groups is to exert forces on the coach body by a deflection of the GRP-springs.
A chassis framework that can again be displaced by three hydraulic cylinders is used to simulate the excitation profile. The pressure in these cylinders is controlled separately by a hydraulic pump and therefore we can disregard the dynamics of the pressure supply and work with constant pressure values for the excitation.
The acceleration of the coach body can be measured by two sensors placed at both sides of the coach body. Furthermore, the relative displacement between the coach body and the chassis framework can be quantified in all three dimensions by two contactless sensors.

3.2 Air Gap Adjustment System (AGAS)

The propulsion system of the RailCab is realized by a doubly-fed asynchronous linear drive. In addition to the force transmission this drive enables a contactless energy transfer into the RailCab. A detailed description of direct drives is located in (Zimmer & Schmidt (2005)) and (Zimmer et al. (2005)). The energy transfer of a doubly-fed linear drive is presented in (Schneider et al. (2009)). The immovable part of the drive in the following named stator is installed in the track bed. The rotor is the movable part of the drive and is connected to the RailCab. The distance between rotor and stator is the so-called air gap δ and the optimization variable of the AGAS. Different influences such as incorrectly laid tracks and setting processes lead to an intense fluctuating air gap (Figure 4).

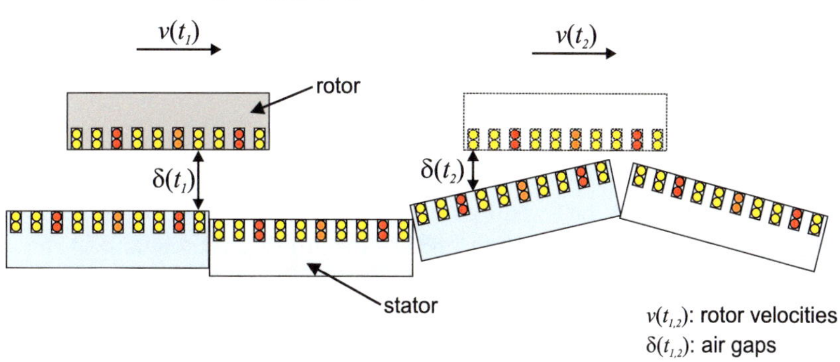

Fig. 4. Air gap over time between rotor and stator

To avoid a collision between rotor and stator a large nominal air gap of 10mm was chosen. This large air gap generates high electrical losses and reduces the efficiency η of the linear drive. The relationship between efficiency and air gap is reciprocal.

$$\eta \sim \frac{1}{\delta} \tag{1}$$

This implies an improvement of the efficiency by minimization of the air gap. The fluctuating air gap at the track motivates a dynamic air gap adjustment.
In order to validate a self-optimizing air gap adjustment a HiL[1] test rig (Figure 5) was developed. In contrast to the RailCab the rotor of the linear drive is mounted at the frame of the test rig and propels the stator elements, which are mounted on the rotating track. The stator elements arranged in a circle allow a continuous track simulation.

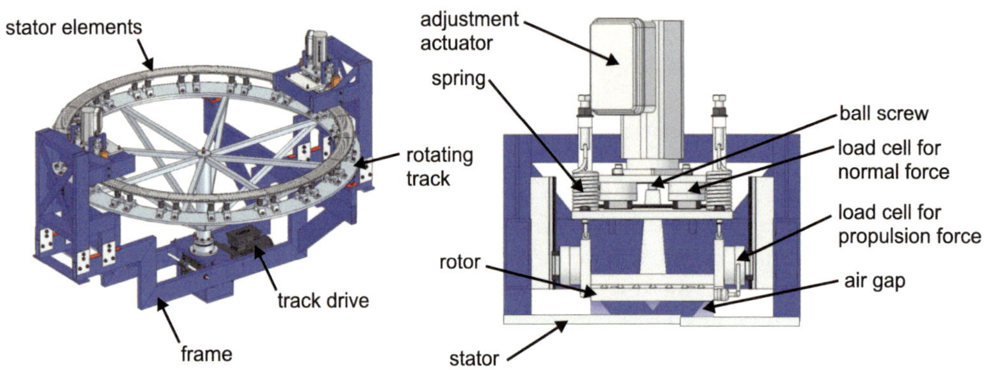

Fig. 5. HiL test rig of the AGAS

The components of the test rig include the adjustment actuators with ball screws. They position the rotor vertically to realize the required air gap. The adjustment actuators work against return springs to minimize the air gap. In case of an actuator failure the linear drive will be raised by the return springs to a safe distance from the stator elements. Hence, a collision between stator and rotor can be avoided. This concept provides the fail safe principle. Load cells measure the normal and propulsion force. They provide the detection of characteristic diagrams for the normal and propulsion force depending on air gap and rotor current. They are used to analyse the system behaviour. The above-mentioned linear drive propels the rotating track, simulating the stator elements laid in the track bed. The stator elements can be adjusted manually in their vertical position to simulate an air gap gradient. The track drive allows to drive or brake the rotating track.

4. Hybrid planning

In this section we will present the hybrid planning architecture. The needed components and information flows are described. We then present the application scenario in the domain of the RailCab in detail and give a step by step description of how the hybrid planning approach is applied.

[1] HiL: Hardware in the Loop

4.1 Overview of the hybrid planning architecture

Planning problems are an integral part of the design and operation of mechatronic systems when it comes to real world problems. Generating a plan that achieves a given objective under a number of constraints and following some global objectives (e.g. save energy, transport in minimal time) appears in most mechatronic systems. Operation needs to be planned along a timeline incorporating several dimensions of continuous physical effects. Under resource constraints like limited energy supply, depending on environmental as well as plan details (e.g. path to travel), consideration of more than the local control problem is needed. A planner needs to take into account these details to generate a plan that is actually executable. This is where hybrid planning is used.

It does not only generate a plan that is feasible under discrete constraints like reaching the target position by traveling along a set of adjacent track sections but it also approximates the continuous behavior of the system while the plan is being built and checks for constraint violations along the way. Doing this with system state updated online with actual data allows for the updating of a plan to reflect the best known approximation of future action's effects. The term hybrid planning tackles an approach to integrate the continuous effects of physical systems without giving up the possibility to build a plan. Discrete planners generally expect the outcome of an action to be deterministic. An action considered to be executed after another can expect a certain state. We extend this scheme by actively approximating the system state that the execution of an action will result in. It bases on discrete planning in that a discrete problem is expected as a basis. Along with the discrete problem, different discrete parameterizations of the planned system can be considered as alternative ways to execute an action.

The way to get the system state including the continuous effects of an action is to run a simulation of the system or the relevant system part. Figure 6 shows the components used to implement such an approach.

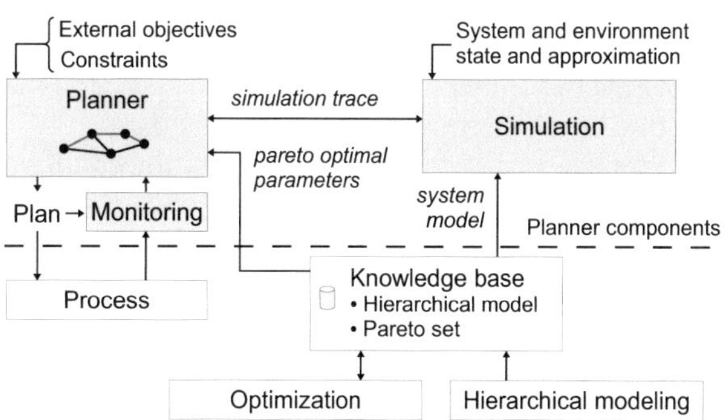

Fig. 6. Components of the hybrid planning architecture

The *planner* is equipped with overall *external objectives* and *constraints* that need to be fulfilled at any time. While creating a plan, the planner initializes a *simulation* with the best known system and environment state for the considered action, parametrizes the simulation and runs it for the projected duration of the planned step. The result of the simulation is a number of continuous value traces that are evaluated according to the constraints and objectives. The constraints are used to rule out an action, e.g. if maximum peak power is too high or comfort value used by the system is too poor during the simulation. Likewise, the global objectives, for

instance the mean comfort or energy consumption of the overall section, are used to evaluate the planned action and to decide upon possible alternatives.
The *environment state* and *approximation* input delivers the necessary information needed for simulation under the expected environmental conditions (e.g. wind, temperature, rainfall). Those parts of the system that no exact physical models are available for (e.g. "'black box"' modules) need to be approximated, too (system approximation). For this purpose we use fuzzy approximation which will be described in more detail and illustrated by an example in Section 6.2.
For system parts where more exact mathematical models exist, the *optimization* and *modeling* components deliver a set of pre-calculated pareto optimal controller configurations and an abstracted model of the system part. The different parameter sets are a discrete dimension of choice to the planner. Due to the applied abstraction the delivered model simulates several times faster than realtime and thus enables a predictive planning for the upcoming track sections. During runtime, delivery of a plan is guaranteed that can be updated to reflect more recent data known about the system state.
During the execution of the plan a *monitoring* component constantly compares the actual system behavior like energy use and damping performance with the predicted behavior. If deviations are detected, replanning may be necessary to choose, for instance, a safer and less energy efficient or less comfortable parametrization. Failing to do this could leave the system running out of energy while in transit, as the models used to simulate system behavior obviously do not reflect the actual system performance.

4.2 Application scenario

As already mentioned in Section 3 we present here a scenario based on some of the subsystems that are being developed around the RailCab: The air gap adjustment system and the active suspension system. They are both parts of the concept of the RailCab and have to work together onboard the vehicle. A central aspect of such mechatronic systems is their energy supply in the form of electrical power that stems from the same system bus. Since we have a mobile system with no overhead contact wire, energy is a scarce, finite resource that needs to be distributed both over time and over the consuming subsystems.
The need to plan ahead of time both distribution of energy is amplified by the fact that settings of one subsystem influence the available amount of energy. We will go on to explain how this works to find out that the scenario actually profits from the use of hybrid planning for the setting of systems.
The RailCab is equipped with a energy storage subsystem that uses capacitors and batteries to bridge times where more energy is consumed onboard that is delivered to it. Obviously this resource is finite. The energy is delivered to the system using the linear motor. In certain operational modes (acceleration and braking) electrical energy can be transferred from the stator elements in the track bed through the magnetic field into the rotor element and can be stored or consumed there. The amount of energy that can be transferred is also dependent on the air gap size between the rotor and stator elements. A larger air gap leads to less energy transfer into the vehicle while a smaller air gap allows the vehicle to extract more energy.
The air gap adjustment system is able to change the height between the rotor element of the linear motor and the nominal axis level. The part that cannot be influenced is the height of the stator element. That has a nominal height relative to the upper rail edge. That level has proven to have a rather large tolerance even on the test track that is hardly used (see the AMV results in (Schmidt, Esau, Adelt & Stern (2008))). An additional factor is the actual

diameter of the wheels that influences the distance between the axis and the rail and so influences the resulting actual air gap between stator and rotor elements. To predict the non-AGAS-originating distance components, the air gap prediction presented in Section 6 was developed. Historical data stored in decentralized track section controls provides best-effort known stator heights measured during earlier runs over the same track section (see Section 3.).

While small air gap sizes are desirable to allow more energy to be transferred, a larger air gap leaves a larger safety area for unexpected stator height deviations from the nominal or last known level. These conflicting goals need to be considered while making a decision upon set points for the AGAS that can then operate to keep the desired level. To make matters worse, every action of the AGAS also needs energy.

The active supension system is the other subsystem considered here. It also draws power from the electrical supply and therefore needs to be conservative with energy usage while providing a maximum comfort for the passengers by damping the coach body from the disturbances introduced by the wheel-rail-contact. This leads to the desire to moderate the available energy usages and objectives between the subsystems.

The solution to these problems presented here as hybrid planning is based on choosing operational modes for the subsystems per section (called configurations) and using simulations of system models to evaluate a plan choice. The scenario focuses on the two subsystems but we will eventually need to take into account all subsystems. Every subsystem is operable with a finite number of configurations, each of them is pareto optimal regarding a set of objective functions. The common objective function here is energy use. Opposing objective functions are for example suspension quality (active suspension system) or safety (AGAS).

Since we don't have a model of the energy storage ready, we will focus on the case of a finite amount of energy onboard the vehicle that must not be depleted by the end of a travel along a network of track sections. Also we will set an upper limit on the peak energy usage that must not be exceeded at any time to keep the electrical system bus from dropping in supply voltage. Under these constraints, we will present the planning of active suspension configurations for a given set of track sections that are consecutively travelled and that exhibit different disturbances to the system. All this under the consideration of maximum peak and mean power usage, validated using system model simulations. At the same time, the air gap prediction emits the air gap that is expected if not influenced by AGAS. The presented scenario emits a plan with optimal suspension performance under the given constraints. An extension to this will be the choice of alternative track sections in the case of constraint violations.

5. Modelling and optimization of the suspension system

For design and analyzing tasks of such complex mechatronic systems as the active suspension system it is convenient to modularize and introduce a hierarchical structure as it is already done in Section 3.1. This allows to focus design activities on an entity or a limited neighborhood of entities that is less complex than the complete system. In the following we give a brief overview of our structuring concept and show how to build a hierarchical model out of such a complex mechatronic system. This model is particularly modified to be used for hierarchical optimization, that is also described in this section. The aim of the hierarchical optimization is to calculate feasible controller configurations that are used for decision making of the hybrid planning as explained in Section 4.2.

5.1 Hierarchical modelling

Often mechatronic systems can be divided into mechatronic subsystems or modules, which can be ordered hierarchically according to their function within the whole system (Lückel et al. (2001)). Such a decoupling is shown for the active suspension system in Figure 3. The information processing of each module can be described by a so called Operator-Controller-Module (OCM, see Figure 7(b)), which is explained in detail in (Hestermeyer & Oberschelp (2004)).

In this work we only need to look at the cognitive operator, which is one part of the OCM. It contains various methods to implement cognitive abilities and usually works under soft real-time constraints. Inside, there is among other things a so called knowledge base, which provides different kinds of knowledge for self-optimizing tasks. A priori there exists only information about the associated module. No information is available about overlying and underlying modules. This encapsulates a single OCM and reduces the dependencies on the remaining OCM. But to fulfill self-optimizing tasks in a hierarchically structured mechatronic system, an OCM has to have information about the behavior of the surrounding modules due to physical couplings. So knowledge has to be exchanged at runtime between the OCMs in the hierarchy.

For hierarchical optimization the knowledge mainly consists of different mathematical models. They are arranged in a hierarchical way, so that they form a hierarchical model. This hierarchical model preserves the hierarchical structure coming from the modularization and achieves the following aims. On the one hand it provides detailed mathematical models for every subsystem in the hierarchy and on the other hand it considers the behavior of the underlying subsystems only in a simplified form. Therefore the complexity and the simulation time is reduced on higher levels.

The modelling task can be divided into two main parts. Firstly a general hierarchical model called base model has to be built. In order to optimize the modelled system, the base model has to be extended by some additional parts, which results in the so-called optimization model. Both models are explained in the following.

In the base model the motion behavior of the corresponding module regarding all relevant dynamical effects is described. The structure of such a base model is shown in Figure 7(a) and consists of detailed, perhaps nonlinear mathematical models of the plant and the corresponding controller. During runtime models from the underlying subsystems are requested, so as to consider their behavior as well. This is necessary because having base models on each level of the hierarchy, which do not know anything about the underlying modules, are inapplicable for hierarchical optimization and hybrid planning. They would produce inaccurate results, as they disregard the physical couplings between the different modules. Conflating all base models to one overall system on higher levels of the hierarchy solves this problem principally, but the resulting model would be very complex and additionally has the disadvantage of increasing simulation time. Hence, only simplified mathematical models, which are generated from the base models, are exchanged between the OCMs. The orders of these simplified models are reduced as much as possible under consideration of retaining all dynamical effects that are relevant for the neighboring subsystems. In our approach we use reduced linear models to transfer knowledge between different OCMs. These reduced models are generated as follows.

The first step comprises of linearizing the base model of an OCM at the current operating point. The resulting linear system is transformed to a minimal realization and afterwards it is simplified by applying a combination of several model reduction techniques. At first the

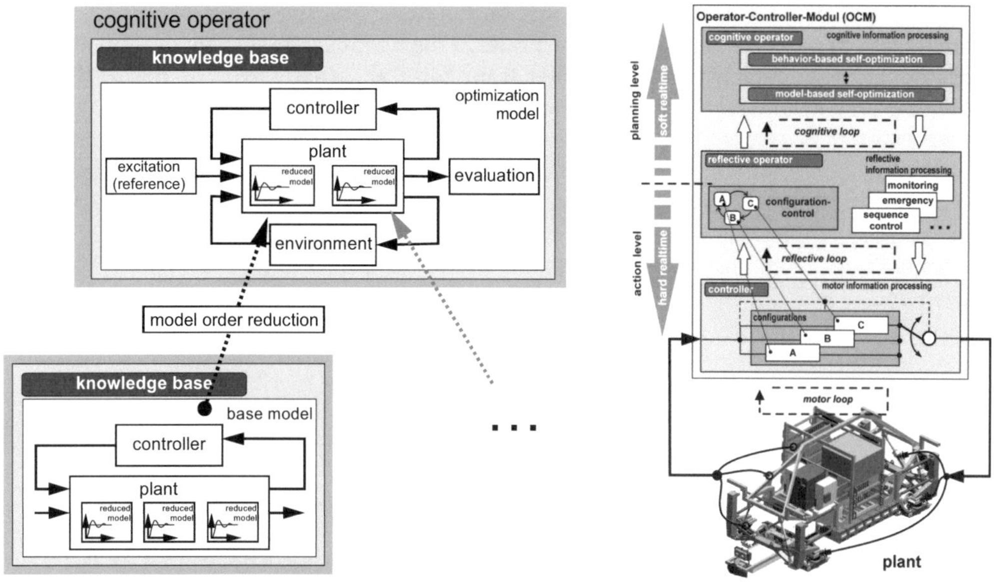

(a) hierarchical model with two levels

(b) Operator-Controller-Module (OCM)

Fig. 7. Information processing for hierarchical modelling

order of the linear system is reduced by *balanced truncation* (Moore (1981)). Afterwards, a *modal model reduction* is used to remove further high eigenvalues systematically. The static error arising from both methods is reduced as much as possible by applying *singular perturbation* (both methods could be found in (Obinata & Anderson (2001)). Currently also methods for *parametric model reduction* are investigated to further improve the hierarchical modelling techniques. These methods go back to a multi-moment-matching firstly suggested in (Daniel et al. (2004)). For more details we refer to (Krüger et al. (2010)).

In order to perform model-based optimization a specific optimization model is needed. Therefore, the base model is extended by an *environment model* as well as an *excitation and evaluation model*. The structure and the interfaces of the resulting model are also shown in Figure 7(a).

The *environment model* emulates the behavior of the surrounding modules in a very simplified manner. It is generated by the overlying OCM and transferred downwards. In contrast to the environment concerning parts in the fuzzy model, presented in Section 6.2, it does not aim to approximate the real environment. It is only used to ensure a correct and reasonable optimization. In fact, this kind of model is necessary, whenever a stand-alone simulation of the base model is not reasonable, e.g. if the base model describes the operating forces of an actuator. This should become clear by looking at the optimization models of the suspension system described subsequently.

The *excitation and evaluation models* emulate excitation or reference signals and calculate the current objective functionals respectively.

A nonlinear base model as well as an optimization model emulating the respective relevant dynamical effects has been built for each module of the HiL test rig of the active suspension

system. Particularly they describe the motion behavior and the hydraulic power. In the following the main elements of each base model in the hierarchy are briefly presented.
hydraulic cylinder: The hydraulic cylinder is modeled as an actuator. The model describes particularly the cylinder and the servo valve including all effects of the control edges. The pressure supply is assumed to act as a constant pressure source and the oil is modeled as compressible. The position of the cylinder is controlled by a PD controller. A single mass, connected with the environment over a spring, works as the environment model, as the base model would be unstable without. A step signal on the input of the controller is used as reference model. Two objectives for *performance* z_{perf} and for *energy consumption* z_{energy} are defined:

$$z_{perf} = \int_{t=0}^{T} |x_{cyl} - x_{ref}| \cdot t \; dt \tag{2}$$

$$z_{energy} = \int_{t=0}^{T} (P_{cyl})^2 \; dt \tag{3}$$

Here P_{cyl} is the hydraulic power, x_{cyl} denotes the cylinder position and x_{ref} the reference position.
actuator group: Each model of an actuator group includes an MBS-model of the guide kinematic and a linear approximation of the GFK-spring. The controller uses an inverse model of the guide kinematic as well as inverse spring characteristics to control the spring forces. There is no feedback controller inside this model. A band limited white noise signal is used as reference force for the actuator group. The objectives are the deviation of the spring forces F_{act} and the reference forces F_{ref}, as well as the energy consumed by the system.

$$z_{perf} = \int_{t=0}^{T} \left| F_{act} - F_{ref} \right|^2 \; dt \tag{4}$$

$$z_{energy} = \int_{t=0}^{T} \sum_{j=1}^{3} P_{cyl,j} \; dt \tag{5}$$

vehicle suspension: The base model contains the coach body and the reduced models of the actuator groups both comprehended as actuators. The controller emulates virtual spring-damper-elements between the coach body and the chassis framework and also the forces of a sky-hook damping. It can be adjusted by nine controller parameters. The calculated forces are referred to the two actuator groups. A band limited white noise signal is used as a track excitation on the chassis. Two objectives for the *comfort* and the *energy consumption* are defined for the entire system.

$$z_{comfort} = \frac{1}{T} \int_{t=0}^{T} |f(a_{coach})| \; dt \tag{6}$$

$$z_{energy} = \frac{1}{T} \int_{t=0}^{T} \sum_{j=1}^{6} P_{cyl,j} \; dt \tag{7}$$

The objective $z_{comfort}$ is calculated from the frequency weighted coach body acceleration a_{coach}. The filter $f(a)$ rates the frequencies of the acceleration signal with respect to the human sensibilities (*VDI 2057 Part 1: "Human exposure to mechanical vibrations – Whole-body vibration"* (2004)).
Due to the model reduction the simulation step size of the base model on the topmost level can be increased by a factor of 5 compared with the nonlinear model; the reduced model on

this level, which is used for hybrid planning, can be simulated more than 56 times faster. More details can be found in (Münch et al. (2008)).
Figure 8 shows a comparison between the hierarchical model and the nonlinear model of the test rig. A controller configuration that leads to moderate power consumption has been chosen and the system has been excited with a vertical step of the railway. At the left the resulting coach body position in vertical direction is depicted. Differences are hardly identifiable in the chosen view, which accounts for the quality of the reduced model. Even the hydraulic power, that is itself a nonlinear function of the system states, can be well approximated as the right part of Figure 8 shows.

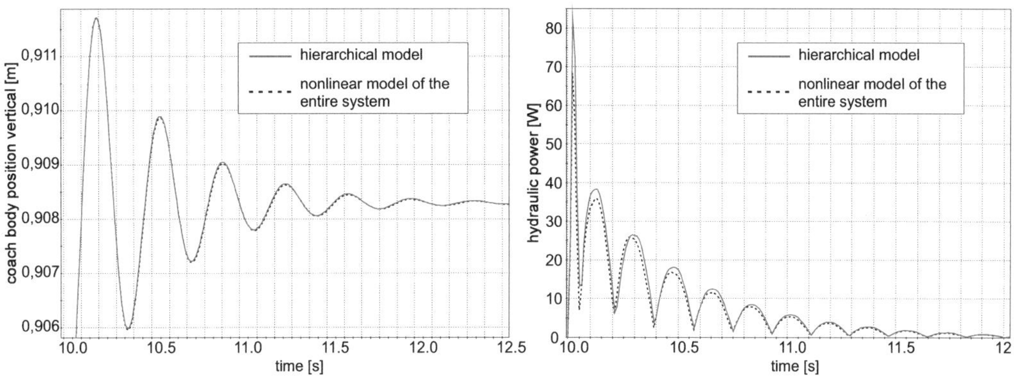

Fig. 8. Hierarchical model of the vehicle suspension compared with a nonlinear model of the entire system

5.2 Hierarchical optimization

A multitude of requirements has to be considered in a mechatronic system. Most of them can be expressed by means of feasible objective functions that can be summarized by the objective vector z. Optimization techniques are used to identify the best suited controller configurations due to the objective functions.
Mathematically, the continuous optimization problem can be formulated as follows:

$$\min_{p} z(p) \tag{8}$$

$$h(p) < 0 \tag{9}$$

Here, $p \in \mathbb{R}^{n_p}$ describes the design parameter space, e.g. the controller parameters and the function $h(p) : \mathbb{R}^{n_p} \mapsto \mathbb{R}^{n_h}$ describes the optimization constraints.
In general, the components of $z \in \mathbb{R}^{n_z}$ do not have a common minimum. Thus there is no unique minimum, but a set of optimal points, the so-called pareto set. Points of this set can be improved in some objectives only by deteriorating others.
There exists a multitude of methods that calculate the pareto set. There are Methods that approximate the whole pareto set, e.g. Dellnitz et al. (2005), as well as methods which compute single pareto points by summarizing the whole objective vector into a single scalar objective function, e.g. Hillermeier (2001). This objective function is then minimized by a conventional optimization method. In this paper we use the *goal attainment method* (Gembicki (1974)) to calculate single points. With this method, a pareto point with a specific objective

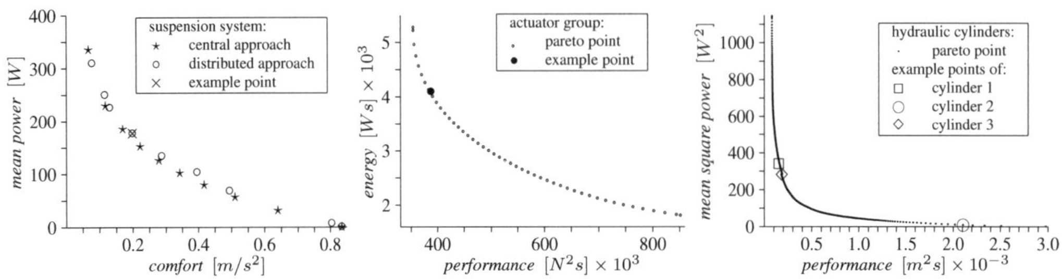

Fig. 9. Pareto sets of the different modules

ratio is selected by a weighting vector. Multiple points of the pareto set are then calculated by repeating the optimization with different weightings.

To optimize the entire hierarchical model, two different approaches are considered: a *central optimization* and a *distributed optimization*.

The central optimization is the simplest approach of optimizing the hierarchical model. Only the objective functionals of the topmost level are considered. Objectives of the lower levels could be regarded as additional optimization constraints if necessary. The optimization, i.e. the controller parameters, are directly transferred from the topmost level of the hierarchy to the underlying modules. Therefore the structure of the whole system has to be accessible in the optimization process.

This contradicts the idea of encapsulated modules that exchange as few information as possible, which has been one aim of the hierarchical modelling. Our second optimization approach, the distributed optimization, overcomes this shortcoming. The main idea of the distributed approach is to decompose the whole optimization problem into several smaller, less complex optimizations. Each OCM performs its own optimization of the optimization model considering only its "local" objectives. It influences the remaining system by setting its own "local" design parameters and also by referring aims to underlying OCMs. Due to the latter the optimizer gets the ability to affect the behavior of the underlying systems. On the basis of its pareto set the underlying system determines the actual controller parameter, adjusts the base model and transfers the new reduced base model back to the overlying OCM. Since the pareto sets of the underlying systems have to be known by the optimizer, the distributed optimization is executed sequentially from the lowest to the topmost level.

As mentioned, the exchange of information between different OCMs is restricted to transferring aims s_i and reduced base models m_i. A direct access to the design parameters or the values of the objective functionals of the underlying systems does not take place. The determination of the actual design parameters is realized in the more abstract way of specifying aims. Thus the exchanged information is highly generic and this leads to a strong encapsulation of the OCM.

The pareto sets of the different modules are plotted in in Figure 9. Due to the symmetry of the two actuator groups their pareto sets are identical and equal aims are used during the distributed optimization process. The pareto sets of the six different hydraulic cylinders are also identical. In the left graph one can see a comparison of both optimization approaches. The central approach performs slightly better than the distributed optimization. This can be put down to the fact that the design parameters of the vehicle suspension are limited to the pareto sets of the underlying system by using the distributed approach.

In order to point out the interrelation between the coach body, the actuator groups and the hydraulic cylinders, an example point of the pareto set of the coach body has been chosen and

the corresponding pareto points of the underlying actuator groups and hydraulic cylinders are marked. Although the three hydraulic cylinders are described by identical mathematical models, the distributed optimization results in different pareto points for each cylinder. This is caused by the different mounting positions of the cylinders. Cylinder 2 has less influence to the comfort, since the optimizer determined a poor *performance* for the benefit of a very low energy consumption.
The pareto set of the topmost level (suspension system) is used for the hybrid planning.

6. Modelling for air gap adjustment and prediction

While the active suspension system in the previous section is a good example for the use of the hierarchical OCM (Operator-Controller-Module) structure, the AGAS subsystem is modelled as a single OCM. It illustrates different kinds of models and their role during the design and operation of self-optimizing mechatronic systems. First Section 6.1 describes a physical model of the AGAS, which is used for system control and for determination of the actual air gap during operation based on measurements of physical parameters, i.e. proper adjustment of the air gap. Such a physical model can hardly consider all parameters influencing a system's behavior or the variety of operating conditions that may arise for instance due to environmental conditions (wear and tear, weather, etc.). These conditions are however important for the hybrid planner to predict the dynamic course of the air gap for future track sections via its simulation component. In such cases a fuzzy model, as described in Section 6.2 for the air gap, is used. It provides a promising alternative to predict the system behavior considering all relevant aspects of such an eventually non-linear system. This fuzzy model is initially based on expert knowledge and may be incrementally improved by measurements gained during operation based on the physical models.

6.1 Physically motivated modelling of the AGAS

To control and optimize the AGAS a physically motivated approach is recommended. Based on (Adelt et al. (2008)) an air gap adjustment system is developed to reach optimal system behavior. It consists of a linear drive and an adjustment actuator wich allows a vertical adjustment of the rotor. The electric drives and the mechanical structure can be described clearly by applying physical methods and techniques.
The optimization of the propulsion system requires the optimal compromise according to the objective functions of the linear drive and the adjustment actuator. The losses of the linear drive P_{ls} and the adjustment power P_{aa} will be minimized. The electrical power P_b fed into the RailCab should be a maximum value. The losses will be reduced by minimization of the air gap and the fed-in power. A small air gap counteracts to the minimization of the adjustment power. This contrast is founded in the return springs located at the rotor. They are working against the adjustment actuator. The optimal compromise between the diverging objective functions will be strived by a multi-objective-optimization. It calculates the optimized vectors for the air gap $\boldsymbol{\delta}_{opt}$, the rotor frequency $\mathbf{f_{r,opt}}$, the RailCab speed $\mathbf{v_{m,opt}}$ and the propulsion Force $\mathbf{F_{m,opt}}$ of the drives. From these vectors the air gap control and the operating point control of the propulsion system computes the setpoint-values for the rotor and stator current i^*_{rq} and i^*_{sd}, the rotor frequency f^*_r of the linear drive, the RailCab speed v^*_m and stator current i^*_{sq} of the adjustment actuator. The rotor speed ω_r and rotation angle ϵ_r of the adjustment actuator will be measured by an integrated resolver. The stator current i_{sq} of the adjustment actuator, the stator and rotor current i_{sd} and i_{rd} of the linear drive will be measured by power electronics. The linear drive temperature ϑ_{ld} will be computed by the drive currents. This

temperature also influences to the multi-objective-optimization because of increasing losses at increasing drive temperature. A direct link exists between the linear drive and the adjustment actuator by the system-inherent normal force F_n (Figure 10). The spring forces F_s are working

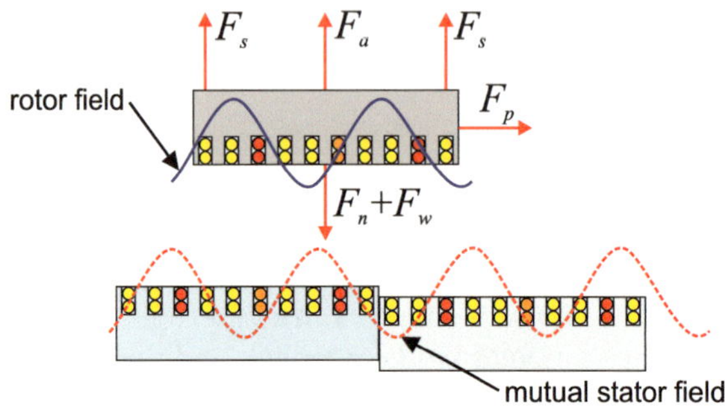

Fig. 10. Forces at the linear drive of the AGAS

against the normal and weight force F_w. Therefore the adjustment force F_a allocated by the adjustment actuator can be reduced. Hall sensors measure the magnetic flux B_s of the stator field (Figure 11). Among other things the air gap δ, rotor frequency f_r and RailCab speed v_m can be determined. A detailed description of air gap and stator frequency measurement will be presented below. The constraints ct of the optimization will be preset by the requested driving profile. Besides these constraints a collision of stator and rotor has to be avoided. The driving profile is composed of the position, the speed and the acceleration which has to be achieved by the RailCab. To avoid a collision of the stator and the rotor the RailCab speed and the maximum of the reachable adjustment acceleration have to be considered. In addition the self-optimizing air gap adjustment system fetches up data from the track section control (TSC) described in Section 3. This information includes inter alia the stator positions of the track section and also have influence on the multi-objective-optimization. The value of the

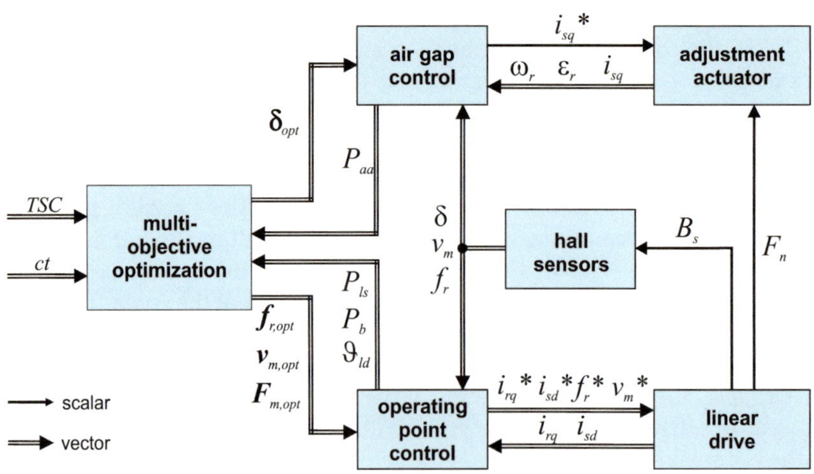

Fig. 11. Block diagram of the AGAS

air gap is also essential for the air gap control. It is possible to model the air gap gradient by a fuzzy-model taking into account enviromental influences as described in Section 6.2. But also a continuous measurement of the air gap is achievable. To measure the air gap of a linear drive different sensors are conceivable. In (Gabel (2009)) the air gap is measured by an inductive distance sensor. Based on (Böcker et al. (2006)) a hall sensor was chosen for the self-optimizing air gap adjustment system. This sensor detects the magnetic flux of the stator field. The arrangement of two sensors with a distance of half a pole pitch allows the detection of the phase angle ϵ_r of the stator field. Therefore the position and the speed of the rotor can be determined in parallel to the air gap. The hall sensor arrangement features a high level of function integration. For measuring the air gap the sinusodial ouput signals will be rectified. The value of the signal will be associated to the air gap by using an appropriate calibration (Figure 12). Air gap gradients will be detected by comparing actual measured data with data of the past.

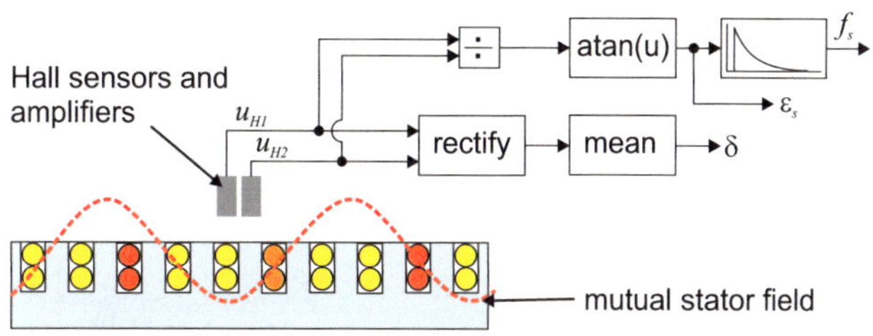

Fig. 12. Hall sensor arrangement

6.2 Fuzzy model for air gap prediction

A fuzzy model of the air gap is used by the simulation component to predict the expected air gap δ_{ag} during system operation given the system and environment state. Two fuzzy models were implemented that reflect the physical system: One model describes the influence of the track and the integrated stator elements on the air gap and a second one describes the influence of the vehicle including the rotor element. Together these models comprise the system's and the environmental conditions that influence the air gap's change.

6.2.1 Fuzzy modelling

The process of modelling was also separated into two steps. During *structure identification* the influencing factors for the shuttle and the track fuzzy model were determined and structured in a so called *influencing factor tree* (IFT, Figure 13). These input variables and the model outputs were fuzzified, and two fuzzy rule systems describing the input-output dependencies for the shuttle, respectively the track fuzzy model, were constructed. The subsequent *parameter adaptation* step is responsible for adapting the parameters of the fuzzy membership functions of the system's input and output variables, for adapting the structure of the IFT and the set of rules and their weights according to new measurement data. Structure identification and parameter adaptation for the AGAS are described below in more detail.

Structure Identification

At first the influencing factors for the two AGAS models shuttle and track are classified and represented as an influencing factor tree. The root node of an IFT represents the model output, i.e. the change of the air gap resulting from the shuttle $\Delta\delta_{shuttle}$ or the track $\Delta\delta_{track}$ in our case. The total change of the air gap $\Delta\delta_{ag}$ can then be calculated as

$$\Delta\delta_{ag} = \Delta\delta_{shuttle} + \Delta\delta_{track} \tag{10}$$

The leaf nodes of the IFT represent the influencing factors, i.e. the models' inputs, and the internal nodes of the tree mirror the classification hierarchy. Figure 13 shows part of the IFT for the shuttle fuzzy model.

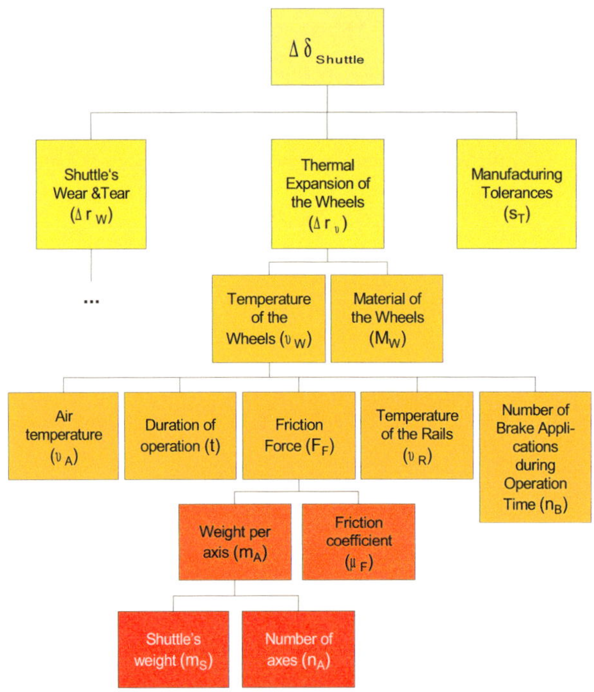

Fig. 13. Influencing factor tree (IFT) of the shuttle fuzzy model

In the shuttle fuzzy model the change of the air gap depends on three classes of influencing factors: *shuttle's wear and tear, thermal expansion of the wheels* and *manufacturing tolerances* (see Figure 13). While *manufacturing tolerances* is a leaf node of the tree corresponding to an influencing factor, *shuttle's wear and tear* and *thermal expansion of the wheels* in turn depend on several classes of factors. As leaf nodes in the sub tree describing *thermal expansion of the wheels* the influencing factors *material of the wheels, air temperature, duration of operation, temperature of the rails, number of brake applications during operation time, friction coefficient, shuttle's weight, number of axes* are distinguished. The classification hierarchy for *shuttle's wear and tear* is omitted due to spatial reasons.

After their classification the influencing factors (leaf nodes of the IFT) first have to be fuzzified. As an initial choice, symmetrical triangular membership functions separating the influencing factors' domains in equal parts are reasonable. The domains and the number of membership functions (ranging between three and seven) were determined by domain experts. The *shuttle's weight*, for instance, may range from 6 tons to 32 tons. It is split into five parts by the membership functions *small, smallmedium, medium, mediumlarge, large*. The *number of axes* for instance, lies between 2 and 8 and is represented by three membership functions *small, medium, large*.

Based on the fuzzification of all influencing factors and factor classes (internal IFT nodes) the fuzzy rule bases for the shuttle and track fuzzy models were developed, relying on the knowledge of domain experts. Each rule base is described by a so called knowledge matrix.

$m_a \setminus \mu_f$	small	medium	large
small	small	small	small
medium	small	medium	medium
large	medium	large	large

Table 1. Knowledge matrix related to friction force

For example, the rule base relating to the *friction force* contains 9 rules, because the influencing factors *weight per axis* (m_a), *friction coefficient* (μ_f) and the *friction force* (F_f) are represented by three membership functions each. The corresponding knowledge matrix is shown in Table 1. The upper row relates to μ_f, the leftmost column to m_a, and the body of the matrix specifies the resulting fuzzy value of F_f.

The knowledge matrix of the *friction force* is completely occupied. But some knowledge matrices were only sparsely occupied. Hence, the initial rule base contained only a few significant rules. For example the rule base of the *temperature of the wheels* has only 17 rules (the complete rule base of the *temperature of the wheels* would contain 2205 rules).

Similar matrices were developed for all leaf and internal nodes of the shuttle and the track IFTs and combined to two knowledge matrices for the complete shuttle and track fuzzy models. The complete rule bases hence reflected the IFT structure in their rule dependencies. The initial knowledge matrix of the shuttle fuzzy model contained 102 rows and the one for the track fuzzy model contained 40 rows. These initial rule bases were tested and adapted during the parameter adaptation phase described below.

Parameter adaptation

The correctness of the initial rule bases was checked by assigning values to the input parameters (influencing factors corresponding to IFT leaf nodes) for which the output (change of the air gap) was known. As expected, the outputs of the two rule bases deviated significantly from the expected ones. In order to find out how parameters e.g. of membership functions should be adapted each rule base was decomposed into sub rule systems corresponding to sub trees in the IFT, that span only two levels of hierarchy and had one root node. Hence, for these sub rule bases the input parameters (corresponding to the leaf nodes of the sub tree) could be directly assigned and their influence on the output (corresponding to the sub tree's root) could be observed, facilitating an adaptation of rules and membership functions. In some cases also new nodes were inserted into the IFT hierarchy. For instance in the sub tree describing the *thermal expansion of the wheels* a new internal node *temperature* was inserted to combine the influencing factors *temperature of the rails* and *air temperature*. After decomposing the models and extending the IFT, the rule bases were extended. For the shuttle fuzzy model 534 rules were created and the track fuzzy model contained 81 rules.

6.2.2 Air gap prediction by the TSC

Starting point for the air gap prediction is the path along track sections selected by the hybrid planner. The driving actions planned for these track sections have to consider several parameters for the AGAS in order to find a tradeoff between safe, energy efficient or energy maximizing operation, that fulfills the given constraints. The track section control (TSC) is responsible for predicting the air gap for a track section that is currently planned to be traveled next. For this purpose it relies on historic data and on the two fuzzy models for the shuttle and the track. This allows arriving RailCabs to request a forecast of the air gap in order to adjust an optimal air gap according to their internal system of objectives. An important aim of the TSC is also to deliver an exact prediction of the air gap's change that depends on many influencing factors like environment conditions, RailCab and track properties. The actual values of these influencing factors are needed as inputs for the *Air Gap Fuzzy Model* and are determined by the *RailCab System*, the *Track System* and the *Environment System*. These calculations have to consider several mutual interdependencies. This structure is depicted in Figure 14.

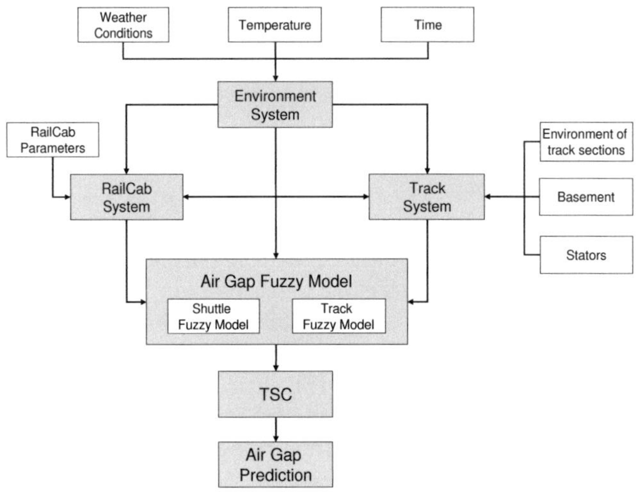

Fig. 14. Principle concept of the air gap prediction by the TSC

The *Track System* determines the factors derived from the track itself, as well as their change, e.g., the wear of the rails. As already mentioned, the *RailCab System* and the *Track System* have also effects on the properties of each other. A track is made up of different section elements like *track switch*, *station*, *bridge* or *normal section*. The track sections are surrounded by one specific environment like *forest*, *ocean*, *shadow*, *normal* and *tunnel*. The basement can be specified to be *gravel* or *concrete*. These parameters influence parts of the fuzzy approximation model for the air gap. Additionally, each section element is characterized by a slope value. A track section consists of a number of stators that have a fixed length and are layed out with a regular spacing. Each stator has *sleeper distance*, *profiling of the rails*, *assembling tolerance* and *manufacturing tolerance* as parameters which are used in the fuzzy approximation model of the air gap.

The *RailCab System* handles all factors which are influenced and determined by a RailCab itself, as for example its weight and the number of its axes. The RailCab parameters for the *Air Gap Fuzzy Model* can be configured as well: *average weight per wheel*, *total distance per wheel*, *total mass* and *number of brake activations*.

The *Environment System* includes weather, temperature and time data which influence on the one hand the *RailCab System* and the *Track System*. For example, rain lowers the friction coefficient between rail and wheel. On the other hand, the *Environment System* provides also the input variables, like e.g. temperature, for both fuzzy models. A forest element, for example, has leaves on the rail in autumn and is generally cooler than open terrain. Through the systems of the *RailCab System*, the *Track System* and the *Environment System* the TSC can adapt the input parameters of the *Air Gap Fuzzy Model* dynamically to the respective circumstances.

7. Evaluation

We are going to evaluate the hybrid planning approach in two parts. The first part will cover the pure simulation based planning results, followed by the results of the air gap modelling based on an example scenario.

7.1 Simulation based planning

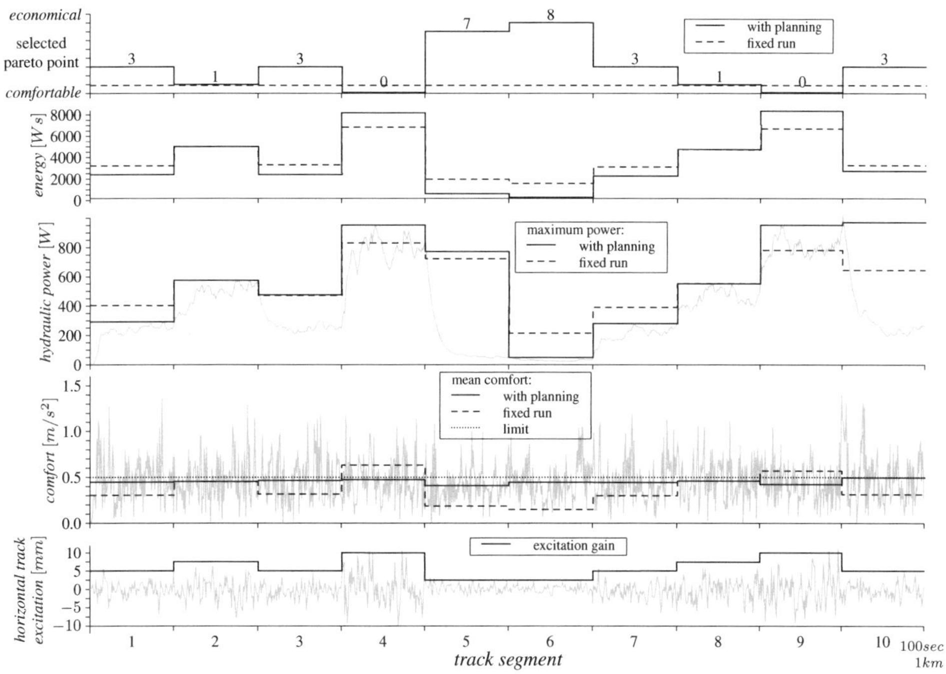

Fig. 15. Energy use (third chart from top) and comfort levels (fourth chart) on planned and static (fixed) configuration. Lower values are better.

Figure 15 shows an example run of a RailCab. It drives with a constant speed over a course with a random but fixed excitation profile. The excitation acts as a disturbance that the active suspension controller needs to damp in order to reach a lower comfort value. The passenger experiences less vibration in a setting with a low comfort value. The system has the choice among ten pareto optimal configurations for the active suspension module resulting from the previous described hierarchical optimization (Figure 9). Each pareto point, numbered from 0 to 9, is a tradeoff between energy use (maximized at *economical*, number 9) and comfort (maximized at *comfortable*, number 0). The topmost chart in Figure 15 illustrates the chosen configuration in the scenario.

The course is divided into ten sections. Each of these sections is configured with a factor that mutiplies the otherwise equally distributed excitation profile, resulting in the excitations plotted in the lowermost chart in Figure 15. This is used to realize different types of rail quality. Larger factors correspond to larger disturbances.

We compare two types of operation. In a fixed run, we choose one configuration for all the sections. The configuration is chosen by maximizing the comfort while not exceeding the energy budget limit. The chosen configuration is pareto point 1.

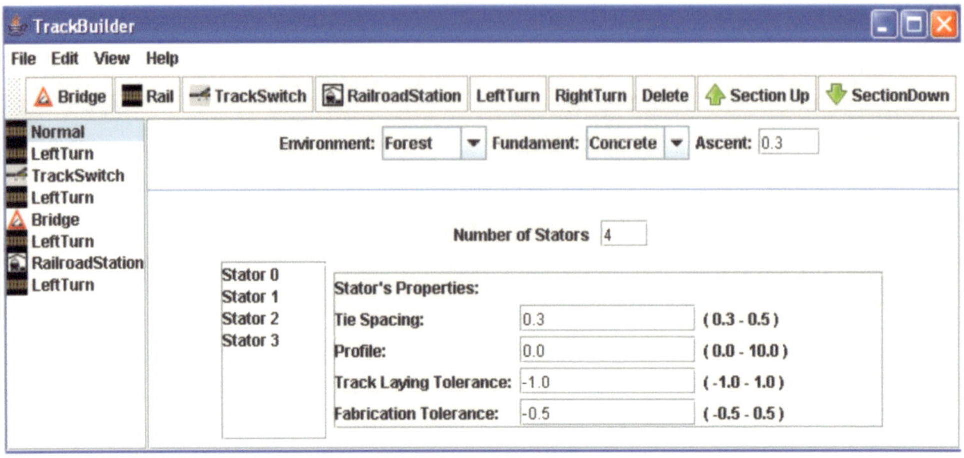

Fig. 16. TrackBuilder interface for building simulated track sections

The other run is the result of the hybrid planner trying to maximize the passenger comfort while staying in all the limits that the fixed configuration had to stay in, too.
The figure shows that the upper limit for the comfort could only be met using the planned version. Using the static configuration, sections 4 and 9 violate the limit. The planned version uses sections with little excitations (1-3 and 5-8) to sacrifice comfort to save some energy. This is then used to improve comfort in the critical sections without violating the energy budget limit. This way, sections 4 and 9 can be travelled with a comfort level well in the set limit. In the planned run the energy use was 6,8% lower. At the same time, the peak energy use was higher.

7.2 Air gap prediction sample scenario

For executing the simulation steps with various parameter values the Track Sectioning Control Simulation System (TSCSS) is realized. The TSCSS is used to test and analyze the air gap system using simulated tracks and simulated environment conditions. The system is split up into two components. The first one called *TrackBuilder* is used to construct a model of track sections (Figure 16). The second component of the TSCSS is the simulation system with the track visualisation of a simulated track network.
The example scenario consists of a course, a RailCab and a sample weather profile. The course consists of 8 track sections with 5 stators each. The sections are a train station, a forest, a bridge, a tunnel, a sea-side section, another train station, a track switch and another forest. These sections have properties, which are influenced by the environment such as a rail temperature. A train station, for example, is roofed, so there is no influence of rain or snow and therefor the friction coefficient between rail and wheel does not lower in this section. In a tunnel there's no influence of solar radiation. As a result, the tunnel conditions do not lead to a rail temperature increase and therefor there is no rail expansion.
We simulate a RailCab travelling over this course and calculate once per stator, what air gap value is predicted. To do this we use properties of the stator, the section, the RailCab and the current air temperature along with the other weather properties. In the simulation the RailCab had to move for 60,000 kilometers. To obtain realistic simulation results we also consider spezific properties of the RailCab like wheel wear. The configuration of the RailCab is that the wheels were profiled each 15,000 kilometers. This leads to a strong increase of a wheel wear.

For each of the 40 stators there is a temperature and an air gap change. In Figure 17 the predicted air gap deviation from the nominal value of 10 mm is shown. The resulting value of the air gap change is between -16 mm and -17.4 mm. So there is a strong deviation from the nominal value because of the high wheel wear.

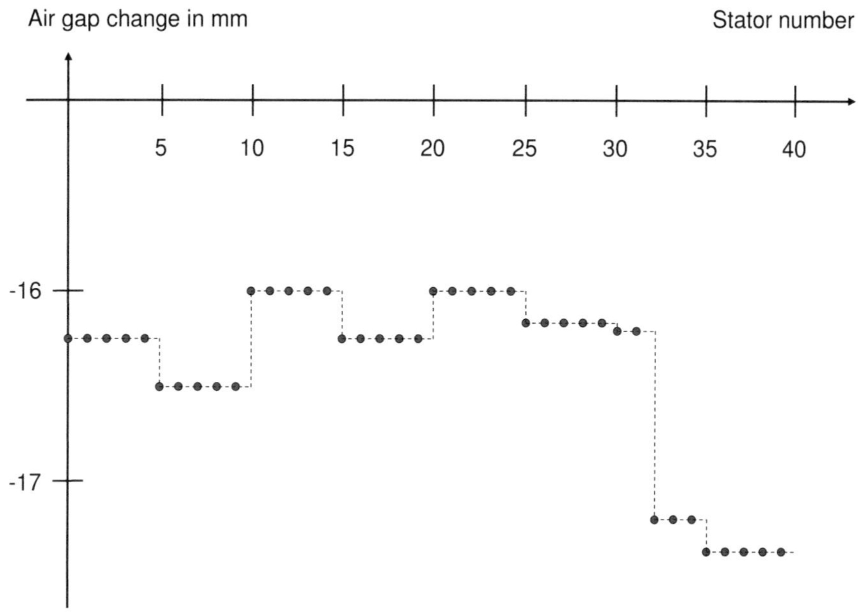

Fig. 17. Predicted air gap change over the simulated course of 40 stators

One noteworthy point is the section with the track switch between stator number 30 and 35. Before and after the track switch, different properties of the stators result in a rather large change in the air gap.
In our presented example, the planner chooses a path along track sections. While considering an action, it considers parameters for the AGAS selecting a preference to maximize the safety. Along with historic data of the TSC, the fuzzy modeled air gap can be simulated for the track section and the actual modeled air gap can be evaluated against safety margins. The prediction we see here could now be used to have the AGAS keep the desired nominal air gap in order to avoid a collision between the stators and vehicle's rotors.

8. Conclusion and outlook

This paper presented a new approach for self optimization based on hybrid (re)planning and its application in a railbound mechatronic system. In contrast to many existing approaches not only discrete actions are planned but also continuous system aspects are taken into account. The planner predicts future system behavior based on mathematical or physical models of system (parts) when possible. However, these are often not available as for instance for environmental conditions. For such cases fuzzy models to predict the behavior of respective system parts - in our example for the AGAS were incorporated. Furthermore, the planner is able to online monitor the execution of a plan and initiate replanning if necessary. Such a replanning can not only change the course of action but also the planning goals which is essential for self-optimization. Evaluations of our approach revealed that hybrid planning with online adjustments to actual system and environment conditions can considerably improve the system performance compared to an a priory fixed plans and fixed parameters from a set of parameter settings determined e. g. by Pareto optimization. Furthermore, it is able to detect impending future violations of system constraints like peak power or comfort level and avoid them by corresponding online parameter and plan adaptations. Ongoing work will try to distribute the planning process along the hierarchical structure of mechatronic systems using the tree-like properties to distribute work load and improve planning results.

9. References

Adelt, P., Donoth, J., Gausemeier, J., Geisler, J., Henkler, S., Kahl, S., Klöpper, B., Krupp, A., Münch, E., Oberthür, S., Paiz, C., Podlogar, H., Porrmann, M., Radkowski, R., Romaus, C., Schmidt, A., Schulz, B., Vöcking, H., Witkowski, U., Witting, K. & Znamenshchykov, O. (2008). *Selbstoptimierende Systeme des Maschinenbaus – Definitionen, Anwendungen, Konzepte.*, Vol. 234, HNI-Verlagsschriftenreihe, Paderborn.

Adelt, P. & Klöpper, B. (2009). Buildings blocks and prototypical implementation of a hybrid planning architecture, *in* B. Klöpper & W. Dangelmaier (eds), *Self-x in Engineering*, MV Verlag, pp. 55–67.

Böcker, J., Henke, C., Schneider, T. & Schulz, B. (2006). Redundante Positionserfassung für ein spurgeführtes linearmotorgetriebenes Bahnfahrzeug, *4. Workshop Entwurf mechatronischer Systeme*, Heinz Nixdorf Institute, Paderborn, Germany.

Chiou, K.-C. & Huang, S.-J. (2007). An adaptive fuzzy controller for 1/2 vehicle active suspension systems, *IEEE International Conference on Systems, Man and Cybernetics (SMC 06), October 08-11, 2006, Taipei*, IEEE Service Center, Piscataway, NJ, pp. 1010–1015.

CRC 614 (2010). University of Paderborn, `http://www.sfb614.de/en/sfb614/`.

Daniel, L., Siong, O. C., Chay, L. S., Lee, K. H. & White, J. (2004). A multiparameter moment-matching model-reduction approach for generating geometrically parameterized interconnected performance models, *IEEE Transactions on Computer-Aided Design of Integrated Circuits and Systems* 23(5): 678–693.

Dearden, R. (2005). Planning and learning in hybride discrete domains - continuous models, *20th International Joint Conference on Artificial Intelligence.*

Dellnitz, M., Schütze, O. & Hestermeyer, T. (2005). Covering pareto sets by multilevel subdivision techniques, *Journal of Optimization, Theory and Applications* 124(1): 113–136.

Esau, N., Adelt, P., Schmidt, A. & Rose, M. (2008). Interval optimization in a hybrid planner context, *Asia International Symposium on Mechatronics.*

Gabel, R. (2009). *Aktive Luftspaltregelung von Antrieben in linearen Transportsystemen*, Cuvillier Verlag Göttingen.

Geisler, J. & Trächtler, A. (2009). Control of the pareto optimality of systems with unknown disturbances, *7th IEEE International Conference on Control & Automation (ICCA 2009)*, Christchurch, New Zealand.

Gembicki, F. W. (1974). *Vector Optimization for Control with Performance and Parameter Sensitivity Indices*, PhD thesis, Case Western Reserve Univ., Cleveland, Ohio.

Ghallab, M. & Traverso, P. (2004). *Automated Planning: Theory and Practice*, Morgan Kaufmann.

Henke, C., Tichy, M., Schneider, T., Böcker, J. & Schäfer, W. (2008). System architecture and risk management for autonomous railway convoys, *IEEE International Systems Conference*, Montreal, Canada.

Hestermeyer, T. & Oberschelp, O. (2004). Structured information processing for self-optimizing mechatronic systems, *1st International Conference on Informatics in Control, Automation and Robotics*, Setubal.

Hillermeier, C. (2001). *Nonlinear multiobjective optimization: a generalized homotopy approach*, Birkhäuser Verlag, Basel.

Jin, Y., Yu, D. & Song, X. (2007). An integrated-error-based adaptive neuron control and its application to vehicle suspension systems, *IEEE International Conference on Control and Automation (ICCA 2007)*, Guangzhou, China, pp. 564–569.

Kambhampati, S., Cutkosky, M. R., Tenenbaum, J. M. & Lee, S. H. (1993). Integrating general purpose planners and specialized reasoners: Case study of a hybrid planning architecture, *IEEE Transactions on Systems, Man and Cybernetics* 23(6): pp.1503–1518.

Kambhampati, S., Mali, A. D. & Srivastava, B. (1998). Hybrid Planning for Partially Hierarchical Domains, *AAAI/IAAI*, pp. 882–888.

Kou, F. & Fang, Z. (2007). An experimental investigation into the design of vehicle fuzzy active suspension, *IEEE International Conference on Automation and Logistics*, Jinan, China, pp. 959–963.

Krüger, M., Scharfenbaum, I. & Trächtler, A. (2010). Parametrische Modellreduktion in hierarchisch modellierten selbstoptimierenden Systemen, *in* J. Gausemeier, F. Rammig, W. Schäfer & A. Trächtler (eds), *7. Paderborner Workshop Entwurf mechatronischer Systeme*, Vol. 272, HNI-Verlagsschriftenreihe, pp. 203–218.

Li, H. & Goodall, R. M. (1999). Linear and non-linear skyhook damping control laws for active railway suspensions, *Control Engineering Practice* 7: pp. 843–850.

Lückel, J., Hestermeyer, T. & Liu-Henke, X. (2001). Generalization of the cascade principle in view of a structured form of mechatronic systems, *IEEE/ASME International Conference on Advanced Intelligent Mechatronics*, Villa Olmo, Como, Italy.

Moore, B. C. (1981). Principal component analysis in linear systems - controllability, observability, and model reduction, *IEEE Transactions on Automatic Control* 26(1): pp. 17–32. IEEE Control System Society, Boston.

Münch, E., Adelt, P., Krüger, M., Kleinjohann, B. & Trächtler, A. (2008). Hybrid planning and hierarchical optimization of mechatronic systems, *International Conference on Control, Automation and Systems*, Seoul, Korea.

Nutt, P. C. (1982). Hybrid Planning Methods, *The Academy of Management Review* 7(3): pp. 442–454.

Obinata, G. & Anderson, B. D. O. (2001). *Model Reduction for Control System Design*, Springer-Verlag, London. ISBN 1-85233-371-5.

RailCab (2010). `http://www.railcab.de/`;.

Russell, S. & Norvig, P. (2003). *Artificial Intelligence: A Modern Approach*, 2nd edn, Prentice-Hall, Englewood Cliffs, NJ.

Schmidt, A., Adelt, P., Esau, N., Kleinjohann, L., Kleinjohann, B. & Rose, M. (2008). Approximation of enviroment models for an air gap adjustment system in a hybrid planning context, *IEEE Multi Conference on Systems and Control*, San Antonio, Texas, USA.

Schmidt, A., Esau, N., Adelt, P. & Stern, C. (2008). Dynamic optimization of track sectioning in mechatronic systems for a hybrid planner, *Proceedings of IEEE International Conference on Mechatronics and Automation (ICMA)*, Takamatsu, Kagawa, Japan, pp. 653–658.

Schneider, T., Schulz, B., Henke, C., Witting, K., Steenken, D. & Böcker, J. (2009). Energy transfer via linear doubly-fed motor in different operating modes, *Electric Machines and Drives Conference (IEMDC2009)*, Miami, Florida, USA.

Stribrsky, A., Hyniova, K., Honcu, J. & Kruczek, A. (2007). Energy recuperation in automotive active suspension systems with linear electric motor, *Mediterranean Conference on Control & Automation (MED 07)*, Athens, Greece.

VDI 2057 Part 1: "Human exposure to mechanical vibrations – Whole-body vibration" (2004). Beuth Verlag, Berlin.

Vöcking, H. & Trächtler, A. (2008). Self-optimization of an active suspension system regarding energy requirements, *International Conference on Control, Automation and Systems*, Seoul, Korea.

Yousefi, A., Akbari, A. & Lohmann, B. (2006). Low order robust controllers for active vehicle suspensions, *IEEE International Conference on Control Applications (CCA 06), Oktober 04-06, 2006, Munich*, IEEE Operations Center, Piscataway, NJ, pp. 693–698.

Zhang, Y., Huang, K., Yu, F., Gu, Y. & Li, D. (2007). Experimental verification of energy-regenerative feasibility for an automotive electrical suspension system, *IEEE International Conference on Vehicular Electronics and Safety (ICVES)*, Beijing, China.

Zimmer, D., Böcker, J., Schmidt, A. & Schulz, B. (2005). Direktantriebe passend ausgewählt, *Antriebstechnik* 02: 2–6.

Zimmer, D. & Schmidt, A. (2005). Linear läuft es schneller, *Antriebstechnik* 02: pp. 2–5.

11

An Evidence Accrual Data Fusion Technique for Situational Assessment

Stephen C. Stubberud[1] and Kathleen A. Kramer[2]
[1]Oakridge Technology
[2]University of San Diego
USA

1. Introduction

Systems have become significantly more complex as more sensors, actuators, and processing can all be contained within a single system. Such systems now process ever increasing amounts of data that can be used to better understand the environment and issues that concern the system. Whether the system is autonomous or involves an operator in the decision loop, processing the data into useful information is of paramount importance. The process of combining this data into a clearer picture is often referred to as data fusion.

These systems are more complex than have been developed in the past. This includes an increase in the number of sensors that monitor not only the external environment of the system but also the current operational state of the system. Systems such as unmanned aerial vehicles (UAVs) and robots may operate according to a set script of actions or be tele-operated but still must have a set of autonomous capabilities to overcome changes in operational conditions, avoid obstacles and other hazards, and be able to identify potential threats or mission criteria.

Often, a single sensor cannot provide all the information that necessary to recognize the issue, event, or target of interest. With multiple sensors, more data is available, but the resulting streams of measurement data produced create a number of issues. The concept of data fusion is applied in these cases to sift through the often voluminous amounts of data, determine which data is important, associate the important data that is related, and combine the data into information that provides an improved understanding of internal or external environment.

Data fusion is the ability to combine information from various sources, e.g., sensor systems, databases, individual perspectives, etc., into a coherent picture that improves the overall understanding of the events of interest. In some cases, such as target tracking or image understanding, specific types of algorithms can be used to fuse the data into a clearer picture. Data in those cases are often of a similar format and can be easily combined. For other problems, such as classification of entities or events, situation assessment, or impact assessment, the data come from a variety of nonrelated sources. In this chapter, a situation assessment data fusion technique for mechatronic systems is developed where the development of relationships of events or objects are determined. The examples that will be addressed with this type of fusion include a condition-based monitoring system

(Brotherton et al., 2002) that looks to relate various reports such as engine temperature and fluid pressures to determine if a correlation between events exists and the tracking of targets to determine if the targets are operating together.

Evidence accrual is one approach to fuse the data for situational assessment. Often, evidence accrual approaches are based on Bayesian taxonomies or networks that attempt to directly relate the probabilities involved. While the underlying probabilistic principles are well understood, implementations frequently use heuristics to overcome some of the practical difficulties, such as uncertainties and the injection of a probability of zero. An alternative approach is an evidence accrual algorithm that uses dynamic systems modeling and a fuzzy Kalman filter (FKF) to propagate information and to incorporate new evidence. The technique has the advantage over typical approaches in that it can provide an uncertainty or quality measure with the results estimate. The technique of the FKF allows for the use of observation or measurement uncertainty to be incorporated into the injection value and produced a level-of-evidence state estimate with an associated uncertainty.

These observations can be direct observations in that they measured the estimated state directly. For example, a pressure sensor can measure the oil pressure in an engine. With a more complex system, however, observations are often indirect. Temperature sensors, pressure sensors, and acoustic sensors can be used to estimate an oil leak and an imminent engine seizure. Also, the observations or measurements may provide only a partial description, such as a bearings-only measurement provides to a target track, or might need to be combined with other measurements to provide a full description of a state element, such as positions over time are used to provide velocity estimates.

In this chapter, this evidence accrual technique is developed and then applied to two systems. The first is an automated situation assessment operation where observations of various objects in the region of interest (ROI) are processed to determine which have the potential to be operating together. The second example looks at the health monitoring capabilities of an autonomous vehicle. The development begins with a detailed summary of data fusion. This is followed by an overview of linear systems theory and the fuzzy Kalman filter. These component technologies are then combined into the evidence accrual system.

2. Situational assessment as a level of data fusion

The Joint Directors Laboratory (JDL) fusion model (Steinberg et al., 1999) decomposes the fusion problem into five subcategories, Level 0 to Level 4. These levels of fusion begin with lowest level, Level 0, that deals with the processing necessary to make measurements from the signals that sensors collect. Level 1 defines the state of the object or event. Levels 2 and 3 are much more complex concepts that build relationships between objects and their impact on other events or objects. Finally, Level 4 is refinement. This the process to reconfigure sensor systems or adapt algorithms to improve performance. As seen in Figure 1, the various levels of fusions not only build on each other but also interact. In this development, Level 1 and Level 2 fusion are the two components that are considered.

In Level 1, the objects are considered targets that are being tracked or classified, obstacles that are detected, events that are identified (such as turbulence or an abrupt change in target behavior), and changes in performance that may indicate a fault or failure in the systems capabilities. These objects are the measurements and the elements that are used to create the Level 2 state.

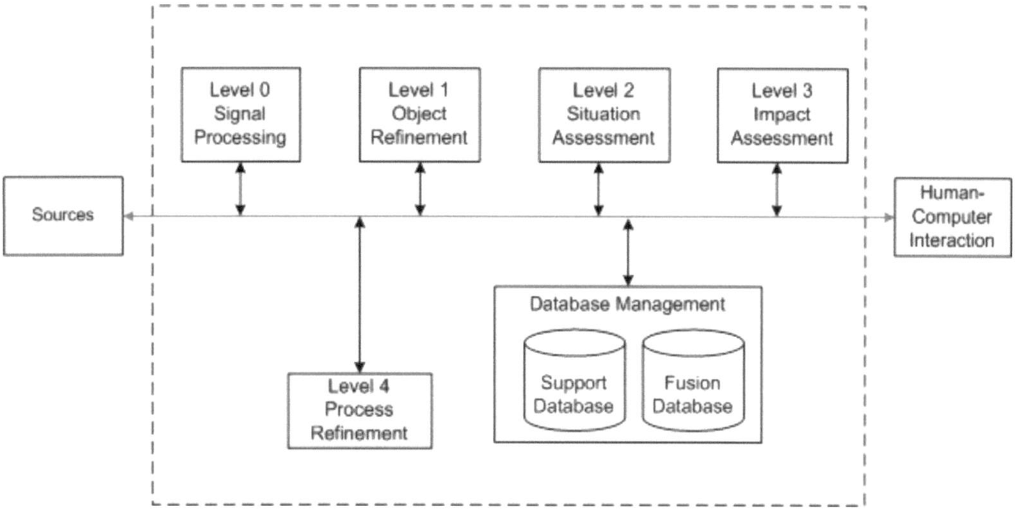

Fig. 1. Situation Assessment, or Level 2 fusion, is a component of the overall data fusion problem.

Level 2 fusion is referred to as Situation Assessment. While there is a significant amount of debate on the definition of Situation Assessment (Hall & Llinas, 2001), for this chapter, Level 2 fusion refers to the development and interpretation of relationships between entities or Level 1 objects. Many approaches have been developed to handle the Level 2 fusion problem. One approach to the Level 2 problem (Stubberud et al., 2003a; Llinas et al., 2004) is an extension of the Bowman model that was developed for Level 1 Fusion. While to some this model implies a Kalman filter implementation, the functional flow can be easily expanded beyond that simplistic viewpoint. As seen in Figure 2, the functional flow of the Level 2 problem is decomposed into a set of functional subcomponents. In this effort, the concentration is on the association component. Unlike Level 1 association that compares measurements to existing tracks as discussed in (Steinberg et al., 1999), Level 2 association must also incorporate interpretation of the relationships.

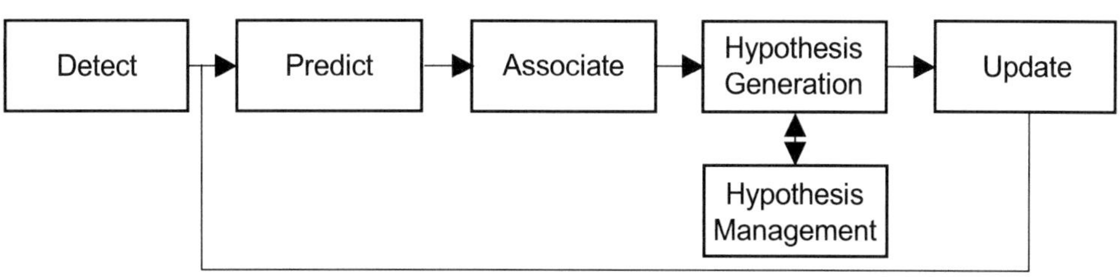

Fig. 2. Level 2 fusion architecture.

The first step to create the relationships is to develop a representation for a Level 2 object. The chosen representation is a state vector that can be comprised of information in a variety of formats, including numeric and linguistic reports. Each state component can be processed using different techniques. For the extension of target tracking, the Level 2 state can be defined as the elements that would define a relationship between targets. No matter the representations or information chosen for a Level 2 object, each subcomponent of the state must have a measure or metric associated with it to develop data association scores with other detected objects and measurements. The basic state structure of a coordinated group would be given as

$$\begin{bmatrix} a & b & c & d \end{bmatrix}^T \tag{1}$$

where

$$a = \begin{bmatrix} x_{unit} \\ y_{unit} \\ x'_{unit} \\ y'_{unit} \end{bmatrix}, \ b = \begin{bmatrix} \#ofclass_1^1 \\ \vdots \\ \#ofclass_1^m \\ \vdots \\ \#ofclass_n^1 \\ \vdots \\ \#ofclass_n^m \end{bmatrix}, \tag{2}$$
$$c = \begin{bmatrix} formation \\ formation\ change \end{bmatrix}, \text{ and } d = \begin{bmatrix} group\ extent1 \\ group\ extent2 \end{bmatrix}$$

This state representation of this group contains four pieces of group information: kinematic, composition, formation, and extent. The extent component represents the physical footprint of the group along with a range of influence which may be the range of weaponry, communications, or the influence of command. In (Stubberud et al., 2003b; Subberud & Shea, 2003), association metrics were developed for each component of the state vector defined in Eqs. (1) and (2). The state vector provides a group representation based on Level 1 information developed for individual targets such as classification and kinematic states. Using other knowledge such as doctrine and inferences such as road structure, the Level 1 objects are combined into Level 2 objects.

Most Level 2 fusion techniques have been developed combine various data including numeric, inference, linguistic, heuristic, and fuzzy (Hall & Llinas, 2001; Chen, 1996). While each of these techniques has been shown to work well as defined, most implementations were developed without consideration of the uncertainty associated with the measurements that created or are fused into them. More effective fusion would provide for means of uncertainty or error estimates in the information provided by the Level 1 object states to influence the determination of the relationship between objects. Poor kinematic estimates as opposed to those with a better pedigree should be weighted less.

The Level 2 fusion approach developed in this chapter is designed to incorporate the uncertainty of the information as well as in the understanding of the process used to fuse the information.

3. Evidence accrual for situation assessment

Evidence accrual is a technique that has been applied often to the Level 2 fusion problem, as in (Buczak & Uhrig, 1995; Hammell & Sudkamp, 1998; Le Hegarat-Mascle, 1997; Stubberud & Kramer, 2007, 2008). The section begins with a discussion of two standard techniques that have been used in the evidence accrual application, Bayesian Taxonomy and Dempster-

Shafer. A newer technique, referred to as Feature Object Exploitation (FOX), can incorporate uncertainty measures and is based on a combination of linear systems theory, fuzzy logic, and Kalman filtering is then developed. The development of the evidence accrual architecture for the FOX approach shows that the data injection is provided by the FKF. The propagation of the data through the architecture is generated using linear systems theory. The FOX approach provides for a variety of input formats, can utilize partial information observations, and can incorporates an uncertainty measure into each state estimate.

3.1 Bayesian taxonomy and Dempster Shafer

Bayesian taxonomy (Pearl, 1988) and the Dempster-Schafer method (Dempster, 1967; Shafer, 1976) are the two classification techniques that are considered standards in the Level 1 fusion community (Blackman & Popoli, 1999).

3.1.1 Bayesian taxonomy

The Bayesian taxonomy is theoretically straightforward in its implementation with the propagation of probabilities from an injection of new information throughout the tree-like structure, as shown in Figures 2a and 2b. The purpose of the Bayesian classification tree is to provide a probability belief in the classification of a target. The architecture selected to create the Bayesian classification tree is the taxonomic hierarchy, also known as a Pearl tree. The Pearl tree is an n-node (as opposed to binary) tree structure whose nodes represent exhaustive and mutually exclusive hypotheses. Each such node is initially assigned an a priori probability, also known as a measure of belief or a score, reflecting the prior probability that the hypothesis is true given all previous evidence. These measures of belief range from 0, indicating no confidence, to 1, indicating complete confidence. The probability at the root node of the tree is always 1, while the measure of belief at each intermediate node is the sum of the scores of its immediate children.

Figure 2a illustrates a simple Pearl tree for target classification. In this example, the numbers included with the node names represent the measures of belief. Clearly, the previous evidence suggests that the target is a hostile aircraft. New evidence can be "injected" into one or more nodes. This evidence takes the form of likelihood ratios, where confirming evidence is greater than 1 and non-confirming evidence is less than 1. The impact of the evidence on each node in the hierarchy is calculated by applying propagation-based updating rules that specify messages to be passed between nodes. Information propagates up and down the tree in such a way that "cutting" across the tree in any way results in a sum of one. For example, confirming evidence with a likelihood of 1.9 is injected into the tree presented in Figure 2a at node "Air Superiority Fighter".

BEL(S), the probability or belief at that node S, is updated via

$BEL^{+}(S) = \alpha_S \lambda_S BEL(S)$, where λ_S is the likelihood ratio and $\alpha_S = \dfrac{1}{1-(1-\lambda_S)BEL(S)}$.

This value is then propagated up and down the tree using Bayesian conditional probabilities. The result of applying the propagation-based updating rules would yield the revised tree shown in Figure 2b. In this example, the injection of the confirming evidence at node "Air Superiority Fighter" caused the shaded nodes' probabilities to increase. The probabilities for all other nodes, with the exception of the never-changing root node, decreased as a result of this confirming evidence.

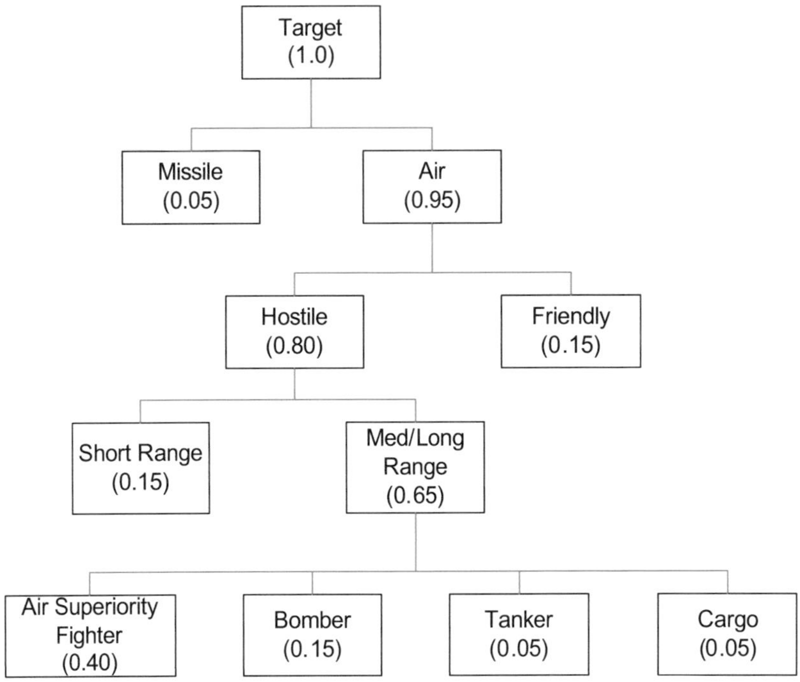

Fig. 2a. Bayesian Network with *A Priori* Probabilities

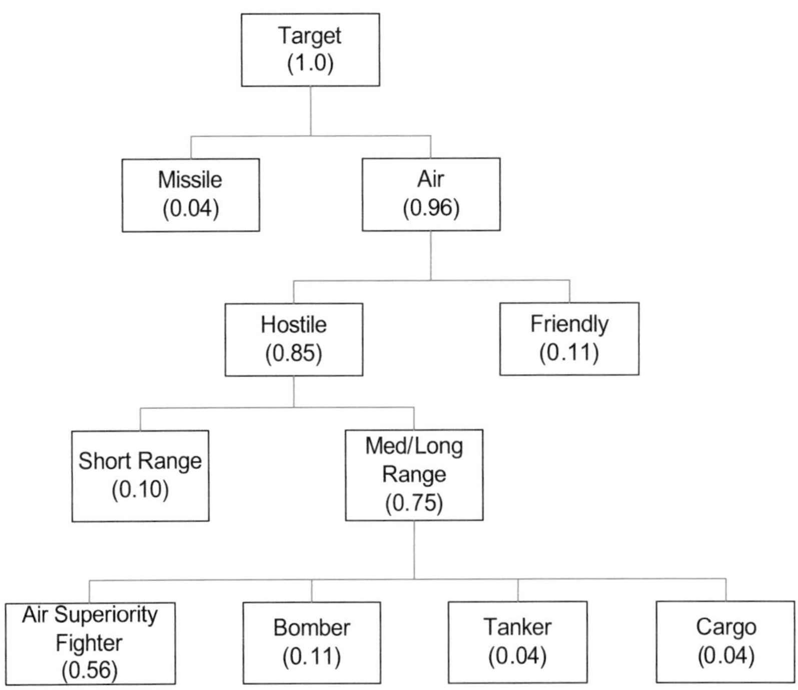

Fig. 2b. Bayesian Network after Processing Confirming Evidence of Type "Air Superiority Fighter".

3.1.2 Dempster-Schafer

The Dempster-Schafer method is a generalized Bayesian inference which can incorporate uncertainty. It is a more complex approach that can incorporate both uncertainty and lack of information into its model. The result provides not only a probability of a given solution, but also a measure of its plausibility. Often, this technique is referred to in the literature as an evidential approach. When the information (input from sensors, sources, and the models used) is independent, the approach is a powerful tool in providing accurate solutions.

Unlike the Bayesian taxonomy, each sensor is allowed to contribute information based on its own capabilities. Assume a set of target types $\{t_1,t_2,...,t_n\}$ that are mutually exclusive and exhaustive. A probability mass is assigned to any element of the set and any possible disjunction, where a disjunction is the event of a subset based on an OR operation. So a disjunction can be the event that Target 1 or Target 2 occurs. Negation indicates the opposite of a proposition. The sum of the probability mass of all propositions, disjunctions, and negations that are defined is equal to 1. The representation of the uncertainty is given as the disjunction of all of the original propositions, θ .

In one example t_1 is defined as a hostile missile, t_2 a neutral launch, and t_3 a friendly missile. A report of an unannounced launch is provided. The mass assignment is given as

$$\mathbf{m}_1 = \begin{bmatrix} m(\theta) = 0.02 \\ m(t_1, t_2) = 0.98 \end{bmatrix}$$

The probability that all three missile types could have occurred is thus 2% and the probability that it is a hostile or neutral is 0.98. The uncertainty in this case is quite small.

A second sensor in this example determines classification based on the plume of the missile. The mass assignment for this sensor is given as

$$\mathbf{m}_2 = \begin{bmatrix} m(\theta) = 0.2 \\ m(t_2, t_3) = 0.4 \\ m(t_1) = 0.4 \end{bmatrix}.$$

Using information from both sensors, the classification becomes

$$\begin{aligned}
m(\theta) &= m_1(\theta) \cdot m_2(\theta) = 0.004 \\
m(t_1) &= m_1(\theta) \cdot m_2(t_1) + m_1(t_1, t_2) \cdot m_2(t_1) = 0.008 + 0.392 = 0.4 \\
m(t_2) &= m_1(t_1, t_2) \cdot m_2(t_2, t_3) = 0.392 \\
m(t_{1,}, t_2) &= m_1(t_1, t_2) \cdot m_2(\theta) = 0.196 \\
m(t_2, t_3) &= m_1(\theta) \cdot m_2(t_2, t_3) = 0.008
\end{aligned}$$

Inconsistencies between reports can be incorporated when using Dempster-Schafer by increasing the uncertainty. This is done by summing the multiplications of all inconsistencies. For example, this would be a sensor m3 that only classifies each individual target. Thus, when fused with Sensor 1, the inconsistency of m(t2) and m(t1,t3) would be computed. This value is then subtracted from 1. The resulting value then divides the other values.

3.1.3 Drawbacks of probabilistic classification algorithms

Both of these probabilistic techniques have drawbacks to their applications. First, when the measurements are not independent, information is repeatedly injected into the system, thus skewing the results. This condition is often a result of improper modeling of sensors or the sensors' relationships to one another. With a Bayesian taxonomy, a concern arises when the probabilistic models of the sensor reports to the potential class are not properly defined or are not consistent across sensors. For example, a given classifier may be developed on the premise that the probability of a class is based on the number of vehicles of each class that would be seen for a given event. With other sensors using the same approach, multiple classes that all have similar characteristics but different weighted probabilities could be adversely affected when the information is combined. Usually, this results from the misunderstanding of the probabilities or from the fact that the models for different sensors are developed by different individuals with different viewpoints of the problem.

Another concern is that the inputs of these probabilistic techniques require numeric information. If the information is symbolic or linguistic, it must be mapped into numeric information in the form of a crisp value with a possible associated uncertainty measure. The Bayesian taxonomy technique has difficulty in incorporating uncertainty about measurements or the quality of the information sources. The solution is to either to spend the time to properly model the various uncertainty conditions into the underlying probability density functions or to use ad hoc methods. The modeling of the uncertainty can be quite complex as environmental and sensor nuances are not always readily understood in a sensor's deployment. When ad hoc techniques are used to correct this problem, the underlying probability foundation is lost. Both the Bayesian taxonomy and the Dempster-Schafer method have the problem that they are confined to the axioms of probability, requiring that the evidence does couple across classes when normalization takes place. Even with Dempster-Schafer, the effects of evidence that only pertains to one class will have an indirect effect on all of the others.

Finally, in both techniques, the formal application of the techniques is computationally complex. The Dempster-Schafer method, in particular, is also known to be computationally intensive and, for large number of classes, the method is avoided.

To overcome these drawbacks, a fuzzy logic approach is developed. According to (Kosko, 1992) and (Watkins, 1994), fuzzy logic can be considered a super set of Bayesian probability. Based on the Bayesian similarities of fuzzy logic developed in (Watkins, 1994), a fuzzy approach to Level 2 fusion based on well known techniques such as Kalman filtering and systems theory is developed.

4. Evidence accrual architecture

To develop the evidence accrual algorithm, a tree structure similar to that of the Bayesian taxonomy (see Figure 3) is developed. Each node is a Level 2 object or an important element of the development of a Level 2 object. The links between nodes represent the functional relationships between the various Level 2 objects and elements. The top node is the highest Level 2 object of interest.

The tree structure allows for a decomposition of the fusion problem into smaller sub-problems that are less complex to develop than creating a single Level 2 object from all of the inputs and databases. The measurements are injected at the different levels using a fuzzy Kalman filter. The FKF provides an ability to incorporate numeric and nonnumeric data as measurements. It also allows for the incorporation of uncertainty measures into the state vector. The data is propagated up the tree structure using systems theory. This provides for both state propagation and uncertainty propagation. Linear, nonlinear, and fuzzy system approaches are all able to be used along the links.

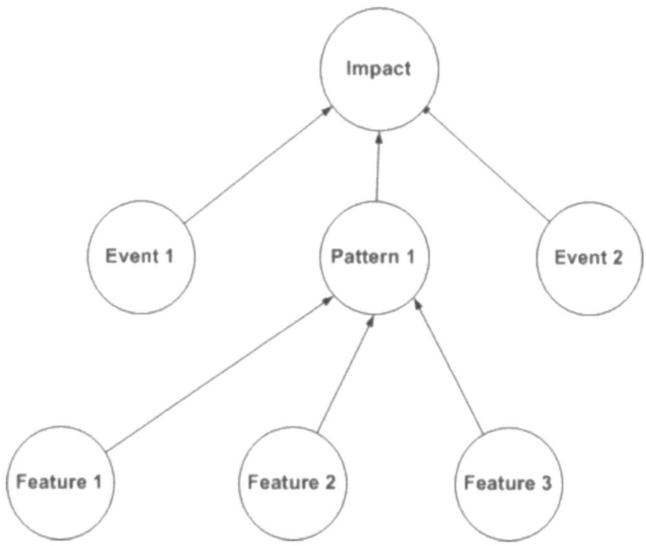

Fig. 3. Tree structure of levels of information provides an impact estimate.

5. Fuzzy Kalman filter

The Kalman filter is the standard estimation technique for dynamic systems. It is a measurement-driven technique that provides an estimate of the state representation of the object being observed and an associated uncertainty to that state. The Kalman filter, if properly designed, can incorporate various measurement types into the state vector as different sensor measurements and observations become known. The use of a fuzzy Kalman filter allows for nonnumeric information, as well as nontraditional information, to be mapped into the impact assessment object state. The ability to inject observations from these various information sources provides the proposed evidence accrual system the capability to determine understanding of impact.

The standard Kalman filter equations are given as

$$K = P_{k|k-1} H^T (H P_{k|k-1} H^T + R)^{-1} \tag{3a}$$

$$P_{k|k} = (I - KH) P_{k|k-1} \tag{3b}$$

$$P_{k|k} = (I - KH) P_{k|k-1} \tag{3c}$$

$$x_{k+1|k} = F x_{k|k} \tag{3d}$$

$$P_{k+1|k} = F P_{k|k} F^T + Q_k \tag{3e}$$

If the measurements are represented as fuzzy membership functions, they cannot directly be incorporated into the Kalman equations. In (Watkins, 1994) it was shown that a fuzzy membership function can be incorporated into a linear vector equation through the relationship

$$\frac{\int (A+Bz)m_{adj}(z)dz}{\int m_{adj}(z)dz} = A + B\frac{\int z\, m_{adj}(z)dz}{\int m_{adj}(z)dz} = A + B\, mom_1\left(m_{adj}\right) \quad (4)$$

where m_{adj} is the informative fuzzy set of the measurement and mom_1 indicates the first moment. From this development, it follows that the fuzzy measurement can be incorporated into the Kalman filter through the update equation, Eq. (3b) as

$$x_{k|k} = x_{k|k-1} + K(mom_1(m_{adj}) - Hx_{k|k-1}) \quad (5)$$

The incorporation of this uncertainty, also referred to as the measurement error covariance, significantly departs from the approach developed in (Watkins, 1994) where the argument is made that only measurements are fuzzy and all other components of a system are crisp. The uncertainty associated with a measurement is represented as a set of fuzzy membership functions. This results from modeling the quality of the reports provided as the lower level classifiers that feed the centralized classifier are incorporated. Often, these classifiers are given a linguistic quality score that must be mapped into a matrix of values to be processed by the Kalman filter. To incorporate the fuzzy membership representation of the uncertainty, it is noted that a linear matrix equation can also be considered a vector equation. The concept of Eq. (4) is again applied it to the Kalman gain equation, Eq. (3a). First, it is noted that

$$(HP_{k|k-1}H^T + R) \quad (6)$$

is a linear vector equation. Therefore, the fuzzy gain equation becomes

$$K = P_{k|k-1}H^T(HP_{k|k-1}H^T + mom_1(m_{adj}(R)))^{-1} \quad (7)$$

With these two methods of incorporating fuzzy logic into the Kalman filter iteration, the evidence accrual system can be developed.

6. Linear systems theory

The evidence accrual technique begins with the development of tree structure, as seen in Figure 3. In this simplified view of an impact assessment problem, the top node represents the final state of interest for the user. The lower nodes indicate the evidence state of information that is used to comprise the lower level data. The state vector of each node would be defined as

$$\mathbf{x} = \begin{bmatrix} xa & x_{s1} & x_{s2} & \dots & x_{sn} \end{bmatrix}^T \quad (8)$$

where xa is the state of the impact object report while xsi are the subcomponents of the impact object that are created from information provided by the subnodes of Figure 3. This information is injected into the system using the fuzzy Kalman filter (FKF). The state elements of the state vector represent the evidence or values of interest of the impact object. The final state, xa, is always a level of evidence that represent the Level 3 fusion object that is reported to the end-user.

The evidence accrual system relies on the concept of first-order and low-order observers (Santina, 1994). In this approach, the state elements are decoupled. Most of the state elements related to the subnodes do not interact. Figure 4 provides a system diagram of this. The state elements can be updated using a scalar equation or first-order equation:

$$x(k+1) = f\,x(k) + \mathbf{g}^T \mathbf{y}(k) + \mathbf{h}^T \mathbf{u}(k) \tag{9}$$

For this evidence accrual application, $\mathbf{u}$(k), the external input, is set to 0. When using the FKF rather than an observer, the gain $\mathbf{g}$ associated with the measurement $\mathbf{y}$ is the Kalman gain. If states are coupled, then a vector-matrix version of Eq. (8) is used. The benefit of this approach is that problem is decomposed into a number of smaller and more computationally simpler problems. The state *xa* is the only coupled state. It has no direct observations. It is updated similar to the velocity state when only position information is available (Blackman, 1986).

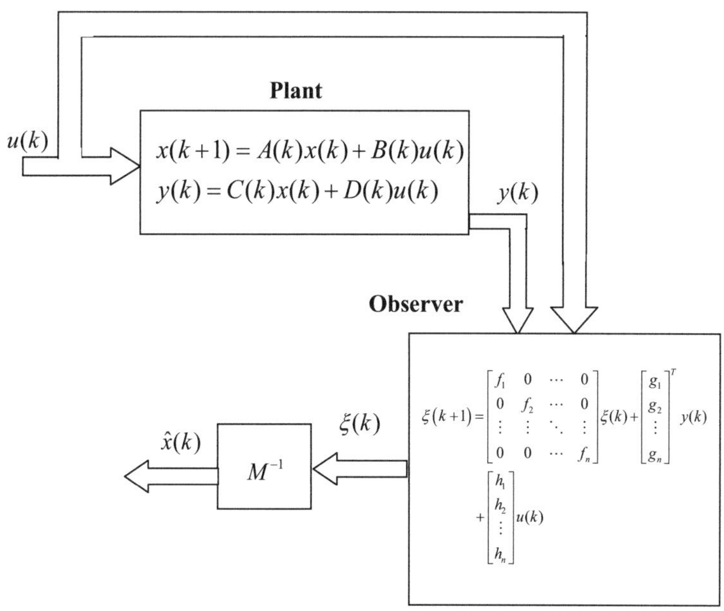

Fig. 4. The evidence accrual approach emulates the concept of first-order observers.

The dynamics of the system that is used to generate a state for a given node is given as

$$\mathbf{x}_{k+1} = \begin{bmatrix} f(xa, x_{s1}, \ldots, x_{sn}) \\ \begin{bmatrix} 0 & f_1 & 0 & \cdots & 0 \\ 0 & 0 & \ddots & \ddots & \vdots \\ \vdots & \vdots & \ddots & \ddots & 0 \\ 0 & 0 & \cdots & 0 & f_n \end{bmatrix} \end{bmatrix} \mathbf{x}_k \tag{10}$$

The benefit of the first-order estimator is that each fusion object can be computed separately based on the available observations. The timestamp k is not an actual time but indicates an event. This approach is based on the implementation of scan reports in target tracking (Buczak & Uhrig, 1995) and event-driven control concepts used in engine control (Skjetne et al., 2002). This allows for the new information to be incorporated into the state vector and the state element xa when it is available.
The significant function of Eq. (10) is $f(.)$. This function maps the observation states into the evidence level for the given class. In this effort, a linear combination is assumed for the evidence combination of Eq. (10) and is defined as

$$x_k = f_a \; xa_{k-1} + f_1 \; x_{s1,k-1} + \cdots + f_n \; x_{sn,k-1} \tag{11}$$

In the covariance equation of the prediction, the relationship becomes

$$P_k = f_a \; P_{a,k-1} + f_1 \; P_{s1,k-1} + \cdots + f_n \; P_{sn,k-1} + q_x \tag{12}$$

If the function f(.) is nonlinear, the covariance equation becomes

$$P_k = \frac{df}{dxa} P_{a,k-1} + \frac{df}{dx_1} P_{s1,k-1} + \cdots + \frac{df}{dx_n} P_{sn,k-1} + q_x \tag{13}$$

The variable q_x indicates the quality of the evidence based on the combination. In other words, the level of understanding of the functional relationship of the various attributes to the classification of a given target type affects the size of the value of q_x. Poorer understanding inflates the value while greater understanding reduces it. The output value at any level can be a level of evidence or an actually computed value. This computed value may be a figure of merit or the output to an equation such as a position or velocity.

7. Evidence accrual for situation assessment of multiple targets

As part of system automation, the ability to understand the local issues is paramount. One part of this understanding is situation assessment of the coordination and independence of non-cooperative entities in one's region of interest. As seen in Figure 5, multiple targets are operating in a region. In operations, coordinated behavior can only be assumed based on observations. In this scenario, five targets are moving together in a formation. Three targets are traversing a road with one target not travelling in conjunction with the other two as its speeds vary. Finally, three targets appear to be converging on a specific point from different angles. The sampled track points are listed in Tables 1a and 1b. The reported classifications are shown in Tables 2a and 2b.
The kinematic Level 1 reports provide target track location and uncertainty as well as a classification with an associated quality, e.g., expressed as a value from 0 to 1 with 1 being absolute confidence. The second functional component will use location and classification metrics defined in (Stubberud et al., 2003b; Subberud & Shea, 2003). These include a classification cardinality metric (Kelley, 1961), a classification Gap metric(Kelley, 1961), and a group distance metric for position and velocity (Stubberud et al., 2003b). The targets are defined as military vehicles including a transporter erector launcher (TEL). The classification of each target can be used to estimate composite groups to determine if the group falls within standard operational units.

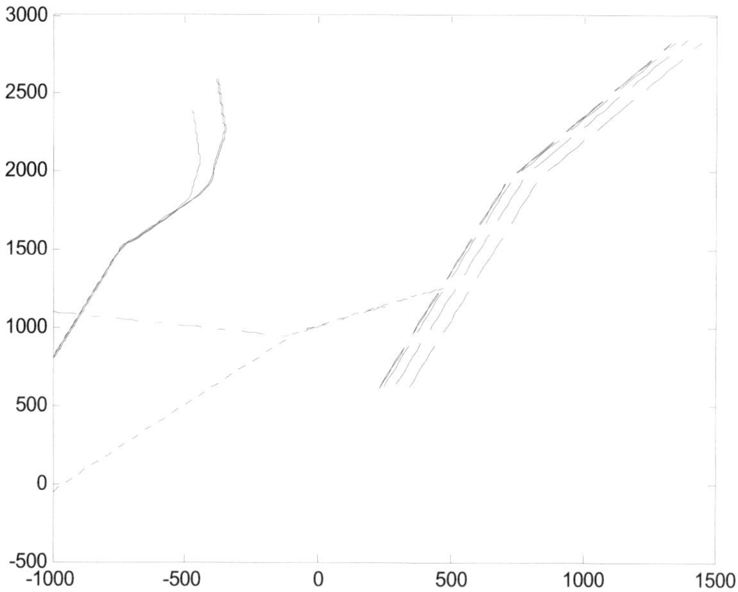

Fig. 5. Scenario of the multiple targets with multiple group relationships

The second component of the report consists of four separate estimates. The first is based on group position distance which was introduced in (Stubberud et al., 2003b):

$$\max\left\{\min_{\mathbf{x}_j \in B}\left\{\left\|\mathbf{x}_i - \mathbf{x}_j\right\|_2 : \forall \mathbf{x}_i \in A\right\},\ \min_{\mathbf{x}_i \in A}\left\{\left\|\mathbf{x}_i - \mathbf{x}_j\right\|_2 : \forall\ \mathbf{x}_i \in B\right\}\right\} \tag{14}$$

This is the distance between two groups of elements. The composite information to generate this measure is all of the position information for each target in each group. Thus, the overall position representation is defined as

$$\mathbf{x}_1(k+1) = \begin{bmatrix} \mathbf{A} \\ \mathbf{x}^{grp\_a} \\ \mathbf{x}^{grp\_b} \end{bmatrix} \tag{15}$$

where

$$\mathbf{A} = \max\left\{\min_{\mathbf{x}_j^{grp\_b} \in B}\left\{\left\|\mathbf{x}_i^{grp\_a} - \mathbf{x}_j^{grp\_b}\right\|_2 : \forall \mathbf{x}_i^{grp\_a} \in A\right\},\ \min_{\mathbf{x}_i^{grp\_a} \in A}\left\{\left\|\mathbf{x}_i^{grp\_a} - \mathbf{x}_j^{grp\_b}\right\|_2 : \forall\ \mathbf{x}_i^{grp\_b} \in B\right\}\right\}$$

with an associated Jacobian

$$\mathbf{F} = \begin{bmatrix} 0 & \dfrac{\partial \mathbf{B}}{\partial x^{grp\_a}} & \dfrac{\partial \mathbf{B}}{\partial y^{grp\_a}} & \dfrac{\partial \mathbf{B}}{\partial x^{grp\_b}} & \dfrac{\partial \mathbf{B}}{\partial y^{grp\_b}} \\ \mathbf{0} & & \mathbf{I} & & \end{bmatrix} \tag{16}$$

where

$$\mathbf{B} = \left\|\mathbf{x}_i^{grp\_a} - \mathbf{x}_j^{grp\_b}\right\|_2 .$$

Sample Time (sec)	Tgt 1	Tgt 2	Tgt 3	Tgt 4	Tgt 5	Tgt 6
0	-1000 800	-1000 804	-995 800	-1000 -50	-1000 -50	-1000 1100
5	-967.94 881.16	-994.97 805.56	-992.6 798.32	-929.39 30.504	-990.04 -31.861	-908.74 1083.1
10	-929.14 1005	-969.96 888.17	-974.9 866.03	-826.91 144.06	-903.38 60.31	-786.5 1064.5
15	-879.59 1121.1	-922.09 1013.1	-926.59 982.93	-720.16 257.78	-797.09 174.06	-660.02 1038.3
20	-838.35 1241.1	-875.77 1138.8	-879.74 1106.1	-620.98 371.09	-685.58 292.31	-537.36 1021.1
25	-792.35 1360.2	-830.91 1267.2	-835.48 1235.3	-517.83 484.78	-578.44 412.14	-414.49 998.34
30	-751.24 1480.3	-788.18 1391.6	-791.35 1363.7	-419.19 597.85	-479.7 526.85	-289.15 975.02
35	-697.14 1589.7	-739.37 1509.3	-749.56 1482.2	-317.17 705.01	-378.1 638.37	-164.82 954.34
40	-584.54 1711.8	-679.95 1607.5	-695.29 1588.7	-218.03 814.42	-275.83 754.74	-52.183 947.48
45	-450.22 1868.3	-588.07 1712.5	-582.82 1713.4	-113.9 928.98	-176.01 868.16	18.583 979.55
50	-361.77 2025.7	-486.21 1823.9	-445.21 1867.1	-22.073 1021.3	-77.271 971.9	75.327 1024.8
55	-330.41 2180.4	-430.93 1933.1	-358.23 2029.5	44.679 1059.6	3.4299 1039.1	132.23 1066.3
60	-326.35 2315.2	-419.91 2031.5	-323.35 2184.8	109.28 1082.8	69.283 1069.8	194.67 1108.9
65	-335.44 2381.7	-423.87 2115	-322.73 2316.4	171.46 1109.5	129.48 1094.1	258.27 1144.5
70	-352.8 2419.4	-432.05 2164.4	-338.43 2380.2	232.29 1142	193.17 1120	325.94 1181.7
75	-365.59 2453.2	-448.6 2212.4	-351.2 2416.4	296.07 1173.2	258.79 1152	390.58 1219.5
80	-372.93 2491.8	-458.18 2253.9	-363.07 2451.2	367.27 1205.3	324.08 1187	457.91 1252.5
85	-378.97 2537.1	-468.78 2299.8	-363.83 2494.7			

Table 1a. Track position reports for targets 1-6

Sample Time (sec)	Tgt 7	Tgt 8	Tgt 9	Tgt 10	Tgt 11
0	295.47 642.01	345.47 622.01	245.47 620.01	231.47 615.01	227.47 613.01
5	347.81 791.44	397.69 769.36	297.02 767.02	283.19 762.85	281.01 764.21
10	429.87 1010.8	476.82 991.62	385.29 980.21	365.23 984.62	363.77 982.36
15	503.7 1228.7	558.18 1207.1	460.68 1200.6	442.07 1198.3	439.14 1197.3
20	585.23 1442.7	636.46 1422	537.35 1416.5	523.16 1410.7	517.11 1414.9
25	664.36 1653.4	712.43 1631.2	615.04 1634.9	601.36 1628.2	593.51 1626.1
30	741.97 1866.8	792.38 1845.5	693.39 1849.6	678.65 1838.8	675.42 1838.6
35	832.38 2076.3	883.61 2060.4	783.71 2061.2	766.23 2057.3	765.41 2054.5
40	982.12 2305.6	1035.6 2286.5	935.04 2285.2	916.08 2281	912.81 2277.5
45	1153.2 2531.5	1201.6 2516.8	1106.3 2512.8	1088.9 2509.2	1083.9 2506.2
50	1327.1 2761.3	1371.7 2741.4	1274.6 2738.1	1256.3 2733.1	1255.6 2733.4

Table 1b. Track position reports for targets 7-11

Sample Time (sec)	Tgt 1	Tgt 2	Tgt 3	Tgt 4	Tgt 5	Tgt 6
5	Truck	Truck	Truck	TEL	Cmd	Fuel
10				TEL	APC	Fuel
20				TEL	Cmd	TEL
31	Truck	Truck	Truck			
35				TEL	Cmd	Fuel
45				TEL	Cmd	Fuel
60	Truck	Truck	Truck			

Table 2a. Classification reports for targets 1-6

Sample Time (sec)	Tgt 7	Tgt 8	Tgt 9	Tgt 10	Tgt 11
5	Tank	Tank	APC	APC	Cmd
10	Truck	Tank	APC	APC	Truck
20	Tank	Tank	Tank	APC	Truck
31					
35	Tank	APC	APC	APC	Fuel
45					
60					

Table 2b. Classification reports for targets 7-11

The Jacobian only has four elements based on the x and y coordinates of the two elements used for the difference. The second estimate is the velocity distance and has a similar state representation and Jacobian.

The composition estimate is similarly developed based on the norm of the cardinality metric

$$card(A) = \begin{cases} 0, & \text{if } A = \varnothing \\ n, & \text{if } A \neq \varnothing \text{ and } n = \#\text{ of elements in } A \end{cases} \tag{17}$$

which measures the difference in the number of elements of a classes in a group and the Gap metric

$$\sin\theta_{\max} \text{ where } \theta = \arccos\left(\frac{x^T y}{\left(x^T x\right)^{1/2}\left(y^T y\right)^{1/2}}\right). \tag{18}$$

which measures the ratio between the number of different classes of targets in a group.

The norm of the cardinality and the Gap compare the potential group compositions with known grouping compositions. The compositions are based on reported classifications of the targets.

For simplicity of implementation, in this example, potential groupings are only allowed when targets are within 300 m of each other. These three second component estimates provide the measurements to the third component estimation, the development of groupings levels of evidence. The third component estimator is a four state model

$$\mathbf{x}_{evidence} = \begin{bmatrix} \left(0.4x_{pos} + 0.3x_{vel} + 0.3x_{comp}\right)\operatorname{sgn}\left(x_{pos}\right) \\ \begin{bmatrix} 1 & 0 & 0 \\ 0 & 1 & 0 \\ 0 & 0 & 1 \end{bmatrix}\begin{bmatrix} x_{pos} \\ x_{vel} \\ x_{comp} \end{bmatrix} \end{bmatrix} \tag{19}$$

where the estimates from position, velocity, and composition are mapped via fuzzy logic to levels of evidence, a value from 0 to 1. The fuzzy map for position distance (antecedent fuzzification function, inference engine, and consequence Defuzzification function) is given in Figure 6.

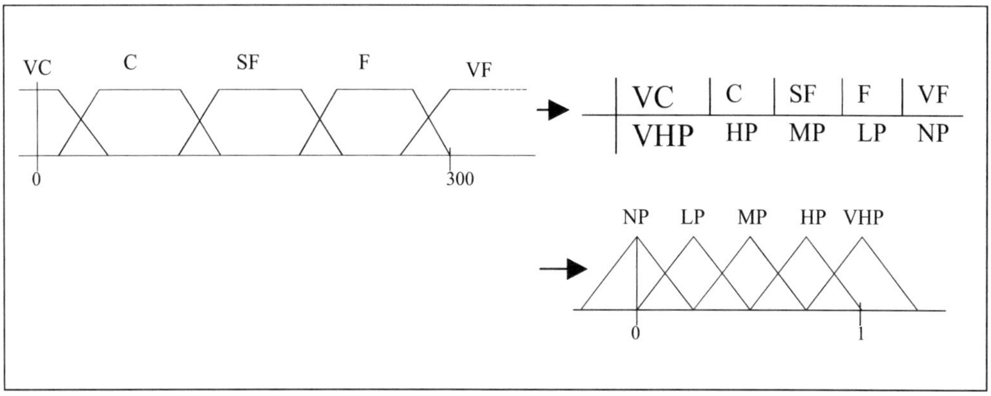

Fig. 6. Fuzzy map for position distance

The notation in the antecedent functions indicate targets are very close (VC), close (C), somewhat far (SF), far (F), and very far (VF). In the consequence function, no probability (NP), low probability (LP), medium probability (MP), high probability (HP), and very high probability (VHP). A similar map is defined for velocity. A two dimensional inference engine that uses the cardinality and the Gap have been defined as well. The initial covariance **P** was set to 10**I**. The process noise was defined as

$$\mathbf{Q} = \begin{bmatrix} 0.1 & 0 & 0 & 0 \\ 0 & 10 & 0 & 0 \\ 0 & 0 & 10 & 0 \\ 0 & 0 & 0 & 10 \end{bmatrix}. \tag{20}$$

The second component estimates are updated with each track report. The third component estimate updates the observations only when a report is provided. The overall estimate is updated when any observation is reported. The initial values for the various groupings of targets are set to 0. If an evidence score is greater than 0.6, then it is reported as a grouping. The largest grouping with such a score is reported.

The results for the Figure 5 scenario are shown in Table 3 and reflect the estimated situation assessment over the course of the scenario. Without velocity information and composition information, no groups can be formed.

Once the velocity information and composition information is provided, the initial groupings can be formed. The group of five targets always uses position and velocity to maintain its relationship while the composition strengthens and weakens the relationship. Target 1 separates early from Target 2 and 3 and the other two do not keep pace. The convoy of truck composition increases the evidence to generate a relationship. Near the end of the scenario, Target 3 closes in and then keeps pace with Target 1 while Target 2 moves away. While the evidence level was close to the threshold, it did not maintain its grouping level. Finally, Targets 4 and 5 almost come together based on composition and position early on but clearly remain apart as the second composition report and velocity are below threshold. After the velocity relationship improved, Targets 4 and 5 form a group. As Target 6 approaches the group, it uses position and composition to create the larger group.

Time	Group 1	Group 2	Group 3	Group 4	Group 5	Group 6
0	No Groups					
5	1,2,3	4	5	6	7,8,9,10,11	
10	2,3	1	4	5	6	7,8,9,10,11
20	1,2,3	4,5	6	7,8,9,10,11		
30	1,2,3	4,5	6	7,8,9,10,11		
35	1,2,3	4,5	6	7,8,9,10,11		
45	2,3	1	4,5	6	7,8,9,10,11	
50	2,3	1	4,5,6	7,8,9,10,11		
60	1,3	2	4,5,6	7,8,9,10,11		
80	1,3	2	4,5,6	7,8,9,10,11		

Table 3. Groupings based on evidence levels of 0.6 trashold.

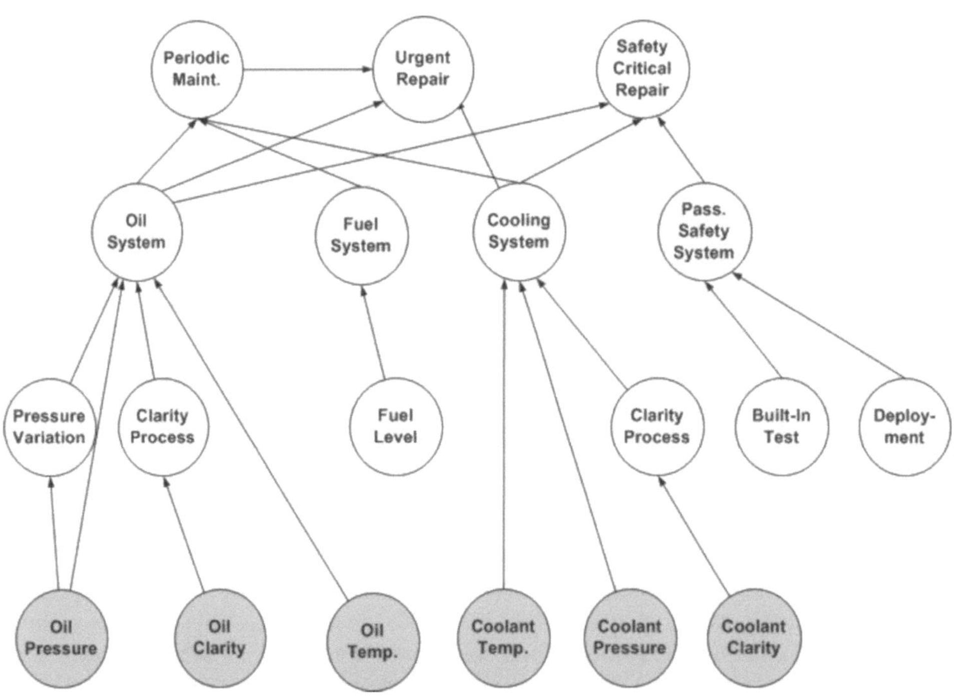

Fig. 7. Basic tree structure for engine health monitoring.

8. Condition-based health monitoring example

Another application of this evidence accrual technique is that of condition based monitoring. As described in (Liggins et al., 2009), many systems, such as helicopters (Brotherton et al., 2002), UAVs, and automobile engines, have various sensors throughout the vehicle to monitor the current operating conditions of the vehicle. By combining the data from these sensors, the health state of the vehicle can not only be determined but also in some instances predicted. In this example, an automobile with sensors on the cooling system, the oil system, the fuel system, and passenger safety system is used. The individual systems are examined to determine their performance to provide the final recommendation of the need for periodic maintenance, urgent repairs, and safety critical repairs. In Figure 7, the basic tree structure is shown. The gray-filled node, indicate basic sensors. Each subsystem of the car creates it own branch of the tree structure.

The oil system branch has three sensors: oil pressure, which is quite accurate, the oil temperature, which is also accurate, and an optical sensor to measure the oil clarity which is not very accurate. The pressure sensor feeds a change in pressure calculation as well as provides a direct measurement to the evidence accrual system. The temperature sensor is a direct input to the evidence level. The clarity sensor is the input to a more complex system that determines the level of particular content in the oil. The oil system state representation is given by

$$\mathbf{x}_{oil\_sys} = \begin{bmatrix} x_{oil\_ch} & x_{oil\_pr} & x_{\mathrm{var}} & x_{clarity2} & x_{clarity} & x_{temp} & x_{press} \end{bmatrix}^T \tag{21}$$

where

$$x_{oil\_ch} = 1 - e^{-x_{clarity2}/1.5}$$
$$x_{oil\_pr} = \max\left(1,\quad 0.6r1 + 0.7r2 + 0.9r3 + 0.9r4\right)$$
$$x_{clarity2} = f\left(x_{clarity}\right)$$

where

$$x_{\mathrm{var}} = \frac{x_{press}(t) - x_{press}\left(t - \Delta t_{\mathrm{sec}}\right)}{\Delta t_{\mathrm{sec}}}$$

and f is system that provides a level of clarity.

The fuel system branch has one sensor, the fuel level gauge which is not very accurate. This sensor provides both a direct measurement of the fuel level and a change in fuel level. The state representation is given as

$$\mathbf{x}_{fuel\_sys} = \begin{bmatrix} x_{fue\_ch} & x_{gauge} \end{bmatrix}^T \tag{22}$$

where

$$x_{fuel\_ch} = \mu\left(-x_{gauge} - 0.25\right).$$

The cooling system has three sensors, a temperature sensor, which is very accurate, a pressure gauge which is accurate, and a clarity sensor similar to that of the oil system. The temperature and pressure sensors provide direct input to the evidence accrual system. The clarity sensor again goes through a more complex system. The state representation is given as

$$\mathbf{x}_{cool\_sys} = \begin{bmatrix} x_{cool\_ch} & x_{cool\_pr} & x_{clarity2} & x_{clarity} & x_{temp} & x_{press} \end{bmatrix}^T \tag{23}$$

where

$$x_{cool\_ch} = 1 - e^{-x_{clarity2}/1.5}$$
$$x_{oil\_pr} = \max(1, \quad 0.6r1 + 0.9r3 + 0.9r4)$$
$$x_{clarity2} = f\left(x_{clarity}\right)$$

where all terms are the same as for the oil system except

$$r1 = r1 + x_{cool\_ch}^3 * \mu\left(\Delta t_{days} - 200\right)$$
$$r4 = \left(1 - \mu\left(x_{press} - 20\right) + \mu\left(x_{press} - 120\right)\right).$$

The passenger safety system has two sensors, a built-in test at start-up to check the that the system has been activated and a simple deployment sensor. The first sensor checks to see if the airbage deployment system is operational. For the latter sensor, if any of the airbags have been deployed without being replaced, then it is set off. The state represenation of the safety system is given as

$$\mathbf{x}_{spass\_afety\_sys} = \begin{bmatrix} x_{safety\_ch} & x_{deployed} & x_{sys\_test} \end{bmatrix}^T \tag{24}$$

where

$$x_{safety\_ch} = \delta\left(\left(1 - x_{deployed}\right)\left(x_{sys\_test}\right)\right).$$

The benefit of the evidence accrual system defined in this chapter is that the reporting evidence can be considered independent or interactive. In this case, the structure allows for periodic maintenance to also provide input into the urgent repairs while the safety issues are completely stand alone. The evidence state for periodic maintenance is defined as

$$x_{per\_mat} = \max\left(1, \quad 0.8x_{oil\_ch} + 0.9x_{fuel\_ch} + 0.8x_{cool\_ch}\right),$$

while the urgent repairs is given as

$$x_{urgent\_repairs} = \max\left(1, \quad 0.8x_{oil\_ch} + 0.9x_{cool\_ch} + 0.3x_{per\_mat}\right).$$

The safety system state is given as

$$x_{safetys} = \max\left(1, \quad 0.5x_{oil\_ch} + 0.8x_{cool\_ch} + x_{pass\_safety\_sys}\right).$$

For the first example will look at an oil system problem. The oil clarity over time is given as seen in Figure 8. The oil pressure is given as described in Figure 9. The clarity clearly over time degrades. This indicates that periodic maintennance is needed. Near the end of the operations, the oil pressure has spikes in the negative direction. While the oil pump and engine maintain pressure, the amount of oil has decreased such that turns induce a drop in the oil pressure indicating a potential major fault.

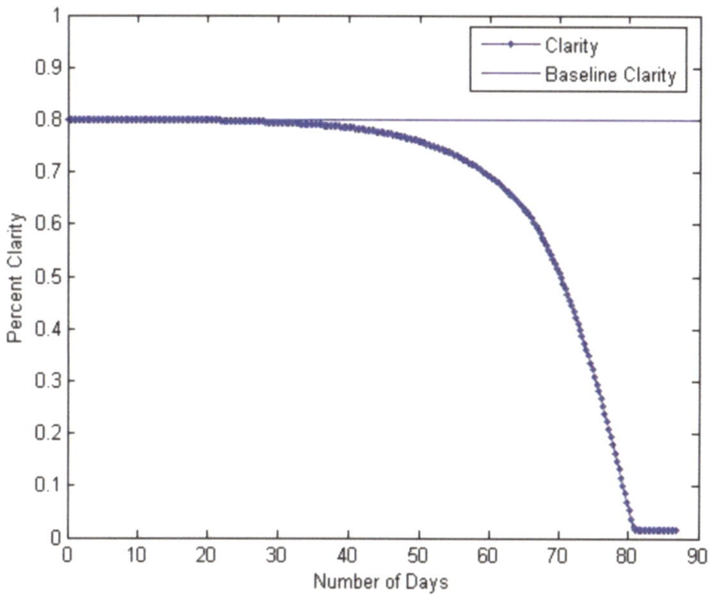

Fig. 8. Oil clarity over time.

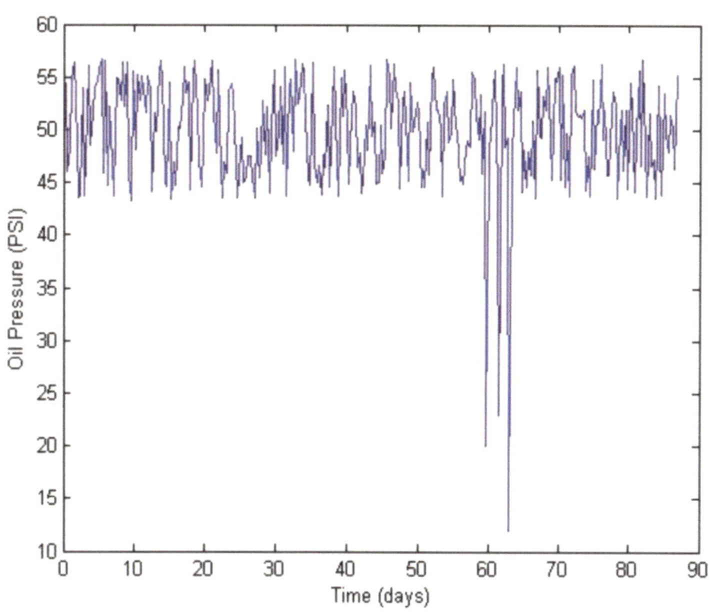

Fig. 9. Oil pressure over time.

Table 4 shows sampled results of the system. This indicates that, at first, the automobile is fine but soon needs periodic maintenance. Over time, the oil pressure problem plus the ignoring of the need for periodic maintenance signals the need for urgent repairs.

	Day 0	Day10	Day 40	Day 59	Day 70
Oil Clarity	0.8	0.8	0.775	0.7	0.4
Oil Clarity 2	0	0	0.09	0.12	0.62
Oil Pressure	54.5	44.7	47.6	26	48.1
Oil Variation	3	2.6	2.8	32	10
Oil Change	0.0	0.0	0.058	0.077	0.339
Oil Problem	0	0	0	1	1
Periodic Maintenance	0	0	0.048	0.061	0.26
Urgent Repair	0	0	0	0.83	0.91

Table 4. Selected results of the oil system.

9. Conclusion

A fuzzy evidence accrual technique for data fusion as applied to mechatronics was developed. The technique incorporates multiple sensor inputs and uncertainty measures.

10. References

Blackman, S. (1986). Multiple-Target Tracking with Radar Applications, Artech House, ISBN 0890061793, Norwood, MA, USA

Blackman, S. & Popoli, R. (1999). Design and Analysis of Modern Tracking Systems, Artech House, ISBN 1580530060, Norwood, MA, USA

Brotherton, T.; Grabill, P.; Wroblewski, D.; Friend, R.; Sotomayer, B. & Berry, J. (2002). A Testbed for Data Fusion for Engine Diagnostics and Prognostics, Proceedings of the 2002 IEEE Aerospace Conference, pp. 6-3029 - 6-3042, ISBN 0-7803-7231-X, Big Sky, MT, USA, March 2002.

Buczak, A.L. & Uhrig, R.E. (1995). Hierarchical Fuzzy Genetic Method for Information Fusion, Intelligent Engineering Systems through Artificial Neural Networks, Vol. 5: Fuzzy Logic and Evolutionary Programming (Proceedings of the Artificial Neural Networks in Engineering Conference), pp. 357-362, ISBN 0791800482, St. Louis, MO, USA, November 1995.

Chen, C.H. (1996). Fuzzy Logic and Neural Network Handbook, McGraw-Hill, ISBN 0-07-011189-8, New York, NY, USA.

Dempster, A. (1967). Upper and Lower Probabilities Induced by a Multivalued Mapping, Annals of Mathematical Statistics, Vol. 38, No. 2, (April 1967), pp. 325 – 339, ISSN 0003-4851

Hall, D. & Llinas, J. (2001). *The Handbook of Multisensor Data Fusion,* CRC Press, ISBN 0-8493-2479-7, Boca Raton, FL, USA.

Hammell, R.J. & Sudkamp, T. (1998). Learning Fuzzy Rules from Data, NATO RTO Meeting Proceedings 3: The Application of Information Technologies (Computer Science) to Mission Systems, pp. 8-1– 8-10, ISBN 92-837-1006-1, Monterey, CA, USA, April 1998.

Kelley, J.L. (1961). General Topology, D. Van Nostrand Company, ISBN 0923891552, Princeton, New Jersey, 1961.

Kosko, B. (1992). Neural Networks and Fuzzy Systems: A Dynamical Systems Approach to Machine Intelligence, Prentice Hall, ISBN 0136114350, Englewood Cliffs, New Jersey, USA

Le Hegarat-Mascle, S.; Bloch, I. & Vidal-Madjar, D. (1997). Application of Dempster-Shafer Evidence Theory to Unsupervised Classification in Multisource Remote Sensing, IEEE Transactions on Geoscience and Remote Sensing, Vol. 35, No. 4, (July 1997), pp. 1018 - 1031, ISSN 0196-2892

Liggins, M.E.; Llinas, J. & Hall, D. L. (2009). Handbook of Multisensor Data Fusion: Theory and Practice, Second Edition, CRC Press, ISBN 1420053086, Boca Raton, FL, USA

Llinas, J.; Bowman C.; Rogova, G.L.; Steinberg, A.; Waltz, E. & White, F. (2004). Revisions and Extensions to the JDL Data Fusion Model II, Proceedings of Fusion 2004, pp.1218-1230, ISBN 91-7056-117-6, Stockholm, Sweden, July 2004.

Pearl, J. (1988). Probabilistic Reasoning in Intelligent Systems: Networks of Plausible Inference, Morgan Kaufmann Publishers, ISBN 1558604790, San Mateo, CA, USA

Santina, M.S.; Stubberud, A.R. & Hostetter, G.H. (1994). Digital Control System Design 2nd Edition, Saunders College Publishing, ISBN 0030760127, Fort Worth, Texas, USA

Shafer, G. (1976). A Mathematical Theory of Evidence, Princeton University Press, ISBN 978-0691100425, Princeton, N.J., USA

Skjetne, R.; Teel, A. R. & Kokotovic, P.V. (2002). Nonlinear Maneuvering with Gradient Optimization, Proceedings of the 41st IEEE Conference on Decision and Control 2002, pp. 3926-3931,ISBN 0-7803-7516-5, Las Vegas, NV, USA, December 2002

Steinberg A.; Bowman, C. & White, F. (1999). Revisions to the JDL Data Fusion Model, Proc. of the SPIE Sensor Fusion: Architectures, Algorithms, and Applications III, pp. 430-441, ISBN 0-8194-3193-1, Orlando, FL, USA, April 1999.

Stewart, G.W. & Sun, J. (1990). Matrix Perturbation Theory, Academic Press, ISBN 0126702306, Boston, MA, USA

Stubberud S. & Kramer, K. (2007). Incorporation of Partially Observable Evidence into an Evidence Accrual Data Fusion Technique, Proceedings of the Third International Conference on Intelligent Sensors, Sensor Networks and Information Processing, pp. 251 - 256, ISBN 978-3-540-38619-3, Melbourne, Australia, December 2007.

Stubberud S. & Kramer, K (2008). Incorporation of Indirect Evidence into an Evidence Accrual Technique for Higher Level Data Fusion, Signal Processing, Sensor Fusion, and Target Recognition XVIII Proceedings of the SPIE, Volume 6968. pp. 696811-696811-12, ISBN: 9780819471598, Orlando, FL, USA, April 2008.

Stubberud, S. & Shea, P.J. (2003). More Metrics for Level Fusion Association, Proceedings of the 16th International Conference on Systems Engineering, pp. 657-663, ISBN 0-905949-91-9, Coventry, UK, September 2003.

Stubberud, S.; Shea, P.J. & Klamer, D. (2003). Data Fusion: A Conceptual Approach to Level 2 Fusion (Situational Awareness), Proceedings of SPIE, Aerosense03, pp. 455-462, ISBN 0-8194-4956-3, Orlando, FL., April 2003.

Stubberud, S.; Shea, P.J. & Klamer, D. (2003). Metrics for Level 2 Fusion Association, Proceedings of Fusion 2003, pp. 180-186, ISBN 0-9721844-4-9, Cairns, Australia, July 2003.

Watkins, F.A. (1994). Fuzzy Engineering, Ph.D. thesis, University of California Irvine, Department of Electrical and Computer Engineering, June 1994

12

Intelligent Mechatronic System for Automatically Evaluating the Training of the Laparoscopic Surgeon

Minor A.[1], Lorias D.[1], Ortiz Simon[2] and Escamirosa F.[1]
[1]*Centro de Investigación y de Estudios Avanzados del IPN*
[2]*Instituto Tecnológico de Nuevo Laredo*
México

1. Introduction

Laparoscopic surgery is seen as the surgery of the future. Its low mortality rate, its fast recuperation rate and the aesthetic advantages it offers the patient make it adequate for application in all of the medical sub-specialisms. This is the main reason why technological innovation for this surgery is accelerated. However, all this technology is insufficient if the specialist surgeon does not also use training and evaluation [Dankelman 2005, Martin et al. 1997, Vassiluo et al. 2005].

Training in the development of the laparoscopic surgeon continues to be the cornerstone of the specialism. It is for this reason that the surgeon should count with training methods that augment their skills and, even more than that, intelligent mechatronic systems that can measure this improvement in their skills. Current training systems are based principally on the time taken to complete tasks and the quality of the development of these tasks. [Peters et al. 2004, Chung & Sackiier 1998, Keyser et al. 2000)].

Time is an objective and impartial factor, but measures of quality are normally subjective. Systems for training and for the evaluation of virtual reality, such as the haptic systems, [MaClusky et al., Sutherland et al. 2006], rate the quality of the training, but the cost of this evaluation is very high. For this reason cheaper mechatronic systems are required, whose philosophy of evaluation is based on the work developed by the user and the evolution of his learning.

In this way it could be possible to objectively evaluate the increase in the skill level achieved by the surgeon. The TrEndo system [Magdalena et al., 2006] is a good example of this effort to complement the measurement of quality of the training, but, even today, no methodology for establishing the clarity of this measurement has been developed. For this reason, we propose a new methodology of intelligent training evaluation, using the analysis in time of the three-dimensional behavior of each instrument during the execution of the task of transfer, based on a mechatronic system.

2. Development

In order to measure the evolution of training, a mechanotronic system was developed. This system carries out the three-dimensional registration of each instrument during the execution of each training task.

Each laparoscopic training task has a particular objective. This could be visuo-spatial adaptation or the choreography of both hands, etc.
The design of the system that digitalizes the temporal position of each instrument during the execution of the task, in conjunction with the practice model, was selected taking three of the most commonly used environments for learning laparoscopic techniques into consideration [Keyser et al., 2000] (the virtual environment, direct practice in the operating theatre and the use of physical trainers). The operating theatre is the most realistic environment for evaluation. In this environment, the equipment and its interaction with the surroundings provide a natural feedback of the effort used in surgery. Unfortunately, it is not possible to carry out the evaluation of a novice surgeon in these conditions. These virtual and haptic systems are expensive and this substantially limits their practical application. The physical trainer is also a realistic and versatile environment [Schijven & Jakimowicz, 2003], which, just like in the operating theatre, has a natural feedback through the use of the equipment. For this reason, a physical practice model was used, and a mechatronic system was designed which, in conjunction with a program of analysis developed in Matlab (the MathWorks, Inc.), evaluates the laparoscopic skills during the execution of a training task, see figure 1.

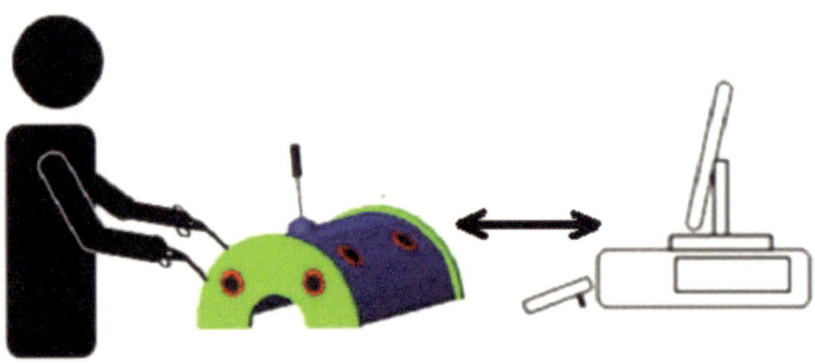

Fig. 1. Schematic diagram of the mechatronic laparoscopic evaluation system

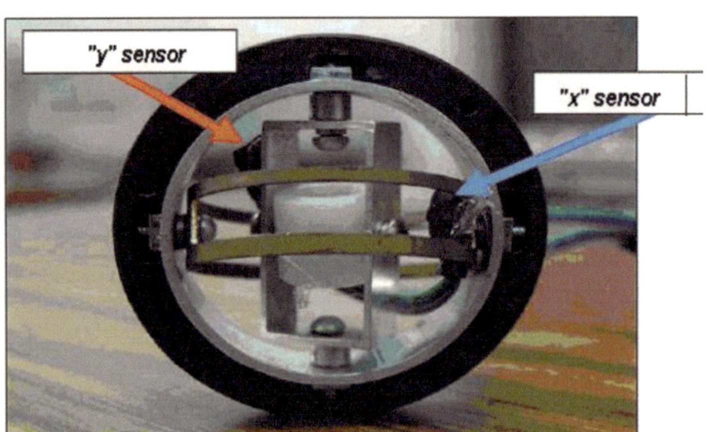

Fig. 2. Gimbal-type structure and assembly of the accelerometers for the registering movement on the x and y axes.

2.1 Three-dimensional registration system

A mechatronic system of movement registration (X,Y,Z) of the active type was designed. This consisted of a pair of accelerometer type sensors, (ADX330) which register the movement in the X y Y axes respectively. The reading of the movement in the z axis is measured using a resistive multiturn sensor. The sensors were joined with a gimbal-type structure and in a commercial trocar of 5mm (see figures 2 and 3)

Fig. 3. Assembly of the trocar in the gimbal arrangement and location of the resistive sensor that measures the variable of movement along the z axis.

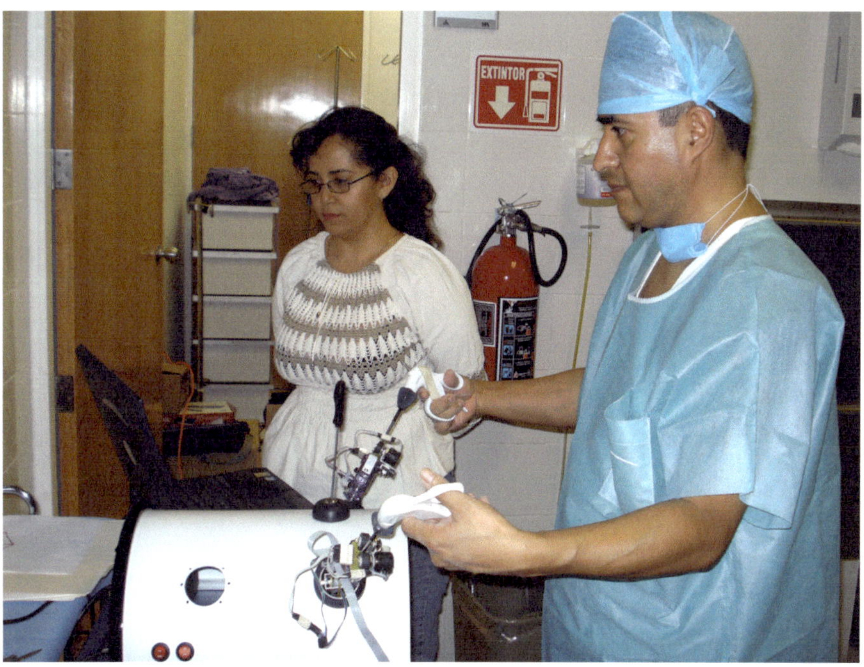

Fig. 4. General set up of the instrumented trochar installed in the trainer. The image shows the arrangement during the training.

The main advantage of this arrangement is that equipment of any make or model can be used without the need for any costly modification of the equipment. In figure 4 the general 'fitting together' of the mechatronic unit can be seen; it is joined together with laparoscopic equipment.

Referring to the sampling frequency, the specifications of the protocol of communication allow a communication of up to 1 Mb/s. Defining a channel of communication of 115.2 kb/s. And the frequency of sampling was adjusted via software to 16 Hz.

Transfer of data

The information provided by the mechatronic system is sent to an electronic unit that conditions the digital data of 8 bits through a series of filters. These data are then sent to the computer through the USB port.

2.2 User interface

A program in the Matlab programming platform was designed that reads the data from the electronic unit. These data are processed in order to calculate different kinematic parameters

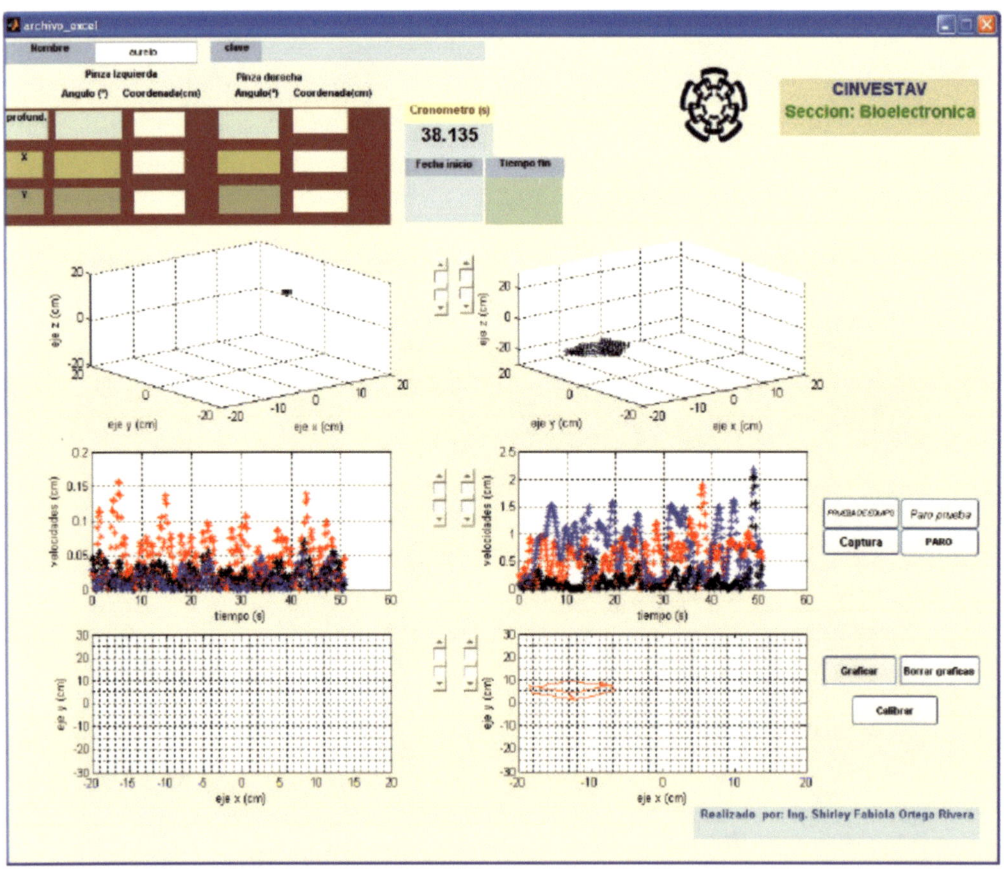

Fig. 5. Graphic interface of communication between the mechatronic system of movement digitalization and the user; Where the kinematic performance for both hands is shown in an independent and personalized form for each task.

during the execution of the task such as: velocity (cm/S), movement (cm), time (s), spatial position of each forcep ($X_{(t)}$,$Y_{(t)}$,$Z_{(t)}$) for total time (t). As well as the angles of inclination (ϕ). These data are processed independently and are presented for each hand, see fig. 5.

2.3 Calibration of the mechatronic evaluation system

Calibration is very important and necessary for ensuring precision in the registration of laparoscopic movements. This calibration consists of adjusting the workspace variable, which is defined as the reach of the equipment. For carrying out this calibration the limits of inclination (ϕ) and the longitude of insertion (L) of the equipment are adjusted. See figure 6. On executing the calibration command, this is carried out automatically. A series of messages appear that tell us the position in which the equipment should be placed. The total time for this routine lasts less than 2 seconds.
Once the calibration of the system is finished, it is possible to graph the trajectory of the laparoscopic equipment of each hand on carrying out a specific task in 3D.

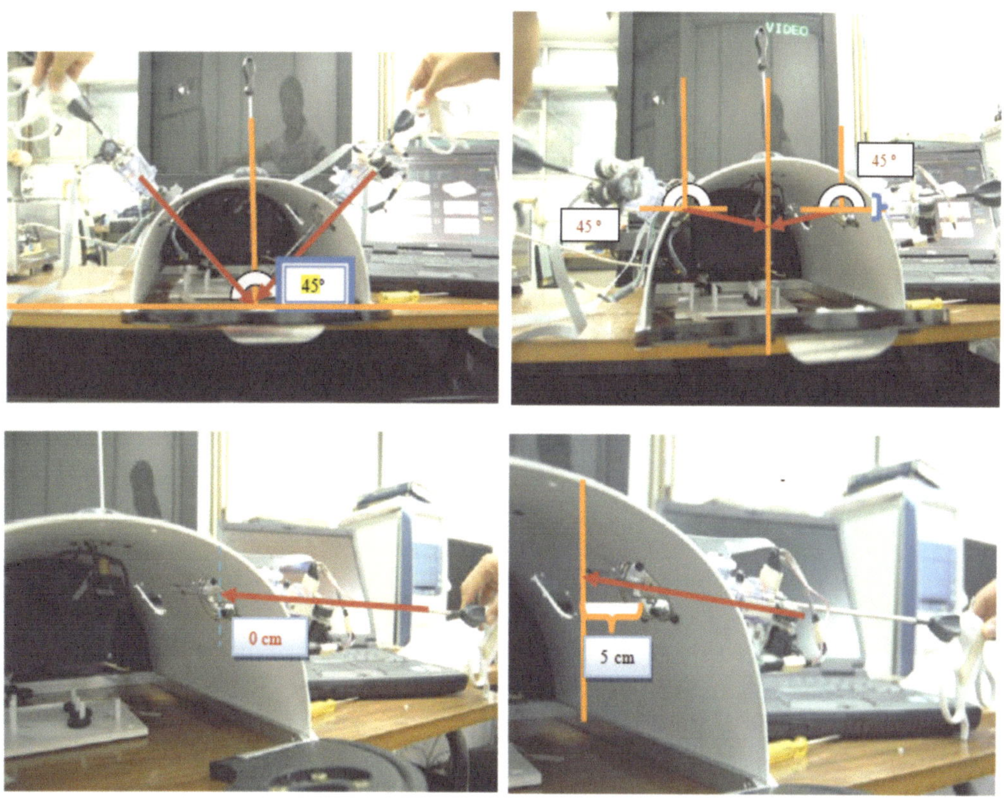

Fig. 6. Calibration of the system during which the workspace variable is adjusted, determined by the inclination (ϕ) and longitude of insertion (L) of the laparoscopic equipment.

The following figure 7 shows a three-dimensional graph of one task that consists of following the contour of a three-dimensional figure, which is placed inside the trainer, with the point of the laparoscopic instrument.

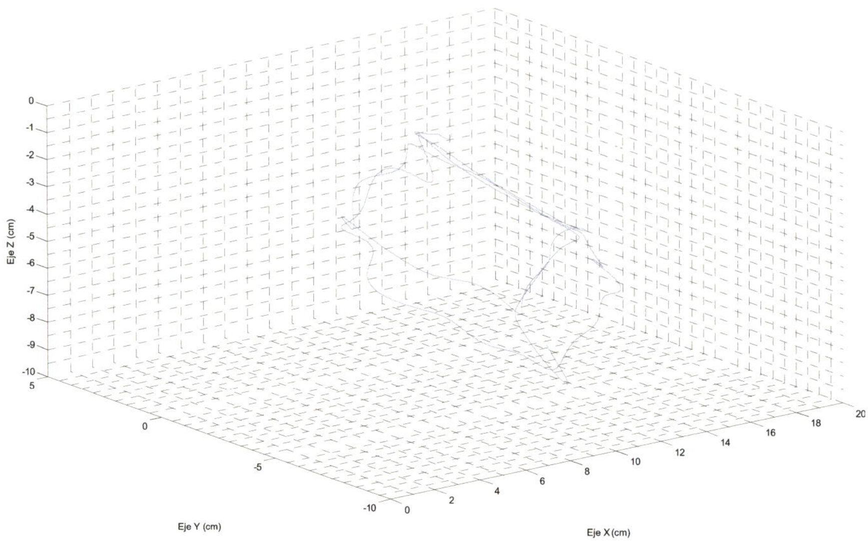

Fig. 7. Three-dimensional presentation of the x,y,z registration of the mechatronic system once it is calibrated

3. Automatic evaluation of laparoscopic training.

The task selected to demonstrate automatic evaluation is the task of transfer of the McGill Inanimate System for Training and Evaluation of Laparoscopic Skills (MISTELS) protocols [Derossis et. al., 1998]. This task has two main objectives: to develop visuo-spatial adaptation and to develop synchronization between both hands in the training.

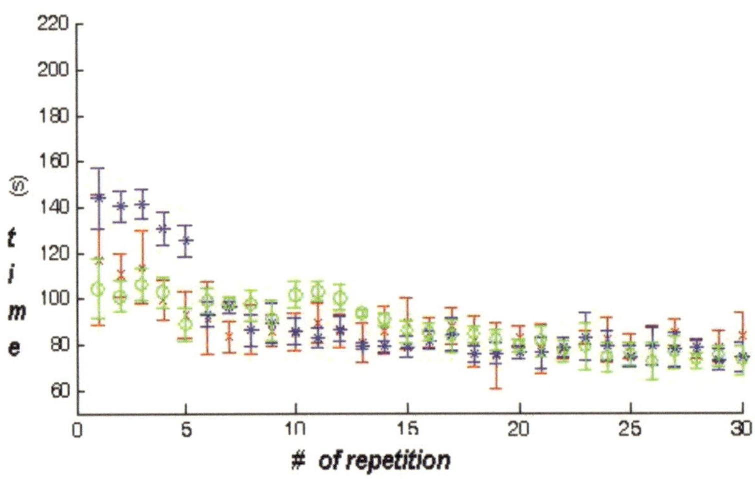

Fig. 8. A typical learning curve of one training task

This graph fig. 8 shows the natural evolution of learning (Grantcharov et al., 2003). As more repetitions of the same task are carried out, the time taken diminishes until it reaches one half, in this moment, the student has dominated the task.

The Mistels transfer task consists of fig (9): The use of a plane with 12 poles, divided in two parts, 6 poles in each section. In one section, each pole is placed in a circle, for this reason it has a section which is full with 6 circles, these circles are transferred in the air from the left laparoscopic instrument to the right laparoscopic instrument, in the middle area of the plane.
Initially, they pass from the left to the right side, and the afterwards from right to left. Registers are taken in time and through this task visuo-spatial adaptation, and the choreography of both hands are evaluated.

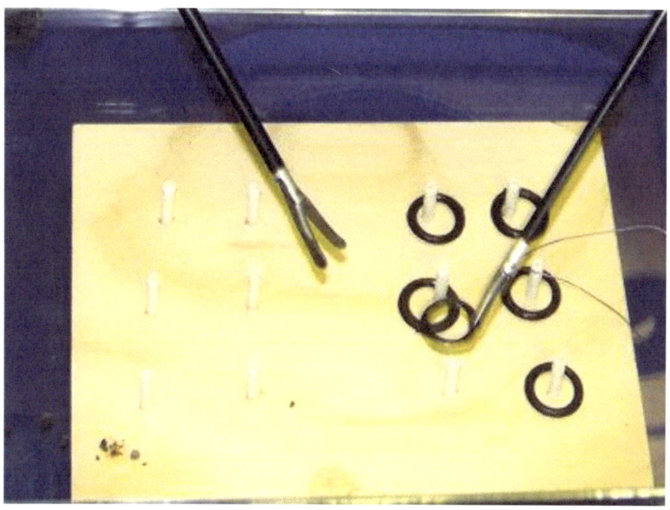

Fig. 9. Mistels Transfer task

The mechatronic system enables the registering of the temporal behavior of each instrument, permits the graphing of the dynamic behavior of each instrument in each plane fig. and produces a statistic evaluation of individual behavior in velocity, acceleration and the time taken to complete the task fig. 10.

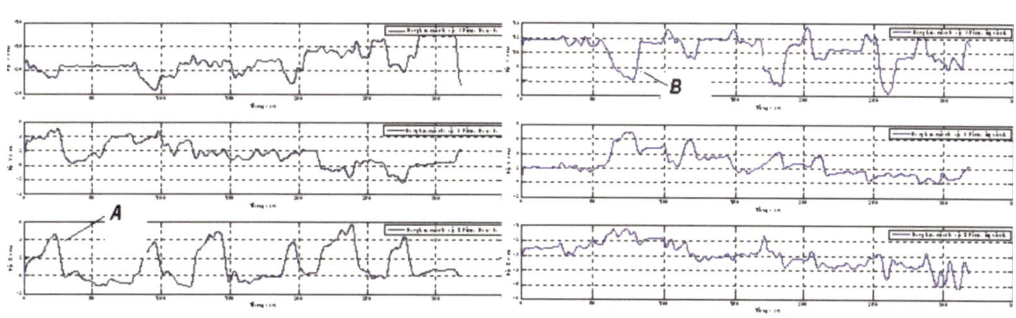

Right Forcep Left Forcep

Fig. 10. In some of these graphs, it is possible to clearly observe, when the circles are taken (A) with an instrument and afterwards, when they are deposited (B) with the other instrument.

From these graphs it is possible to calculate the low area of the curve. The low area of the curve represents the work, or effort, invested in carrying out the task.

$$W = \int_{x_i}^{x_f} F_x(x)dx \quad (1)$$

We propose the calculation of the work (W), or effort, as an "intelligent" factor in evaluation of learning, as long as the work, or effort, invested is directly related with the development of the task:
The work in discreet form is:

$$W \approx \sum F_x(\tilde{x})\Delta x \quad (2)$$

Our hypothesis is: The resident expends less effort in carrying out a laparoscopic training task at the end of this. And if this effort can be calculated, then it is possible to use this value as an immediate objective factor that is capable of evaluating if a laparoscopic surgeon has dominated each laparoscopic training task.
If the task is carried out by a resident at the beginning of the training we have unwanted movements because of the lack of visuo-spatial adaptation, and the lack of hand-eye coordination. This is translated into the excessive time taken in developing the task and in abrupt movements during its execution.
If the same resident carries out the task after practicing the same task, he can complete the task in less time, and without such abrupt movements.
This value can be compared in any moment with itself or against a group. In this way, it is possible to evaluate the level of skill of each laparoscopic surgeon, whether the surgeon is novice, intermediate or experienced.
For the evaluation, 10 resident laparoscopes in the first year of their specialism were selected, without a close relationship with the handling of the laparoscopic equipment and with the dominant right hand. The transfer task of the Mistels protocol was selected, using optics of zero degrees.

4. Results

Data was registered and the average low area of the registration curve was calculated in the three planes x,y, and z, of the right forcep fig. 11.

5. Discussion

A new automatic factor of evaluation of laparoscopic training was designed using the transfer task of the MISTELS protocol and a mechatronic system for the dynamic registration of the equipment.
To objectively measure the evolution of laparoscopic training is a technological challenge that, in being resolved, would permit the establishment of new parameters of immediate evaluation of the surgeon, whether he is novice, intermediate or expert. For this reason we propose the "work" as a new factor for evaluating laparoscopic training.
The results of the calculation of the "work" at the beginning of training, demonstrate that the surgeon invests greater energy in both hands. When visuo-spatial adaptation and synchronization between both hands are better, the effort invested starts to diminish substantially, until it reaches the average value.
According to the results, before the training of the surgeon uses more energy on not having enough ability to adapt himself visually, and through coordination, to the resolution of the

task. On the other hand, on developing the skill, the surgeon invests less work in developing the same task in less time. One important characteristic is that this parameter can be calculated automatically from the registered data and can give an objective parameter of evaluation of the training for each hand.

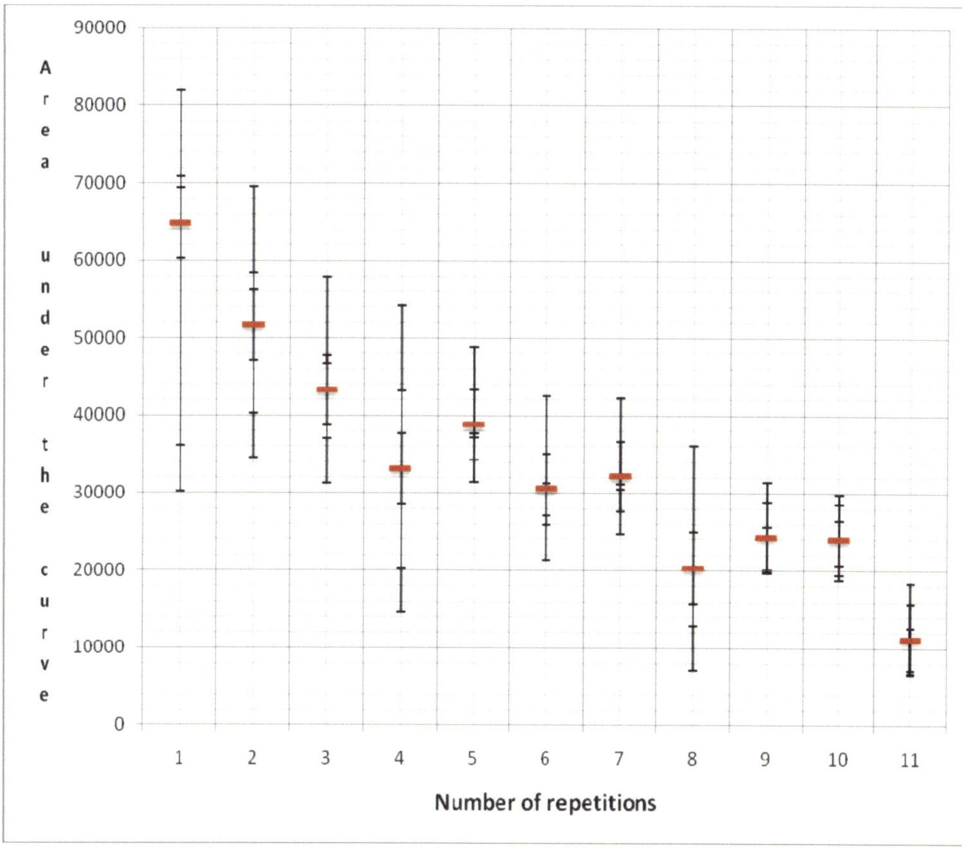

Fig. 11. "Work" developed during the transfer task

6. Conclusions

A new factor for the evaluation of laparoscopic training was developed. The method used 3D registers of each instrument during the execution of a specific task, using a mechatronic system. The adequate analysis of the three-dimensional signal during the execution of a task, demonstrates that after the training, the "work" invested in developing this task diminishes and is a factor that is measurable, and can be objectively reproduced for rating the skills of a laparoscopic surgeon.

7. References

Cotin S, Stylopoulos N, Ottensmeyer M, Neumann P, Rattner D, Dawson S. Metrics for Laparoscopic Skills Trainers: The Weakest Link! Lect Notes Comput Sc LNCS 2002;2488:35–43.

Chmarra MK, Grimbergen CA, Dankelman J. Systems for tracking minimally invasive surgical instruments. Minim Invasiv Ther & Allied Technol. 2007;16: 328–40.

Derossis AM, Fried GM, Abrahamowicz M, Sigman HH, Barkun JS, Meakins JL. Development of a model for training and evaluation of laparoscopic skills. Am J Surg 1998; 175: pp 482-487

Dankelman J, Chmarra MK, Verdaasdonk EGG, Stassen LPS, Grimbergen CA. Fundamental aspects of learning minimally invasive surgical skills. Minim Invasiv Ther & Allied Technol. 2005; 14:247–56.

Martin JA, Regehr G, Reznick R, MacRae H, Murnaghan J, Hutchison C, et al.. Objective structured assessment of technical skills (OSATS) for surgical residents. Brit J Surg. 1997; 84:273–8.

Moore MJ, Bennett CL. The learning curve for laparoscopic cholecystectomy.Am J Surg. 1995; 170:550–9.

McClusky DAM, van Sickle K, Gallagher AG. Relationship between motion analysis, time, accuracy, and errors during performance of a laparoscopic suturing task on an augmented reality simulator. 12th International Congress of the European Association

Sutherland LM, Middleton PF, Anthony A, Hamdorf J, Cregan P, Scott D, et al.. Surgical simulation. A systematic review. Ann Surg. 2006; 243:291–300.

Schijven M, Jakimowicz J. Virtual reality surgical laparoscopic simulators. How to choose. Surg Endoscopy. 2003;17:1943–50.

Vassiliou MC, Feldman LS, Andrew CG, Bergman S, Leffondre K, Stanbridge D, et al. A global assessment tool for evaluation of intraoperative laparoscopic skills. Am J Surg. 2005;190:107–13.

Peters JH, Fried GM, Swanstrom LL, Soper NJ, Sillin LF, Schirmer B, et al. Development and validation of a comprehensive program of education and assessment of the basic fundamentals of laparoscopic surgery. Surgery 2004; 135: 21–27.

13

Reliability of Authenticated Key Establishment Protocols in a Complex Sensor System

Kalvinder Singh[1] and Vallipuram Muthukkumarasamy[2]
[1]*IBM and Griffith University*
[2]*Griffith University*
Australia

1. Introduction

Wireless sensor and actuator networks are an important component in mechatronics. As the sensors permeate the environment they can monitor objects, space and the interaction of objects within a space. Sensors can monitor a wide range of diverse phenomena by collecting information such as vibrations, temperature, sound, and light. Different sensors have different associated costs. For example, a sensor simply detecting light will have different costs to a sensor recording sound. However, less costly sensors can be used to detect a phenomenon before alerting the more costly sensors to start their monitoring. As the number of heterogeneous sensors increases, so will the amount of interactions between the sensors. In an instrumented system sensors are becoming more interconnected, interactive, intelligent, and interdependent.
There are many different types of sensor environments, ranging from large areas covered by sensors, to many sensors in a small area.Different environments have a wide range of different characteristics. The different environments may include :

Transportation The management of traffic and the transportation of goods. This can include traffic lights, traffic controllers to manage traffic flow, or shock sensors, impact indicators for fragile goods.

Smart Buildings A distributed control system to efficiently manage building automation. For instance, smart buildings allows people to visualize and manage air–conditioning, lighting, or other appliances in the building.

Environmental Monitoring The monitoring of the characteristics and quality of the environment. This can include continuous monitoring a river system for pH level, dissolved oxygen, conductivity, turbidity and colour

Health and Well–Being The monitoring and delivery of health to a patient. This can include wearable wireless health sensors for remote bio-monitoring. Enabling patients to live independently.

Water The management of distribution of water to the populace. Water management includes the planning, developing, and distributing water resources.

Each environment can be considered to be a distributed control system with similar requirements. Figure 1 is a common framework used by each environment. The *real world* signifies the environment that needs to be managed. For instance, *real world* could be a building, a patient, water ways, toll way, or a rainforest. To manage each system, there needs to be a *data and measurement platform*. The platform is used to collect in situ continuous sensor data. The data is then sent to the *analysis platform*. The analysis of the data involves the assimilation, interpolation and explanation of the data. By the end of the analysis cycle the system should have high–quality trusted data. The high–quality trusted data is then sent to the *decision model*, where simulations are run and predictions are made. The outcomes or final decisions are sent to the actuators in the system. Actuators can include air-conditioning, lights or even a person.

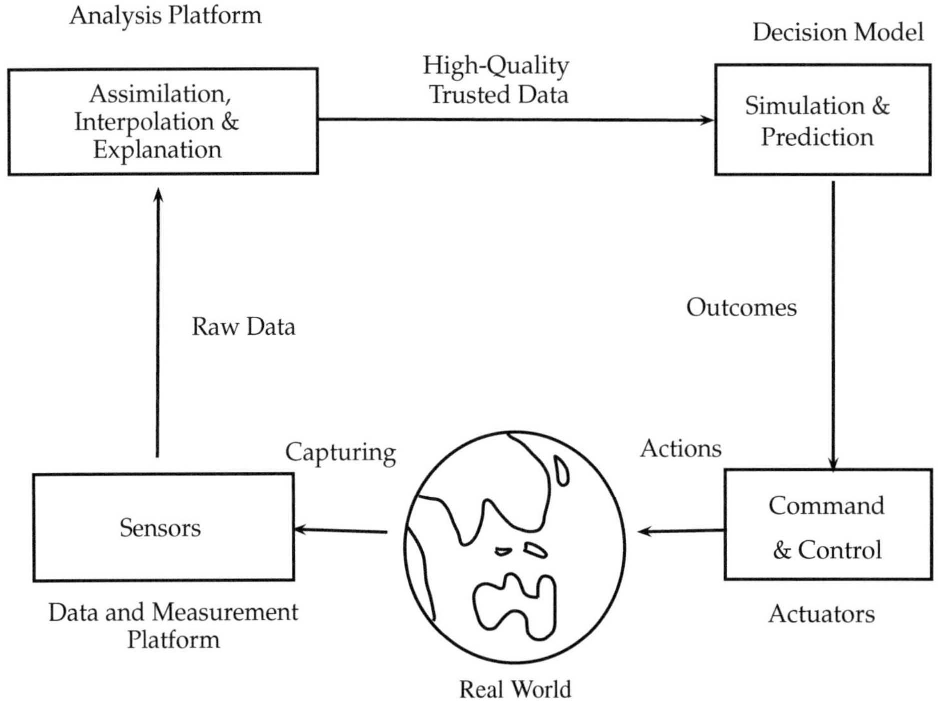

Fig. 1. Logic System Flow Diagram

The requirement for high–quality trusted data in the system, places the need for security, information assurance and reliability into the system. The future networks are increasingly becoming heterogeneous and highly dynamic. Cooperative networking paradigm envisages advanced wireless networks to cost effectively provide multimedia services and applications (incorporating various sensors) anywhere, anytime. Self-organising, opportunistic networking tries to exploit resources (bandwidth, memory, computing power, etc) available in various devices and nodes in the system. This approach of moving away from exclusively relying on infrastructure networks raises serious challenges to overall reliability of the system.

The complex nature of operation of sensor network applications involve use of different communication technologies, ranging from Cellular, WLAN, WiMax, WPN, WMN, WSN. Although much work has been done on the inter-operability between these networks, the authentication of nodes and the level of security available for services in such situations are still needs innovative solutions.

The system may not require encryption, but may require authentication. For instance, a sensor detecting the amount of light in a publicly available room supplies no or little information to an adversary. However, the information the light sensor sends to another device should be authenticated before any actions are taken.
One of the security vulnerabilities investigated in this chapter is the use of small keys. For performance reasons, sensors use small keys, however, small keys give an attacker a greater opportunity to compromise a sensor node. Thus, small keys will require frequent refreshing. Old keys can be used to generate new keys. But, if the old key is compromised then the new key can easily be compromised. A secure, efficient and scalable mechanism to freshen the keys between the sensors nodes is discussed.
When sensors are deployed in battlefields or developed to monitor homeland security, they have a likelihood of becoming the target of adversaries. In an open environment, an individual server or an intermediate sensor node may not be completely and permanently trustworthy. Several existing protocols (Chen et al., 1995; Gong, 1993) found in traditional network protocols handle the shortcomings of untrusted servers. We, however, show that those existing protocols have an $O(n^2)$ complexity, and are therefore not suitable to a resource constrained environment.
In this paper, we describe five protocols to address the problems described above. These protocols use information from the sensor nodes and the servers to generate a new key. Some of the protocols do not rely on the sensor nodes to generate cryptographically good pseudo–random numbers. We will show the sensor nodes can prove that the new key is fresh, and will demonstrate how key confirmation ensures that the nodes are guaranteed to be using the same key.
Section 2 shows the notations used throughout this paper. Section 3 provides a background to sensor networks, and the unique security challenges. Section 4 describes the multiple server protocols found in traditional networks, and the problems with using those protocols in sensor networks. Section 5 describes three multiple server protocols for sensor networks. Section 5.3 provides a detailed analysis of the multiple server protocols. Section 6 compares the protocols in different sensor network environments. Finally, a conclusion is provided in Section 7.

2. Notation and assumptions

Table 1 gives a list of notations, which are used when describing authentication and key establishment protocols in this paper.
In protocols using a multiplicity of servers an assumption is made that A and B do not trust any individual server. TinyOS is used as the development environment, with the following restrictions on the size of the data structures. The key size is 64 bits, the nonce size is 1 byte, the packet size is 29 bytes, and finally the location size is 2 bytes (which allows 64K of nodes in a sensor network).

3. Background

We refer to a sensor network as a heterogeneous system combining small, smart, and cheap, sensing devices (sensors) with general–purpose computing elements. A sensor network consists of a potentially large number of sensors; there may also be a few control nodes, which may have more resources. The functions of the control nodes include: connecting the sensor network to an external network; aggregating results before passing them on; controlling the sensor nodes; providing services not available to a resource constrained environment.

Notation	Description
A and B	The two nodes who wish to share a new session key.
S	A trusted server.
S_i	A server in a set of servers $S_1, \ldots, S_n$, where n is the numbers of servers.
N_A	A nonce generated by A.
$\{M\}_K$	Encryption of message M with key K to provide confidentiality and integrity.
$[[M]]_K$	Encryption of message M with key K to provide confidentiality.
$[M]_K$	One-way transformation of message M with key K to provide integrity.
K_{AB}	The long-term key initially shared by A and B.
K'_{AB}	The value of the new session key.
K_{AS}, K_{BS}	Long-term keys initially shared by A and S, and by B and S for centralized authentication server.
K_{AS_i}, K_{BS_i}	Long-term keys initially shared by A and S_i, and by B and S_i, for each $i \in 1, \ldots, n$.
X, Y	The result of the concatenation of data strings X and Y.
$A \rightarrow B : m$	Denotes that A sends a message m to B.
$A \rightarrow S_i : m$	Denotes that A sends a message m to each server.
$S_i \rightarrow A : m$	Denotes that each server sends a message m to A.
$X \oplus Y$	Exclusive–or operation with X and Y.

Table 1. Notations

Sensor network applications (Bulusu, 2005) include tracking bushfires, monitoring wildlife, conducting military surveillance, and monitoring public exposure to contaminants.
Some sensor nodes are resource constrained, such as the Mica mote (Crossbow, 2006). The Mica motes contain a 4 MHz processor with 512 KB flash memory and 4 KB of data memory. A Mica mote also has a separate 512 KB flash memory unit accessed through a low-speed serial peripheral interface. The RF communication transfer rate is approximately 40 kbps. The maximum transmission range is approximately 100 meters in open space. Communication is the most expensive operation in sensor networks.
Other components in the sensor network may have more computation power and memory. Examples are the Stargate platform (Crossbow, 2006), the GNOME platform (GNOME, 2006), the Medusa MK–2 platform (Medusa, 2006), and the MANTIS platform (MANTIS, 2006). These platforms may use other higher–level operating systems such as the Linux operating system. The platforms themselves may have additional communication mechanisms. For instance, the GNOME platform also has an Ethernet connection.

3.1 Key establishment in wireless sensors

Security in sensor environments (where there are sensors with low resources) differs in many ways from that in other systems. Efficient cryptographic ciphers must still be used with care. Security protocols should use a minimal amount of RAM. Communication is extremely expensive; any increase in message size caused by security mechanisms comes at a significant cost. Energy is the most important resource, as each additional instruction or bit transmitted means the sensor node is a little closer to becoming nonfunctional. Nearly every aspect of sensor networks is designed with extreme power conservation.
There are many aspects to WSN security (Deng et al., 2005); ranging from data fusion security, location aware security, to the lower level security primitives such as cryptography,

authentication and key establishment protocols. We will not cover all aspects of WSN security in this paper, instead only concentrate on some of the lower level primitives: authentication and key establishment protocols.

Several cryptography libraries using symmetric keys (Karlof et al., 2004; Perrig et al., 2001) have been proposed. Recent work has shown that even asymmetric keys may be used in WSNs (Malan et al., 2004; Watro et al., 2004). Singh et al. (Singh et al., 2006) has proposed an efficient key establishment protocol using elliptic curves. However, they still consume considerably more resources than the symmetric counterparts.

Key establishment protocols are used to exchange and set up shared secrets between sensor nodes. Asymmetric cryptography is unsuitable for most sensor architectures because of the higher computational overhead, and energy and memory consumption. When using symmetric keys, we can classify the key establishment protocols in WSNs into three main categories: Pair–wise schemes; Random key predistribution schemes; Key Distribution Center (KDC) schemes.

The simplest is the full pair–wise scheme (Chan et al., 2003), where each node in a network of total of n nodes shares a unique pairwise key with every other node in the network. The memory overhead for every sensor node is $(n-1)$ cryptographic keys. Other pair–wise schemes (Blundo et al., 1993; Leighton & Micali, 1994) also have $O(n)$ memory cost. In a pair–wise scheme, the sensor network is not compromised even if a fraction of the sensors are compromised.

Random key predistribution schemes are the second category (Chan et al., 2003). This is a major class of key establishment protocols for sensor networks. They rely on the fact that a random graph is connected with high probability if the average degree of its nodes is above a threshold. After the connected secure network is formed, the protected links can be further used for agreeing on new keys, called *path–keys*. One of the main problems with random key predistribution schemes is that if a certain number of sensor nodes become compromised, then the entire sensor network can be compromised.

The third category is the KDC scheme. If two entities sharing no previous secret want to communicate securely with each other, they can receive assistance from a third party. In WSNs the two entities are typically resource–constrained sensor nodes, and the third party is a resource–heavy base station. However, in a multi-tiered environment, such as a military battlefield, the third party may be a resource heavy camera. Typically, the base station provides an authentication service that distributes a secure session key to the sensor nodes. The level of security of a typical key distribution protocol depends on the assumption that the third party is trustworthy (Boyd, 1996).

KDC schemes use the least amount of memory compared with the other two categories, and has an extra advantage of providing authentication for the sensor nodes. Examples of KDCs in WSNs were first proposed in SPINS (Perrig et al., 2001). However, the SPINS protocol may not be suited for every WSN topology. For instance, it does not easily scale to a large WSN, since the non–uniform communication will focus the load onto the KDC. This may cause the battery life of the network nodes to diminish considerably. However, a KDC mechanism is suitable for a surveillance sensor environment.

Hybrid schemes can also by created by combining different key establishment categories. The PIKE scheme (Chan & Perrig, 2005) is such a scheme, it combines a pair–wise scheme with the KDC scheme, where one or more sensor nodes act as a trusted intermediary to facilitate the key establishment. The scheme was developed to limit the amount of memory used by the pair-wise and random key predistribution schemes, and also to limit the communication load because of the KDC schemes. However, the difficulties in using a sensor node as the trusted third party are: the trusted intermediary can easily become compromised; key sizes in sensors nodes are not large; sensor networks may only accept authenticated messages, so may not have access to an encryption algorithm.

It should be noted that none of the above sensor key establishment schemes can handle authentication when the third party is compromised. Also, if the KDC is a sensor or controller node, then there is a higher likelihood that it can be compromised.

3.2 Security for the KDC

Several researchers have addressed security of the controller nodes and/or the base station. SIA (Przydatek et al., 2003) addresses the issue of compromised nodes by using statistical techniques and interactive proofs, ensuring the aggregated result reported by the base station is a good approximation to the true value, even if a small number of sensor nodes and the aggregation node may have been compromised. However, the communication overhead between sensor nodes and the base station is high. Other works have shown that some of the statistical methods used are not resilient to a group of malicious sensor nodes, and the end user should be aware of which statistical methods are easily cheated (Wagner, 2004). Another way to protect results is to use a witness node mechanism (Zhu et al., 2003).

A different approach is to protect the base station location. Routing mechanisms to protect the location and disguise the identity of the base station have been proposed (Deng et al., 2004). Hop-by-hop re-encryption of each packet's header and data fields is designed to change the presentation of a packet so that it cannot be used to trace the direction toward or away from the base station. Uniform rate control is advised so that traffic volume nearer the base station is undifferentiated from traffic farther from the base station. Time decorrelation between packet arrivals and departures further increases the difficulty of tracing packets.

However, ensuring that the authentication services are not hindered by a compromised or broken controller node or base station presents different challengers. A simple approach is to replicate the authentication services of the server so that any one of several servers can perform authentication. However, this approach reduces the level of security; if one server is compromised, security for every replicated server is compromised.

3.3 Limitations and concerns

When WSNs are deployed in battlefields or developed to monitor national security, they have a likelihood of becoming the target of adversaries. In an open environment, an individual server or an intermediate sensor node may not be completely and permanently trustworthy. To make a key distribution protocol work in an environment where sensor nodes do not trust an individual base station, an authentication scheme, which can be used with limited resources and can reduce the requirement for trusting servers, needs to be found.

For performance reasons, sensors use small keys. However, small keys give an attacker a greater opportunity to compromise a sensor node. Thus, small keys will require frequent refreshing. Old keys can be used to generate new keys. But, if the old key is compromised than the new key can easily be compromised. Many existing sensor protocols concentrate on the initial key distribution but do not have a secure mechanism available to update the keys. Keys in sensor networks are usually 64 bits in size, and they may become easily compromised. Sensors that are short–lived may not require that cryptographic keys be updated, however any sensors that exist for an extended amount of time will require updating of keys.

4. Traditional reliable key establishment protocols

Many key establishment protocols assume that server is never compromised. In a sensor environment, this may not be always true. A method alleviate the problem with a compromised server compromising the entire system is to use multiple server key establishment protocols. Boyd and Mathuria have produced a survey of key establishment protocols using multiple servers (Boyd & Mathuria, 2003) in traditional networks. In their survey, two multiple server protocols were listed: Gong's multiple server protocol (Gong, 1993), and the Chen–Gollmann–Mitchell protocol (Chen et al., 1995). However, this survey did not take into account the unique nature of a sensor environment. The main goals of using multiple servers in a sensor network are:

- even if one or more servers become unavailable, it may be possible for the sensor nodes to establish a session key.
- even if one or more servers are untrustworthy, the sensor nodes may still be able to establish a good key.

Protocol 1 Gong's simplified multi-server protocol

$M1$	$A \to B:$	$A, B, N_A, \{A, B, x_1, cc(x)\}_{K_{A1}}, \ldots, \{A, B, x_n, cc(x)\}_{K_{An}}$
$M2$	$B \to S_i:$	$A, B, N_A, N_B, \{A, B, x_i, cc(x)\}_{K_{Ai}}, \{B, A, y_i, cc(y)\}_{K_{Bi}}$
$M3$	$S_i \to B:$	$\{B, N_A, y_i, cc_i(y)\}_{K_{Ai}}, \{A, N_B, x_i, cc_i(x)\}_{K_{Bi}}$
$M4$	$B \to A:$	$\{B, N_A, y_1, cc_1(y)\}_{K_{A1}}, \ldots, \{B, N_A, y_n, cc_n(y)\}_{K_{An}}, \{N_A\}_{K_{AB}}, N_B$
$M5$	$A \to B:$	$\{N_B\}_{K_{AB}}$

A simplified version of Gong's original multiple server protocol is described in (Boyd & Mathuria, 2003). We describe this version as shown in *Protocol 1*. One of the main features of this protocol is that the nodes, A and B, choose the keying material while the n servers, $S_1, S_2, \ldots, S_n$, act as key translation centers that allow keying material from one node to be made available to the other. Initially A shares a long-term key K_{Ai} with each server S_i, and similarly B shares K_{Bi} with S_i. Node A has split the key x into $x_1, x_2, \ldots, x_n$ and node B has split the key y into $y_1, y_2, \ldots, y_n$. The session key is defined as $K_{AB} = h(x, y)$ where h is a one–way function. The protocol sends a total of $2n + 3$ messages.
To prevent compromised servers from disrupting the protocol, A and B form a cross-checksum for all the shares. The cross-checksum for x is shown in Equation (1).

$$cc(x) = (h(x_1), h(x_2), \ldots, h(x_n)) \tag{1}$$

The $cc_i(x)$ (should be equal to $cc(x)$) and $cc_i(y)$ (should be equal to $cc(y)$) are the cross–checksums returned by server S_i. The node will give a credit point to the servers if their cross–checksum values are the same as the values obtained from the majority of servers. When all the checks are complete, B retains the value x_j with the most credit points.
The major problem with this protocol is the size of the messages. The message sizes of $M1$ and $M4$ in the Gong multi–server protocol are of $O(n^2)$. Message $M5$ is $O(1)$, while $M2$ and $M3$ are of $O(n)$. A message size of $O(n^2)$ is not desirable in a sensor network. Another problem is that the size of the output of the one–way function will have to be reasonably large (otherwise a malicious server can quickly calculate the possible values for x and y). So for small values of n, the message sizes themselves will be very large for a sensor network.
The second multiple server protocol we consider is the Chen et al. multiple server protocol as shown in *Protocol 2*. One of the main features of this protocol is that the servers, rather than

the sensor nodes, choose the keying material. Both nodes employ a cross-checksum to decide which servers have given valid inputs. The protocol sends a total of $2n + 4$ messages.

Protocol 2 Chen-Gollmann-Mitchell multi-server protocol

$M1$	$A \rightarrow B:$	A, B, N_A
$M2$	$B \rightarrow S_i:$	A, B, N_A, N_B
$M3$	$S_i \rightarrow B:$	$\{B, N_A, K_i\}_{K_{Ai}}, \{A, N_B, K_i\}_{K_{Bi}}$
$M4$	$B \rightarrow A:$	$\{B, N_A, K_1,\}_{K_{A1}}, \ldots, \{B, N_A, K_n\}_{K_{An}}, cc_B(1), \ldots, cc_B(n)$
$M5$	$A \rightarrow B:$	$cc_A(1), \ldots, cc_A(n), \{B, N_B, N'_A\}_{K'_{AB}}$
$M6$	$B \rightarrow A:$	$\{A, N'_A, N_B\}_{K'_{AB}}$

The cross-checksum used in this protocol is different from the one used in the Gong multi-server protocol. The sensor node B calculates the $cc_B(i)$ as shown in Equation(2).

$$cc_B(i) = \{h(K_1), h(K_2), \ldots, h(K_n)\}_{K_i}, \ \forall i \in (1, \ldots, n) \tag{2}$$

To prevent A or B imposing the session key, the choice of $h()$ is limited; for example, it cannot be an exclusive–or–operation. If B doesn't receive any message from server S_j then $cc_B(j)$ is an error message, and $h(K_j)$ is replaced by an error message in the calculation of the other $cc_B(i)$ values. When A receives the checksums, A will first decrypt the values and compares the values with its own calculations of the cross-checksums. The valid K_i secrets are retained for the majority of i values and others are discarded. The session key K_{AB} is defined to be the hash of all the good K_i values concatenated, as shown in Equation (3).

$$K_{AB} = h(K_j, \ldots, K_m) \tag{3}$$

The messages $M4$ and $M5$ in the Chen et al. multi–server protocol have a computational complexity of $O(n^2)$. While messages $M2$ and $M3$ have $O(n)$, messages $M1$ and $M6$ have $O(1)$ computational complexity.

This protocol encounters similar problems as the Gong multi–server protocol with regard to the size of the messages. Once again, several messages are of $O(n^2)$ in size. With the cross–checksums containing the outputs of a one–way function where the inputs are key values, once again the size of the output will need to be large. The cross-checksums in this protocol are encrypted instead of only requiring a hash algorithm. However, the Chen et al. multi–server protocol is considerably more efficient with regard to the size of the messages.

Another aspect of the multiple server protocols is the creation of the new key K_{AB}. The nodes A and B retrieve the new key by using a secret sharing mechanism such as the one defined in (Shamir, 1979). Secret sharing is a mechanism allowing the owner of a secret to distribute *shares* amongst a group. Individual shares or a small number of shares are no help in recovering the secret. The n shares are distributed, such that any set of t (or more) shares is sufficient to obtain the secret. The most well-known threshold scheme uses polynomial interpolation.

When polynomial interpolation is used in cryptographic applications the field is typically $\mathbb{Z}_p$, the field of integers modulo p, for some prime p. To share a secret $s \in \mathbb{Z}_p$ in the (t, n) threshold scheme, the dealer generates a polynomial of degree $t - 1$:

$$f(x) = a_0 + a_1 x + \cdots + a_{t-1} x^{t-1} \tag{4}$$

The coefficients are randomly chosen in $\mathbb{Z}_p$ except for $a_0 = s$. The shares are values of $f(x)$ with $1 \leq x \leq n$. If t shares are known s can be recovered. For example, if $f(1), f(2), \ldots, f(t)$ are known then:

$$s = \sum_{i=1}^{t} f(i) \prod_{i<j\leq t} \frac{j}{j-i} \tag{5}$$

Given any t points on the polynomial (excluding the value of 0), the value for $f(0)$ can be obtained.

The time complexity to compute n shares is $O(nt)$. The time complexity to recover a_0 is $O(t \log_2 t)$ (Gong, 1993). Having such a scheme or similar scheme in a sensor node will consume significant amount of resources in an already resource constrained environment.

The existing multiple server protocols are therefore not suitable for a sensor network environment. An ideal solution with the desired characteristics requires innovative multiple server protocols, specifically designed for the sensor environment.

5. Multiple server protocols for sensor environments

Three multiple server protocols have been developed specifically for the sensor environment (Singh & Muthukkumarasamy, 2008). In this chapter we will label the three Multiple Server Protocols (MSP) as Singh et al. MSP1, Singh et al. MSP2, and Singh et al. MSP3. The multiple authentication server protocols Singh et al. have developed are based on the concept of the Boyd four–pass protocol (Boyd & Mathuria, 2003). These will maintain similar security characteristics as that of the centralized authentication protocol, as shown in the Boyd four–pass protocol. Because of the severe resource constraints which exist in sensor nodes, a multiple authentication server protocol should have low computational complexity in both time and space. A multiple server authentication protocol in a sensor environment should have the following characteristics:

- Small computation overhead,
- Minimal number of messages,
- Sensor nodes should not be relied upon to generate good randomness,
- The new key should be fresh,
- There should be key confirmation by both nodes.

The next several sections will go into detail about the three different sensor multiple server protocols.

5.1 Singh et al. Multiple Server Protocol 1

The Singh et al. MSP1 is an efficient multiple server protocol specifies n servers. The protocol has the following message flows. The sensor node A sends the first message, A, B, N_A, to each of the servers. Each server sends their message to both sensor nodes A and B. Sensor node B sends N_B, the keying data, and the cross–checksums created by B. It is important to note at this stage that K_S is unknown, so unlike the original protocol, B is not able to send $[N_A]_{K_{AB}}$. When sensor node A receives the next message, it will calculate its own cross checksums, and compare them against the cross–checksums created by B. At this stage, the keys K_S and K_{AB} are created. Sensor node A sends its cross–checksums to B, so B can create K_{AB}. The final message completes the key confirmation between A and B, as shown in *Protocol 3* (Singh & Muthukkumarasamy, 2006).

The Singh et al. MSP1 provides key authentication, key freshness and key confirmation, using multiple authentication servers. In the Singh et al. MSP1 protocol, the following constructs are used:

$AUTH_{Ai} = [A, B, K_i]_{K_{AS_i}}$
$MASK_{Ai} = [[AUTH_{Ai}]]_{K_{AS_i}}$
$AUTH_{Bi} = [A, B, K_i]_{K_{BS_i}}$
$MASK_{Bi} = [[AUTH_{Bi}]]_{K_{BS_i}}$

Protocol 3 Singh et al. MSP1

$M1$	$A \rightarrow S_i :$	A, B, N_A
$M2$	$S_i \rightarrow B :$	$N_A, A, S_i, AUTH_{Bi}, MASK_{Bi} \oplus K_i$
$M2'$	$S_i \rightarrow A :$	$S_i, AUTH_{Ai}, MASK_{Ai} \oplus K_i$
$M3$	$B \rightarrow A :$	$cc_B(1), \ldots, cc_B(n), N_B$
$M4$	$A \rightarrow B :$	$cc_A(1), \ldots, cc_A(n), [N_B]_{K_{AB}}$
$M5$	$B \rightarrow A :$	$[N_A]_{K_{AB}}$

Both of the sensor nodes and the servers contribute to the key value. The values N_A and N_B are generated by A and B respectively as input to the MAC function, that determines the session key. The key used with the MAC function is generated by the servers. Both A and B compute the session key as $K_{AB} = [N_A, N_B]_{K_S}$. The nodes should have a minimum number of servers returning valid results before confirming that the key is valid. Node B will calculate $cc_B(i)\ \forall i \in 1, \ldots, n$.

$$cc_B(i) = \begin{cases} [K_i]_{K_i} & \text{if valid,} \\ EM & \text{otherwise} \end{cases} \tag{6}$$

Where EM is an error message; an example will be the value zero. There is a remote chance a valid case may be zero. If the valid value is zero, the server needs to be considered a compromised server (even though it is not a malicious server).

Node A will calculate $cc_A(i)$, and compare it with $cc_B(i)$. If they are the same, then the server S_i is valid. Below is a way the nodes compare the cross checksum for $cc_A(i)$ and $cc_B(i)$.

$$cc_A(i) = \begin{cases} cc_B(i) = [K_i]_{K_i} & \text{if valid,} \\ EM & \text{otherwise} \end{cases} \tag{7}$$

After the comparison of the entire cross checksums, a set of valid keys $V_1, \ldots, V_m$ should remain. The creation of K_S is defined as follows.

$$K_S = V_1 \oplus \ldots \oplus V_m \tag{8}$$

Where V_i is the ith valid key given by a server, and m is the total number of valid servers $t \leq m \leq n$, where t is the minimal number of trusted servers. However, unlike the existing multiple server protocols, the trusted servers will not be able to calculate K_S. The calculated $cc_A(i)$ values are returned to B, where B performs similar checks as A and calculates K_S.

5.1.1 Analysis of singh et al. MSP1

The Singh et al. MSP1 has a number of advantages, one of which is that the nodes do not need good random number generators to create the nonces. The nodes could even safely use a counter for their nonce values. Another advantage is that if a server or a number of servers are unavailable, the authentication service itself still exists through the other servers. The servers and the sensor nodes have different keys; even if one or more servers become compromised, the authentication service or the security of the system is not compromised.

The protocol only encrypts random information. If the encryption cipher uses an IV value (such as RC5 and SKIPJACK currently used in TinyOS (TinyOS, 2007)) then we can use a constant IV value. However, the constant IV value chosen for Singh et al. MSP1 protocol must only be used to encrypt the random data and should never be used to encrypt other information. Also, a wide variation of different ciphers can safely be used.

Some MACs have vulnerabilities when the message sizes are variable. All of the message sizes are of constant value, allowing us to safely use a wider range of MACs than previously available. The size of the MACs can be lower than that of conventional protocols. The integrity checking is performed by the sensor nodes. If x is the size of the MAC in bits, then an adversary has 1 in 2^x chance in blindly forging a valid MAC for a particular message. The adversary should be able to succeed in 2^{x-1} tries. Because of the low bandwidth of sensor nodes, a 4 byte MAC, requiring 2^{31} packets, will take years to complete. If an adversary did attempt this attack, the sensor node would be non–functional within that period. In addition, an adversary will need to forge $2t$ MACs; t MACs to A and t MACs to B, and stop traffic from the other base stations before they can determine the value of K_{AB}.

In order to study the performance impacts of each of the multiple server protocols, we first define the following symbols. The size of location indicator is a_0, the nonce size is a_1, the key size is a_2, the hash size is a_3, and the number of servers is n. The following equations are used to define how many bytes are sent for each message: $M1_i = 2a_0 + a_1$, $M2_i = 2a_0 + a_1 + a_2 + a_3$, $M2'_i = a_0 + a_2 + a_3$, $M3 = a_1 + na_3$, $M4 = (n+1)a_3$, and $M5 = a_3$. However, there are n messages of type $M1$, $M2$ and $M2'$.

5.2 Singh et al. Multiple Server Protocol 2

The Singh et al. MSP1 protocol was investigated and extended further to reduce the large number of messages sent through the WSN. Thereby the Singh et al. MSP2 was developed as shown in *Protocol 4.* The assumption is made that the servers can communicate through a different network, other than the low bandwidth WSN used by the sensor nodes. For instance, the GNOME platform (GNOME, 2006) also has an Ethernet connection it can use as a high speed backbone network to communicate with other GNOME machines. In the Singh et al. MSP2 protocol, the sensor node A only sends one message to a server, denoted as S_1. Server S_1 then gathers all the required information from the other servers through the server network (rather than the sensor network). Server S_1 concatenates the information and sends it to sensor node B. The list of servers may either be a static list, known by the sensor nodes, or it may be a list based on the trustworthiness of the servers.

5.2.1 Security analysis

If S_1 becomes malicious, there are a number attacks that the server can try. The simplest attack is a denial of service, where the server will not gather any extra information from other servers, or doesn't respond to the sensor node B. If sensor node A does not receive a response from B in a required amount of time, then it should try S_2.

Protocol 4 Singh et al. MSP2 protocol

$M1$	$A \rightarrow S_1:$	A, B, N_A
$M2$	$S_1 \rightarrow S_i:$	A, B
$M3$	$S_i \rightarrow S_1:$	$MASK_{Bi}, AUTH_{Bi} \oplus K_i, MASK_{Ai}, AUTH_{Ai} \oplus K_i$
$M4$	$S_1 \rightarrow B:$	$S_1, MASK_{B1}, AUTH_{B1} \oplus K_1, \ldots, S_n, MASK_{Bn}, AUTH_{Bn} \oplus K_n, N_A, A$
$M4'$	$S_1 \rightarrow A:$	$MASK_{A1}, AUTH_{A1} \oplus K_1, \ldots, MASK_{An}, AUTH_{An} \oplus K_n$
$M5$	$B \rightarrow A:$	$cc_B(1), \ldots, cc_B(n), N_B$
$M6$	$A \rightarrow B:$	$cc_A(1), \ldots, cc_A(n), [N_B]_{K_{AB}}$
$M7$	$B \rightarrow A:$	$[N_A]_{K_{AB}}$

Another possible attack S_1 can try is to forge the MACs to create legitimate messages from the other servers. As discussed earlier, it is unreasonable to assume a server can forge a message, let alone $2t$ messages.

If there are t malicious servers, then S_1 can contact those servers and not involve the trusted servers. However, the Singh et al. MSP1 is also vulnerable to t malicious servers.

If contacting only one server is still a concern, then a higher level reputation based framework (Ganeriwal & Srivastava, 2004) on top of existing authentication protocols to make sure the nodes collaborate with only trustworthy base stations. Another solution is that A can send the first message to p servers, where $p < n$, and each server is allocated servers from which to obtain information (however, this will require more code and logic within the sensor applications).

5.2.2 Cost/complexity analysis

The Singh et al. MSP2 decreases the number of packets sent by A. The following equations are used to define how many bytes are sent for each message, $M1 = 2a_0 + a_1$, $M2_i = 2a_0$, $M3_i = 2a_2 + 2a_3$, $M4 = a_0 + a_1 + na_2 + na_3$, $M4' = na_2 + na_3$, $M5 = a_1 + na_3$, $M6 = (n+1)a_3$, and $M7 = a_3$.

Although the Singh et al. MSP2 decreased the number of messages sent by A, it is not as reliable as Singh et al. MSP1. If S_1 is down, either A will need to detect this and try S_2, or S_1 itself will need to be a clustered system. The drawback of a true replicated clustered system is that it is more likely that the system can be compromised, since there are more machines available for an adversary to attack. Another problem if S_1 becomes malicious is that it may not return any information from the other servers. Once again, A can try other servers.

5.3 Singh et al. Multiple Server Protocol 3

The third multiple server protocol is the Singh et al. MSP3. The third protocol was designed for the case where if the key K_S becomes compromised, and the long–term keys K_{AS} and K_{BS} have not changed. The key K_S is important because it can be used in the future to create or renew a session key between A and B. The multiple server scenarios have strengthened the security between the nodes and the KDC. Compromised keys between a node and one (or more) servers, does not affect the security of the protocol. An adversary can replay previous (portion of) messages to force the sensor nodes to use the same K_S as before.

The Singh et al. MSP3 protocol was designed to remove this problem. The advantage of the Singh et al. MSP3 over Singh et al. MSP2 is that if K_S is ever compromised and the long term keys K_{AS} and K_{BS} have not changed, then the protocol can be run again safely. The calculation for $AUTH$ was also done so that a replay attack is not possible.

$AUTH_{Ai} = [A, B, N_A, N_B]_{K_{Ai}}$
$AUTH_{Bi} = [A, B, N_A, N_B]_{K_{Bi}}$

Protocol 5 Singh et al. MSP3

$M1$	$A \to B:$	A, N_A
$M2$	$B \to S_1:$	A, B, N_A, N_B
$M3$	$S_1 \to S_i:$	A, B, N_A, N_B
$M4$	$S_i \to S_1:$	$AUTH_{Bi}, MASK_{Bi} \oplus K_i, AUTH_{Ai}, MASK_{Ai} \oplus K_i$
$M5$	$S_1 \to B:$	$AUTH_{B1}, MASK_{B1} \oplus K_1, \ldots, AUTH_{Bn}, MASK_{Bn} \oplus K_n$
$M5'$	$S_1 \to A:$	$AUTH_{A1}, MASK_{A1} \oplus K_1, \ldots, AUTH_{An}, MASK_{An} \oplus K_n$
$M6$	$B \to A:$	$cc_B(1), \ldots, cc_B(n), N_B$
$M7$	$A \to B:$	$cc_A(1), \ldots, cc_A(n), [N_B]_{K_{AB}}$
$M8$	$B \to A:$	$[N_A]_{K_{AB}}$

The Singh et al. MSP3 has one more message than Singh et al. MSP2, however, Singh et al. MSP2 has a larger message. This is because of the need to send A and N_A in $M4$, whereas in Singh et al. MSP3 the message $M5$ does not need to send this extra information. The maximum size message in Singh et al. MSP2 can be decreased if another message $M1'$ is sent as shown in Equation (9).

$$A \to B : A, N_A \tag{9}$$

5.4 Performance analysis

When looking at authentication algorithms, a number of different aspects need to be taken into consideration. Apart from the security properties, such as key establishment, key freshness, and key confirmation, a number of performance aspects should also be looked at. In the past, symmetric key algorithm performance was categorized by the number of messages and the number of rounds. However, in sensor networks these are not true indicators of the performance of an algorithm. Other measures such as computational costs, number of bytes and packets sent and received, the amount of memory consumed are also important in a sensor environment.

The computational costs to the scheme arises because of the encryption, decryption and integrity checking of the keys generated by the servers. There is also the generation of the K_{AB} by both A and B. An inefficient aspect of the Boyd four–pass protocol is the need to decrypt the message before the integrity check is done. The protocols have a similar restriction if bits were changed in the $MASK$ then it will not be known until both $MASK$ and $AUTH$ were calculated. Bogus messages injected by a third party cannot be detected using any computationally efficient methods. The Bellare–Rogaway uses the encrypted messages when performing the integrity check, rather than the decrypted message. A very simple modification to the Boyd four–pass protocol removes this minor limitation, as shown in the *Protocol 6.*

Protocol 6 Boyd protocol integrity change

$M1$	$A \to S:$	A, B, N_A
$M2$	$S \to B:$	$[[K_S]]_{K_{AS}}, [A, B, [[K_S]]_{K_{AS}}], [[K_S]]_{K_{BS}}, [A, B, [[K_S]]_{K_{BS}}]_{K_{BS}}, N_A, A$
$M3$	$B \to A:$	$[[K_S]]_{K_{AS}}, [A, B, [[K_S]]_{K_{AS}}]_{K_{AS}}, [N_A]_{K_{AB}}, N_B$
$M4$	$A \to B:$	$[N_B]_{K_{AB}}$

The size of the messages does not change, and there is the added benefit of the integrity check performed before the decryption of the message. This technique was also extended to be used in Singh et al. protocols (however, encryption algorithms are then required to be implemented on the sensor nodes). However, it has been shown that the communication costs are almost an order of magnitude more than the computational costs in security systems (Perrig et al., 2001).

The communication costs will be heavily dependent on the topology of the network. Also, different protocols have different communication overheads for the sensor nodes, and the servers. The communication overheads of the different protocol is examined in detail in Section 6.

6. Comparison

We now compare Singh et al. protocols with the Chen et al. protocol and the Gong protocol. The computational complexity of the existing multiple server protocols is greater since the key is constructed using a key–sharing mechanism. This has two major drawbacks in a sensor network environment. The first is the amount of extra code (and therefore memory overhead) needed when creating the new key. The second is the additional computation (and therefore extra energy) required when creating the new key. If we use a threshold scheme such as Shamir's scheme (Shamir, 1979) to recover the key, the computational complexity will be $O(t \log_2 t)$. Whereas Singh et al. protocols use a simple exclusive–or function (as describe in Equation (8)) to recover the key, which has a computational complexity of only $O(n)$. Not only does this save on computational cost, but also has the added benefit of requiring less code than a full–blown key–sharing threshold scheme. The advantage of using a full–blown key–share threshold scheme, is that t servers can calculate the new session key between A and B if they are the only servers involved in the protocol. However, for performance reasons we have not placed the same restriction on Singh et al. multiple server protocols.

Another comparison is the communication cost of Singh et al. protocols compared with the Chen et al. and Gong protocols. For simplicity, we will assume that the output of the one–way function used to calculate the cross–checksums in the existing multiple server protocols is the same size as the integrity function used in Singh et al. protocols. Although, as described earlier, for security reasons the one–way function in the existing multiple server protocols will need to be larger.

We will compare the total number of bytes sent by sensor node A for each of the existing multiple server protocols, and Singh et al. MSP1 and Singh et al. MSP2. However, similar calculations as the ones shown here can be used to calculate the impact on the base stations and sensor B.

The total number of bytes sent by sensor A in the Gong multi–server protocol is $n^2 a_3 + 2na_0 + na_2 + 2a_0 + 2a_1 + a_3$, which has a complexity of $O(n^2)$. The number of bytes sent by sensor A in the Chen et al. multi–server protocol is $n^2 a_3 + na_3 + 3a_0 + 3a_1 + a_3$. The above two cases have a complexity of $O(n^2)$. However, the number of bytes sent by sensor A in Singh et al. MSP1 and Singh et al. MSP2 are $2na_0 + na_1 + (n+1)a_3$ and $2a_0 + a_1 + (n+1)a_3$, respectively. Both of these messages have a complexity of $O(n)$.

Figure 2 compares the total number of bits sent out in the entire network, relative to the number of servers used. The Gong protocol is the most expensive, followed by the Chen protocol. Also, the protocols consider for this graph send all their messages over the sensor network, rather than sending some of the messages over a faster backbone network.

The existing multiple server protocols have message sizes of size $O(n^2)$, whereas Singh et al. protocols are $O(n)$. It should be noted that the existing multiple server protocols do have added functionality, where the trusted servers are able to calculate the key K_{AB}. However, for

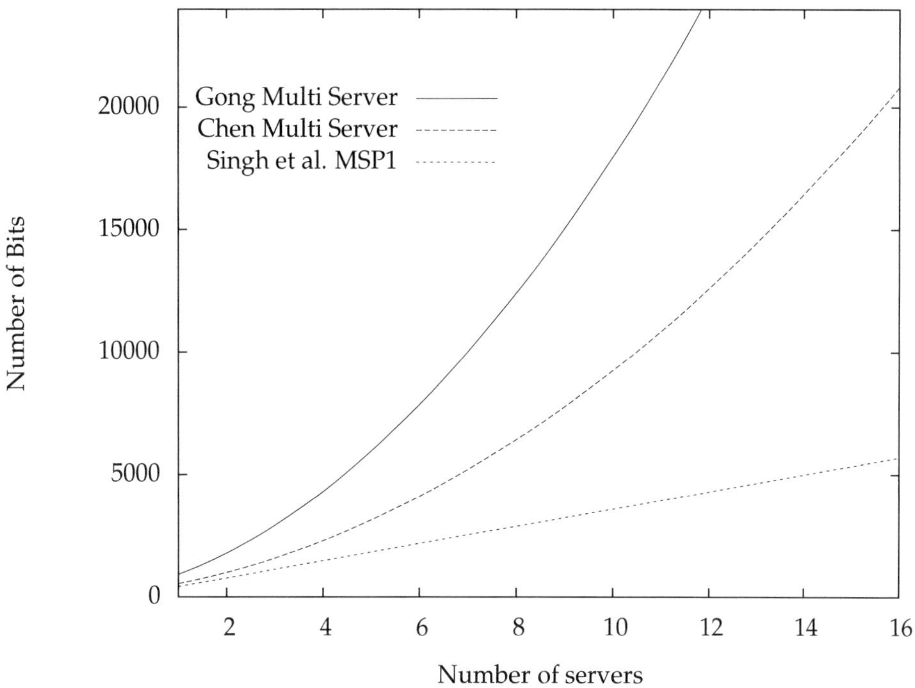

Fig. 2. Total Number of bits sent by different Protocols

performance reasons we have not placed the same restriction on Singh et al. multiple server protocols.

Table 2 describes the costs of the Singh et al. MSP1, where n is the number of servers. The table shows that in the protocol sensor node A transmits the most amount of messages. However, even though sensor node B only transmits two messages, the size of one of its messages is based on the number of servers. Hence, the number of bits sent by sensor node B is similar to the number of bits sent by sensor node A. The messages and size of messages transmitted by a server S_i is constant and independent on the number of servers in the protocol. The number of messages received and the number of bits received is very similar when comparing sensor node A and sensor node B. The message and number of bits received by a server S_i is insignificant compared to all the other messages.

Node	A	B	S_1	S_i
Msgs Tx	$n+1$	2	2	2
Msgs Rx	$n+2$	$n+1$	1	1
Bits Tx	$72n+32$	$32n+40$	248	248
Bits Rx	$144n+40$	$168n+32$	40	40

Table 2. Performance Figures for Singh et al. MSP1

Table 3 describes the costs of the Singh et al. MSP2, where n is the number of servers. The table shows that in the protocol the server S_1 transmits the most amount of messages. However, even though sensor nodes A and B only transmits several messages, some of their message sizes is based on the number of servers. Hence, the number of bits sent by sensor nodes A

and B is still linear. The messages and size of messages transmitted by a server S_i (except S_1) is constant and independent on the number of servers in the protocol. The number of messages received and the number of bits received is very similar when comparing sensor node A and sensor node B. The message and number of bits received by a server S_i (except S_1) is insignificant compared to all the other messages. The number of bits received by S_1 is linear in nature.

Node	A	B	S_1	S_i
Msgs Tx	2	2	$n+1$	1
Msgs Rx	3	2	n	1
Bits Tx	$32n+72$	$32n+40$	$232n-40$	192
Bits Rx	$128n+40$	$128n+56$	$192n-152$	40

Table 3. Performance Figures for Singh et al. MSP2

Table 4 describes the costs of the Singh et al. MSP3, where n is the number of servers. The protocol has many of the same features as Singh et al. MSP2. For instance, the server S_1 transmits the most amount of messages. The sensor nodes A and B only transmits several messages, some of their message sizes is based on the number of servers. Hence, the number of bits sent by sensor nodes A and B is still linear. However, the message sizes are slightly larger when compared against the previous protocol because of the extra security. The messages and size of messages transmitted by a server S_i (except S_1) is constant and independent on the number of servers in the protocol. The number of messages received and the number of bits received is very similar when comparing sensor node A and sensor node B. As in the previous protocol, the message and number of bits received by a server S_i (except S_1) is insignificant, however, it is slightly larger when compared to the previous protocol. As in the previous protocol, the number of bits received by S_1 is linear in nature.

Node	A	B	S_1	S_i
Msgs Tx	2	3	$n+1$	1
Msgs Rx	3	3	n	1
Bits Tx	$32n+72$	$32n+88$	$240n-48$	192
Bits Rx	$128n+40$	$128n+72$	$192n-144$	48

Table 4. Performance Figures for Singh et al. MSP3

The Singh et al. MSP1 puts a larger emphasis on communication costs on sensor node A. The other two protocols put greater emphasis on the server nodes. The extra message in Singh et al. MSP3 does not affect sensor node A, and for a large number of servers it is insignificant.
The number of cryptographic operations performed by each node in the three protocols shown in the table, is the same for each of the protocols. The number of $AUTH$ calculations is $2n+4$ for both A and B. The number of $MASK$ calculations is $2n$ for both A and B. For S_i the number of $AUTH$ and $MASK$ calculations is two. The only nodes that need to create good random numbers are each of the servers S_i and they only create one for each run through of the protocol. Both N_A and N_B can be a simple counter.
Another comparison is the number of packets sent by each of Singh et al. multiple server protocols, as shown in Figure 3. The comparison is done over two different network topologies. The servers either communicate over the sensor network or they have a separate network to communicate over. The Singh et al. MSP1 has the same cost over both network topologies, since every message interacts with a sensor node, and the servers do not need to

communicate with one another. Because of the larger number of messages sent by the Singh et al. MSP2 and Singh et al. MSP3, it is natural for the number of packets to be more than Singh et al. MSP1, when all the messages have to be sent on the same sensor network. As the number of servers increases, the difference (when on the same network) between Singh et al. MSP1 and the other two protocols does not dramatically increase. The other two protocols have the advantage of concatenating messages, which the original multiple server protocol does not have. If the servers can communicate over a different network, the number of packets sent by Singh et al. MSP2 and Singh et al. MSP3 stay very close to one another, and sometimes they are the same. This is because of the first protocol having larger message sizes, and the messages need to be segmented sooner than they do in the other protocol. The number of packets sent on the energy–constrained sensor network is significantly larger if using Singh et al. MSP1, when the servers are on different networks.

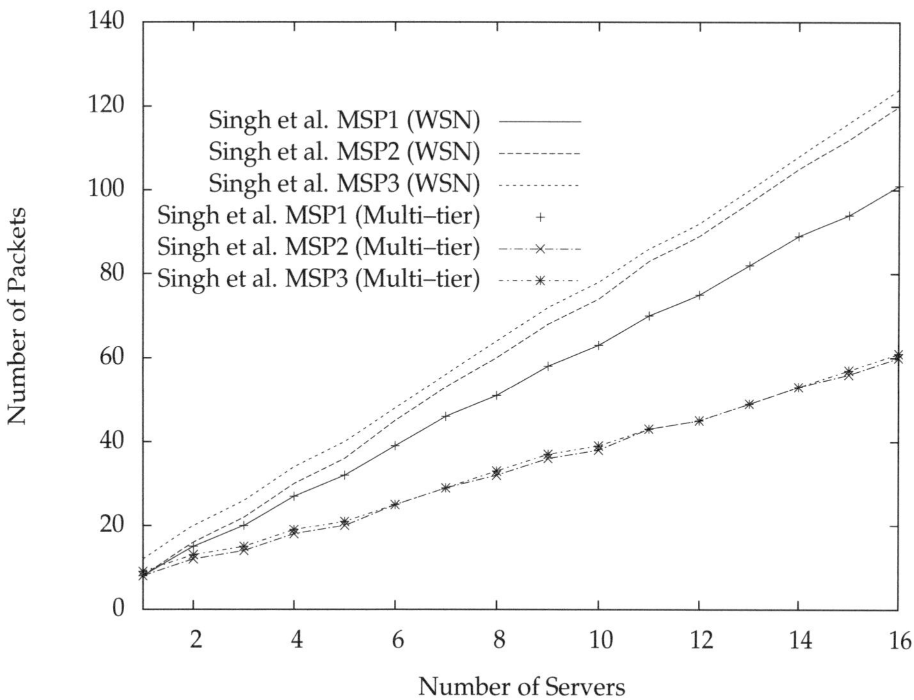

Fig. 3. Number of packets sent by different Protocols

TinyOS has a seven byte overhead when sending a single packet. When comparing the bits sent by each of the protocols, we included the packet overhead. We once again cover the case of two network topologies, as shown in Figure 4. The Singh et al. MSP1 has the same cost over both networks. When comparing the protocols, if the servers can communicate over a different network, there is virtually no difference between Singh et al. MSP2 and Singh et al. MSP3. However, the extra overhead of sending data to the server S_1, which then sends the same data down to the sensor nodes, causes both protocols to have significantly more overhead when run over the same network.

Table 5 compares the existing multiple server protocols with some of Singh et al. protocols. In the table we look at the protocol reliability, and compare the protocols when the servers use

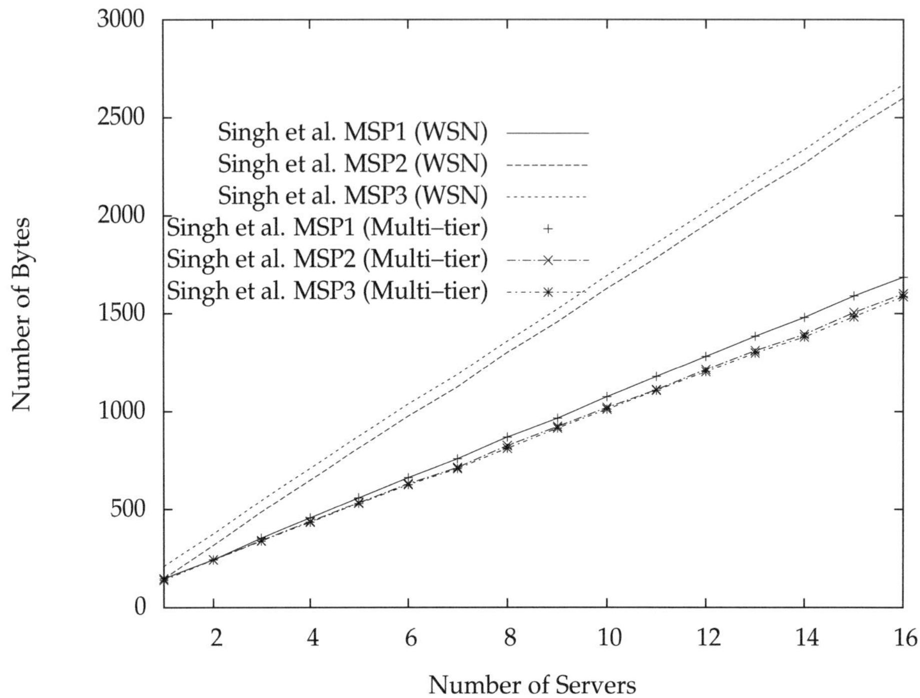

Fig. 4. Number of bytes sent by different Protocols

the WSN or a different network. We also evaluate the robustness of the protocols based on security attacks in any environment.

	Properties of the Protocol			
Protocol	Protocol Reliability	Efficiency of WSN	Efficiency of multi–tiered	Security
Gong	Excellent	Bad	Bad	Excellent
Chen et al.	Excellent	Bad	Bad	Excellent
Singh et al. MSP1	Excellent	Good	Average	Good
Singh et al. MSP2	Average	Average	Excellent	Good
Singh et al. MSP3	Average	Average	Excellent	Excellent

Table 5. Comparison of Multiple Server Protocols

After the detailed analysis of the protocols and the comparison with existing protocols, we conclude that Singh et al. MSP1 should be used in situations where the environment is concerned with reliability. If reliability isn't a major concern and if the servers can communicate over a different network then Singh et al. MSP2 should be used. If the environment is not multi–tiered then the most reliable protocol (Singh et al. MSP1) is also the most efficient. If concerned about the security of K_S then Singh et al. MSP3 should be used. In a complex environment the need for reliability and security puts extra costs on the protocols.

7. Conclusions

Key establishment protocols, without the assumption of trusting an individual authentication server, are needed in sensor environments where clients cannot trust individual servers. We examined the existing multiple server protocols developed for traditional networks, and investigated the problems using those protocols for WSNs. We then three multiple server protocols for sensor networks and a complex sensor system containing multi–tiered networks, and provided a detailed analysis and comparison of each the protocols.

The multiple server protocols designed for a sensor environment removed the requirement for the servers to know the keys between each of the sensor nodes, and thus helping to limit the message sizes to $O(n)$. The protocols have the added benefit of not solely relying on the sensor nodes to generate cryptographically sound pseudo–random numbers, but still using information from each of the sensors to generate the new key. We have shown that our protocols are flexible enough for them to be used with almost any cryptographic primitive and in a range of environments.

8. References

Blundo, C., Santis, A. D., Herzberg, A., Kutten, S., Vaccaro, U. & Yung, M. (1993). Perfectly–secure key distribution for dynamic conferences, *Lecture Notes in Computer Science* 740: 471–486.
URL: *citeseer.ist.psu.edu/blundo95perfectlysecure.html*

Boyd, C. (1996). A class of flexible and efficient key management protocols, *CSFW '96: Proceedings of the Ninth IEEE Computer Security Foundations Workshop*, IEEE Computer Society, Washington, DC, USA, p. 2.

Boyd, C. & Mathuria, A. (2003). *Protocols for Authentication and Key Establishment*, Springer Berlin / Heidelberg.

Bulusu, N. (2005). Introduction to wireless sensor networks, *in* N. Bulusu & S. Jha (eds), *Wireless Sensor Networks: A Systems Perspective*, Artech House.

Chan, H. & Perrig, A. (2005). PIKE: Peer intermediaries for key establishment in sensor networks, *Proceedings of IEEE Infocom*, IEEE Computer Society Press.

Chan, H., Perrig, A. & Song, D. (2003). Random key predistribution schemes for sensor networks, *SP '03: Proceedings of the 2003 IEEE Symposium on Security and Privacy*, IEEE Computer Society, Washington, DC, USA, p. 197.

Chen, L., Gollmann, D. & Mitchell, C. J. (1995). Key distribution without individual trusted authentication servers, *8th IEEE Computer Security Foundations Workshop*, IEEE Computer Society Press, pp. 30–36.

Crossbow (2006). Crossbow, URL: *http://www.xbow.com/*.

Deng, J., Han, R. & Mishra, S. (2004). Intrusion tolerance and anti-traffic analysis strategies in wireless sensor networks, *Dependable Systems and Networks, 2004 International Conference on*, IEEE, pp. 637–646.

Deng, J., Han, R. & Mishra, S. (2005). Sensor–network security, privacy, and fault tolerance, *in* N. Bulusu & S. Jha (eds), *Wireless Sensor Networks: A Systems Perspective*, Artech House.

Ganeriwal, S. & Srivastava, M. B. (2004). Reputation-based framework for high integrity sensor networks, *SASN '04: Proceedings of the 2nd ACM workshop on Security of ad hoc and sensor networks*, ACM Press, New York, NY, USA, pp. 66–77.

GNOME (2006). A controller node, URL: *http://cmlab.rice.edu/projects/sensors*.

Gong, L. (1993). Increasing availability and security of an authentication service, *IEEE Journal on Selected Areas in Communications* 11(5): 657–662.

Karlof, C., Sastry, N. & Wagner, D. (2004). Tinysec: a link layer security architecture for wireless sensor networks, *SenSys '04: Proceedings of the 2nd international conference on Embedded networked sensor systems*, ACM Press, New York, NY, USA, pp. 162–175.

Leighton, F. T. & Micali, S. (1994). Secret-key agreement without public-key cryptography, *CRYPTO '93: Proceedings of the 13th Annual International Cryptology Conference on Advances in Cryptology*, Springer-Verlag, London, UK, pp. 456–479.

Malan, D. J., Welsh, M. & Smith, M. D. (2004). A public–key infrastructure for key distribution in tinyos based on elliptic curve cryptography, *Proc. 1st IEEE Communications Society Conference on Sensor and Ah Hoc Communications and Networks (SECON '04)*, IEEE Computer Society Press, Santa Clara, CA, USA, pp. 71–80.

MANTIS (2006). The mantis contoller node, URL: *http://mantis.cs.colorado.edu/*.

Medusa (2006). The medusa mk–2 controller node, URL: *http://nesl.ee.ucla.edu/*.

Perrig, A., Szewczyk, R., Wen, V., Culler, D. & Tygar, J. D. (2001). SPINS: Security protocols for sensor networks, *Seventh Annual International Conference on Mobile Computing and Networks (MobiCOM 2001)*, Rome, Italy.

Przydatek, B., Song, D. & Perrig, A. (2003). Sia: secure information aggregation in sensor networks, *SenSys '03: Proceedings of the 1st international conference on Embedded networked sensor systems*, ACM Press, New York, NY, USA, pp. 255–265.

Shamir, A. (1979). How to share a secret, *Commun. ACM* 22(11): 612–613.

Singh, K., Bhatt, K. & Muthukkumarasamy, V. (2006). Protecting small keys in authentication protocols for wireless sensor networks, *Proceedings of the Australian Telecommunication Networks and Applications Conference*, Melbourne, Australia, pp. 31–35.

Singh, K. & Muthukkumarasamy, V. (2006). A minimal protocol for authenticated key distribution in wireless sensor networks, *ICISIP '06: Proceedings of the 4th International Conference on Intelligent Sensing and Information Processing*, IEEE Press, Bangalore, India, pp. 78–83.

Singh, K. & Muthukkumarasamy, V. (2008). Performance analysis of proposed key establishment protocols in multi–tiered sensor networks, *Journal of Networks* 3(6).

TinyOS (2007). An operating system for sensor motes, URL: *http://www.tinyos.net/*.

Wagner, D. (2004). Resilient aggregation in sensor networks, *SASN '04: Proceedings of the 2nd ACM workshop on Security of ad hoc and sensor networks*, ACM Press, New York, NY, USA, pp. 78–87.

Watro, R., Kong, D., fen Cuti, S., Gardiner, C., Lynn, C. & Kruus, P. (2004). Tinypk: securing sensor networks with public key technology, *SASN '04: Proceedings of the 2nd ACM workshop on Security of ad hoc and sensor networks*, ACM Press, New York, NY, USA, pp. 59–64.

Zhu, S., Setia, S. & Jajodia, S. (2003). Leap: efficient security mechanisms for large-scale distributed sensor networks, *SenSys '03: Proceedings of the 1st international conference on Embedded networked sensor systems*, ACM Press, New York, NY, USA, pp. 308–309.